T0182993

Annals of Scientific Society for Assembly, Handling and Industrial Robotics 2021

Thorsten Schüppstuhl
Kirsten Tracht · Annika Raatz
Editors

Annals of Scientific Society for Assembly, Handling and Industrial Robotics 2021

 Springer

Editors
Thorsten Schüppstuhl
Aircraft Production Technology (IFPT)
Hamburg University of Technology
Hamburg, Germany

Annika Raatz
Institute of Assembly Technology
University of Hannover
Garbsen, Germany

Kirsten Tracht
Institute for Mechanical Engineering (bime)
University of Bremen
Bremen, Germany

ISBN 978-3-030-74031-3   ISBN 978-3-030-74032-0   (eBook)
https://doi.org/10.1007/978-3-030-74032-0

This Springer imprint is published by the registered company Springer Nature Switzerland AG
The registered company address is: Gewerbestrasse 11, 6330 Cham, Switzerland

Contents

Assembly Planning

Generic Modeling Technique for Flexible and Highly Available Assembly Systems

Rainer Müller and Martin Karkowski

Abstract

To face the major challenges posed by the increasing product variants with shortening life cycles and fluctuating market conditions, modular and adaptable assembly systems are used. Their strong dependency on software creates a new void in the planning and implementation processes of these assembly systems. Usually, the programmer fills this void based on his knowledge, which leads to frequent and large adaptations of the code base. This is rather counter-productive. To address this challenge, we developed a generic user-friendly graphical API (Application Programming Interface) for a process owner in our previous work (Müller et al, Proc CIRP 81:730–735, 2019, [1]). This API can describe any assembly system and the associated task and is used to get the boilerplate code needed to execute the process on a programmable logic controller (PLC)—the standard hardware used in the industry. In this paper, the virtual description of an assembly system used by the API is extended to include a goal-oriented task description by defining the process and the structure. We believe that this extension provides the proper abstraction needed by the process owner. In addition, this extension significantly reduces the modeling effort.

Keywords

Petri nets · Agents · Planning · Process control · PLC · AI

R. Müller · M. Karkowski (✉)
Zentrum für Mechatronik und Automatisierungtechnik gemeinützige GmbH, Eschberger Weg 46, 66121 Saarbrücken, Germany
E-mail: m.karkowski@zema.de
URL: http://www.zema.de

T. Schüppstuhl et al. (eds.), *Annals of Scientific Society for Assembly, Handling and Industrial Robotics 2021*,
https://doi.org/10.1007/978-3-030-74032-0_1

1 Introduction

Integration of new production technologies—such as HRC (Human Robot Collaboration)—into assembly systems and requirements for mass customization lead to frequent adjustments of the assembly systems. For this reason, it is desirable to use flexible and versatile assembly systems [2].

However, constraints such as usage of off-the-shelf hardware (in the form of PLCs), assurance of higher availability of systems, deployment of different recovery strategies after an equipment's malfunction, presence of various analyzing and debugging capabilities, etc., lead to difficulty in the development of reusable, maintenance-friendly and well-structured software implementations [3].

Current methods often perform fundamental modifications to the software to ensure compatibility with the used equipment. This often leads to increased cost (in terms of software development time), slower deployment and increased time-to-manufacture, which contradicts the requirements of modern assembly systems [4].

In this paper, an extension to a comprehensive behavioral description of an assembly system based on Petri nets is presented. Our extension allows a goal-oriented task description of the assembly system and satisfies the aforementioned constraints while also satisfying the requirements of the software-defined modern assembly systems.

The text is structured as follows. Section 2 presents related approaches found in literature that aim to simplify the implementation of assembly systems. Section 3 presents previous fundamental work done on our system. Section 4 details the various building blocks used in our approach. The extension of our behavioral description is presented in Sect. 5. Section 6 presents a modified Refinement Action Engine used in our System.

2 Related Works

Plug and Produce represents a vision to adapt the structure of automated systems with a minimal manual reconfiguration of the software of the system [5]. Various techniques such as high-level programming [5, 6], Automatic Programming [7, 8] etc., form the components of any *Plug and Produce* system.

SMACH—a planning and execution system for ROS (Robot Operating System)—relies on a set of hierarchical state machines where the states are mapped to the execution of user-defined code to interact with ROS nodes [9]. SMACH is not designed for implementing low-level systems.

SkiROS is developed to control robots by efficiently linking semantically described tasks. Tasks are defined as sequences of skills. They comprise of atomic actions, like grabbing, transporting or placing, which are described semantically by an ontology. The actual task is planned by a modular task planner based on PDDL (Planning Domain Definition Language) [10].

These systems presented above don't satisfy one or more of the constraints mentioned in the previous section. For example, SkiROS assumes a deterministic planning domain and assigns the tasks by utilizing the matched production equipment. Usage of hierarchical state machines (like in [9]) in SMACH complicates the aggregation of an assembly system using construction kits that are modular in nature. Stateless description of the assembly process (like in [7, 8] or [10]) in *Plug and Produce*-type systems often forgo the modeling of relevant aspects for industrial applications such as errors or disturbances in processes. Other self-organizing approaches based on multi-agent systems are not suitable for industrial applications because of their unpredictable behavior [11] and high software complexity. Furthermore, a common recurring observation is that approaches lacking a graphical (e.g. [10]) task definition prevents integrating the knowledge of process owners at early stages of the development process.

3 Previous Works

In our previous works (see [1, 12]), a modeling technique based on Petri nets is developed. In order to obtain a virtual model of the assembly system, detailed product analysis is performed. Based on that analysis, a generic description of the process is manually defined graphically. The description uses Petri nets to express the logical sequence of the process. It models the assembly operations as skills which are used to derive the resources of an assembly system. These resources provide services that are linked to their capabilities via a taxonomy and are invoked at runtime. The transitions of Petri nets can be linked with services provided by the resources and the entire structure of the petrinet specifies when the services are called.

The mapping between resources and services can be derived automatically as long as the selected resources provide unique services and the mapping between the used skills and the services are unique. Alternatively, the mapping can be done manually. When deriving resources from the defined skills, it may be necessary to integrate additional skills. This leads to an iterative design process until appropriate resources are identified. The aggregation of the these resources along with its manually modeled relations, restrictions and requirements denotes the model of our system's behavior. The complete model is obtained by adding the mechanical and electrical structure of the system as well as the flow of information. An exemplary application of the modeling technique is given in [1].

To perform the assembly task, the virtual model is transferred into the standardized control system (i.e. *Refinement Action Engine*). To optimize and validate the model, the standardized control system can be coupled to a simulation environment (see [13] for more details). This enables a fast adaptation and validation of the system behavior before the mechanical setup.

The expressiveness of the modeling technique is also its weakness since a complete representation of the entire process is very exhausting. This work addresses the following question: *How can the modeling of an assembly task be simplified?*

4   Theoretical Background of the Work

The following sections briefly describe the different building blocks used in our system. First, *Perti nets*, a de facto standard used for modeling discrete-event systems are presented. In our approach, Petri nets are used to model the base behavior of individual resources, the behavior of the assembly system, and the goal-oriented task-description. Second, *marking graphs* which enable exploration of the state space of the system are introduced. These graphs are derived from the Petri nets and they form the basis for automatic planning. The marking graph will be explored to determine plans for the goal-oriented task description. Afterward, the concept of a *Refinement Action Engine (RAE)*, which is used to execute model of an assembly system automatically, is presented. Integrating *planners* into a RAE enables automatic generation and adaptation of a plan. This approach is called *Automatic Planning* in the related literature.

4.1   Petri Nets

Petri nets comprises of *places*, *transitions*, and *edges* represented by circles, rectangles and arcs respectively. They are used for modeling discrete-event dynamic systems graphically. States of a system or subsystem are represented by places, whereas state changes, e.g. by actions or events, are represented by transitions. Active states are expressed by *marking* a place with tokens (filled circles), which are consumed and produced by transitions. Tokens represent abstract conditions or elements of the modeled system. A Petri net is defined by its places, transitions and their relations, modeled by edges that define the flow of tokens. The distribution of the tokens in the network is denoted as marking of the net [14].

The states of the Petri net are modeled explicitly, whereas other event-based modeling techniques define their states only implicitly. This is helpful because explicitly modeled states simplify the description and analysis of the modeled system. Besides Petri nets have a formal mathematical description and thence, they support simple methods for model-checking and simulation [15].

4.2   Marking Graphs

A marking M of the Petri net is reachable if there is a sequence of transitions that transforms the initial marking M_0 to the desired marking M. All reachable markings can be transformed into the marking graph or reachability graph of the Petri net, where markings are expressed as nodes and the steps between the markings as edges. The Marking Graph defines the potential state space of the modeled system. It is generally infinitely large and often used as a starting point for an automated analysis [14].

4.3 Refinement Action Engine

A Refinement Action Engine (RAE) uses *events*, *facts*, *tasks* and *interactions* to influence an environment. The state of the system is described by facts, which are used to select refinement methods to fulfill the desired task. Refinement methods describe the effect on the environment and the conditions to be met. Each refinement method provides commands which are executed by an execution platform. The interaction of the platform with the surrounding environment results in events. These events are forwarded to the RAE and may cause new tasks. Tasks can be provided by additional external components, like users or planners [16].

4.4 Planners

AI planning systems—although called *planners* in this paper—are used to generate problem-specific solutions to fulfill a task. A plan itself comprises of actions, which can only be executed under certain conditions. A planning system usually develops a plan by either following some heuristics or learning on the fly. Please refer to [16] to gain a more in-depth understanding of different planning models.

4.5 Automated Planning and Acting

To check the correctness and validity of the plan, it must be executed or *acted* and in order to be able to perform an **automated acting**, the planning must be integrated into the *acting* of the system. In a RAE, this could be achieved by refinement planning. Ghallab et al. [16] provide a good overview of refinement planning methods. Our system uses a classical AI planning approach, utilizing the marking graph as search space. The planner generates full plans that are repaired immediately after unexpected events occur.

5   Extension of the Modeling Technique

In order to reduce the modeling effort, we propose an integration of a goal-based task description into our modeling technique and an integration of a planner into our control system.

For this purpose, the behavioral model of the resources is extended. The extension of the behavior description allows to distinguish between event-based state changes and planned state changes. Both are expressed via transitions. Event-based state changes occur in an uncontrolled manner and cannot be influenced, such as a human entering the workspace of a robotic system. They are called **events**. Planned state changes, also called **actions**—such as the movement of the robot from position A to B—occur intentionally. In addition, the target state of the execution of an action must be specified via special edges (see red edges in Fig. 1), which, however, is not necessarily achieved. As an example, the state "Robot at target" represents the target state of the action "Command robot to target", whereas, for example, the event of a "Collision detected" leads to the state "Error" (not shown in the graphics).

The goal-oriented task description can then be automatically derived using the taxonomy. The taxonomy decouples the process description from the implementation by the resources. For this purpose, the linked actions of the transitions, which implement the skills used in the process description, are determined automatically and the entire target system state is derived based on their target states. Furthermore, the use of special edges enables the manual definition of desired system states. The following constraints can be modeled:

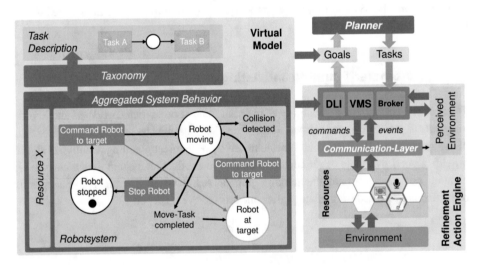

Fig. 1 Model to implement industrial applications connected to the standardized control system

– A resource must be in the defined state. (For this purpose, the marked places of the resource can be specified.)
– A resource must not be in the defined state. (For this purpose, the marked places of the resource, which must be avoided, can be specified.)

To perform the assembly task, the virtual model is transferred to the non-real-time standardized Control System (RAE) and IEC-61131-3-compatible PLC code is generated and transferred to the hardware used in production. This eliminates the need to manually translate the model into control code and improves the efficiency of the commissioning process. To simplify the PLC code, the generated code uses the RAE's broker to perform non-real-time critical tasks.

The manually and automatically defined goals are extracted and forward to a planner during runtime. The aggregated system behavior is utilized by the planner to determine the search space for plan generation by dynamically exploring and expanding the marking graph of underlying Petri net. Thus, a generic process is used to derive the currently active goals of the planner. Based on the current system state and active goal, plans are derived. Changing the system's state or achieving a goal, results in checking the planner's goals and plan. The plan will be updated or repaired if necessary. To optimize the plan generation different aspects like occurrence probabilities of events are learned while performing the assembly task. Different policies of the planner allow optimizing the plan in terms of different criteria.

6   Developed Control System

To perform the defined assembly task, the RAE uses five base elements (see Fig. 1). The defined logical sequence of the assembly task is dynamically interpreted by a dynamic logic interpreter (**DLI**), whereas the flow of data is implemented via a volatile memory system (**VMS**). The dynamic logic interpreter extracts the required assembly subtasks described in the virtual model and forwards them to a planner. The planner uses the virtual model to determine an appropriate solution. The plan is transferred to the dynamic logic interpreter, which ensures its execution. A **broker** delegates the tasks defined by the logic to the resources by utilizing the corresponding **communication layer** of the resource. The **resources** implement the skills as services. A detailed overview of components of the RAE is given in [12]. In the following sections, we present the developed planner and an approach to dynamically optimize the planner which enables dynamic adaptions of the assembly system to changes in its environment.

6.1   Planner

The Planner uses the aggregated behavior model of the system (see Fig. 2 left) in the form
of a search space generated using the Petri net. For this purpose, the marking graph of the
provided Petri net is dynamically expanded. A partial marking graph of the Petri net is shown
on the right of Fig. 2, where the aggregated system states are represented by nodes and the
transitions by edges.

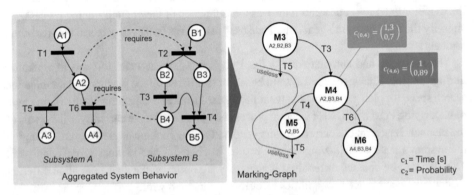

Fig. 2 Aggregated system behavior and its corresponding marking graph

The underlying algorithm, essentially a modified A^* algorithm, is used to determine
an optimal path to the desired goal state based on user-defined criteria (for e.g. shortest
time and probability). Before executing the algorithm, useless transitions are automatically
determined. These transitions are pruned to reduce the search space of the algorithm and
guide it towards the desired goal (see Fig. 2 red line). The psuedo code of the algorithm is
presented in Algorithm 1.

In each iteration, the algorithm needs to determine which of its paths to extend, where a
path $p_{(0,n)}$ represents the traversal from the start node 0 to the node n. The cost $\vec{c}_{(0,n)}$ of a
path $p_{(0,n)}$ and an estimate of the cost $f_h(n)$ required to extend the path to the desired goal
are used to implement a search strategy where paths are explored in increasing order of their
rating calculated by a user-defined rating function f_r. The values of the cost vector $\vec{c}_{(0,n)}$
represent the user-defined criteria. After selection, the corresponding node n of the selected
path $p_{(0,n)}$ is removed from the open list $nodesToTest$ and the costs to reach its successors
are determined.

Thus, every edge of the marking graph contains a cost vector $\vec{c}_{(i,k)}$ (see $\vec{c}_{(4,6)}$ in Fig. 2 as
example) which holds the additional required costs for traversal between marking M_i and
M_k. To determine the cost vector $\vec{c}_{(0,k)}$, every criterion provides a function $f_c := \mathbb{R}^n \mapsto \mathbb{R}^n$
to combine two cost values. These functions apply to the corresponding cost vectors as
follows:

$$\vec{c}_{0,k} := f_c(\vec{c}_{(0,i)}, \vec{c}_{(i,k)}) \tag{1}$$

Algorithm 1 Search Algorithm

1: $nodesToTest = [0]$ where $0 := startNode$
2: $\vec{c}_{(0,0)}$ = user-defined value
3: **repeat**
4:     $currentNode = n$ with min $f_r(f_c(\vec{c}_{(0,n)}, f_h(n)) \cdot \vec{w}$ where $n \in nodesToTest$
5:     **if** $currentNode$ is goal **then**
6:         Reconstruct the path from 0 to $currentNode$
7:     **end if**
8:     remove $currentNode$ from $nodesToTest$
9:     **for** each $edge, successor$ of $currentNode$ **do**
10:         **if** $edge$ not $useless$ **and** $successor$!= $currentNode$ **then**
11:             $\vec{c}_{temp} := f_c(\vec{c}_{(0,currentNode)} \cdot \vec{c}_{(currentNode,successor)})$
12:             **if** $f_r(\vec{c}_{temp}) \cdot \vec{w} < f_r(\vec{c}_{(0,successor)}) \cdot \vec{w}$ **or** $\vec{c}_{(0,successor)}$ is unknown **then**
13:                 $\vec{c}_{(0,successor)} = \vec{c}_{temp}$
14:                 put $successor$ in $nodesToTest$
15:                 parent($successor$) = $currentNode$
16:             **end if**
17:         **end if**
18:     **end for**
19: **until** length of $nodesToTest = 0$

If the required cost to reach the successor node are lower than the previously determind cost or the successor has not been discovered, the newly found path $p_{(0,k)}$ is added to the open list. The algorithm terminates when the open list is empty or when the selected path leads to the desired goal.

To rate a node and select the most promising one for graph expansion, every criterion is rated by rating function $f_r := \mathbb{R}^n \mapsto \mathbb{R}^n$. The aggregated rating is calculated by combing the individual ratings with its user-defined weighting factor w_i:

$$r_k := f_r(\vec{c}_{(0,k)}) \cdot \vec{w} \tag{2}$$

with $\sum_{i=1}^{n} w_i = 1$ and $\mathbf{0} \leq f_r(c_{(0,k)})) \leq \mathbf{1}$. This enables finding an optimized solution for mixed criteria.

In our approach, a delete relaxation of the Petri net (see [17] for more details) is used as heuristic f_h to calculate the cost estimate required to extend a path to the desired goal. The relaxation assumes that transitions only require tokens in their preset places instead of consuming them when fired. A place in the relaxed Petri net, once marked, will never be unmarked. This implies that any reachable marking is reachable by a short sequence of transitions.

6.2   Optimizing the Generated Plans

To optimize the generated plans, the system can learn the cost vectors $\vec{c}_{(i,k)}$ to get more accurate information like an occurrence probability of an event. For this purpose, the events of the system and trackable cost values are recorded. The difference of the tracked values between two markings is used to update the cost vectors of the marking graph.

In our case, the system stores the differences of the tracked costs in a limited ring buffer. During updating the cost vectors of the marking graph, we generate the average value of the ring buffer. Thus, the size of the buffer defines the agility of the system. The system adaptability can be varied by varying the size of the buffer.

7   Conclusion and Outlook

To reduce the implementation effort of flexible assembly systems, a graphical modeling method in combination with a generic modular control system has been extended with aspects of planning. The graphical modeling technique and goal based description of the assembly task allow considering the know-how of the process owners in the early stages of the development process and reduces iterative optimizations during commissioning.

The actual process sequence of the assembly task is determined at runtime by using the behavioral model of the system and a generic description of the assembly task which further reduces the implementation effort. Additionally, we present an approach to automatically optimize the generation of process sequences. This approach adapts the system to changes in its environment during the runtime.

After successful validation of the behavioral model, generated code based on the model can be transferred to common hardware used by the industry. Code generation considerably simplifies and shortens the commissioning of the assembly system.

We are currently developing a physical demonstrator to validate our approach. In future research, aspects of automatic assembly operation optimization will be considered.

Acknowledgements This article is written within the project Mittelstand 4.0-Kompetenzzentrum Saarbrücken, as part of the Support Initiative "Mittelstand-Digital" of the BMWi. The nationally funded competence centers provide information to small and medium-sized companies about the opportunities and challenges of digitization. The authors are responsible for the content of the publication.

References

1. Müller, R., Scholer, M., Karkowski, M.: Generic automation task description for flexible assembly systems. Proc. CIRP **81**, 730–735 (2019). https://doi.org/10.1016/j.procir.2019.03.185
2. Michalos, G., Markis, S., Papakotas, N., Mourtzis, D., Chryssolouris, G.: Automotive assembly technologies review: challenges and outlook for a flexible and adaptive approach. CIRP J. Manuf. Sci. Technol. **2**(2), 81–91

3. Schlick, J., Stephan, P., Loskyll, M., Lappe, D.: Industrie 4.0 in der praktischen Anwendung. In: M. Hompel, B. Vogel-Heuser, T. Bauernhansl (eds.) Handbuch Industrie 4.0, Springer Reference Technik, 2., erweiterte und bearbeitete auflage edn. (2016)
4. Selig, A.: Informationsmodell Zur Funktionalen Typisierung von Automatisierungsgeräte. ISW/IPA-Forschung Und -Praxis, vol. 180. Jost-Jetter (2011)
5. Zoitl, A.: AutoPnP - Plug& Play für Automatisierungssysteme: Schlussbericht - Konsortial-bericht (2014)
6. Antzoulatos, N., Castro, E., Scrimieri, D., Ratchev, S.: A multi-agent architecture for plug and produce on an industrial assembly platform. Prod. Eng. **8**(6), 773–781 (2014). https://doi.org/10.1007/s11740-014-0571-x
7. Danny, P., Ferreira, P., Lohse, N., Guedes, M.: An automation ML model for plug-and-produce assembly systems. In: 2017 IEEE 15th International Conference on Industrial Informatics (INDIN): University of Applied Science Emden/Leer, Emden, Germany, 24-26 2017 : Proceedings, pp. 849–854. IEEE (2017). https://doi.org/10.1109/INDIN.2017.8104883
8. Anandan, P., Ferreira, P., Dorofeev, K., Lohse, N.: An event-based automationml model for the process execution of "plug-and-produce" assembly systems (2018). https://doi.org/10.1109/INDIN.2018.8471955
9. Bohren, J., Cousins, S.: The smach high-level executive. IEEE Robot. Autom. Mag. **17**, 18–20 (2011). https://doi.org/10.1109/MRA.2010.938836
10. Rovida, F., Crosby, M., Holz, D., Polydoros, A.S., Großmann, B., Petrick, R., Krüger, V.: SkiROS—a skill-based robot control platform on top of ROS. In: ROBOT OPERATING SYSTEM: The Complete Reference, pp. 121–160. Springer (2017)
11. Wooldridge, M.J.: An Introduction to Multiagent Systems, 2nd edn. Wiley (2009)
12. Müller, R., Scholer, M., Karkowski, M.: Increasing the flexibility of customized assembly systems with a modular control system. In: 2018 Fifth International Conference on Internet of Things: Systems, Management and Security, pp. 46–53 (2018). https://doi.org/10.1109/IoTSMS.2018.8554528
13. Illmer, B., Karkowski, M., Vielhaber, M.: Petri net controlled virtual commissioning – a virtual design-loop approach. In: Enhancing Design Through the 4th Industrial Revolution Thinking, p. 6. Elsevier B.V. (2020)
14. Reisig, W.: Understanding Petri Nets. Springer, Berlin (2013). https://doi.org/10.1007/978-3-642-33278-4
15. Aalst, W.: Discovering Petri Nets: A Personal Journey, pp. 3–9 (2019). https://doi.org/10.1007/978-3-319-96154-5_1
16. Ghallab, M., Nau, D., Traverso, P.: Automated Planning and Acting. Cambridge University Press (2016)
17. Bonet, B., Haslum, P., Hickmott, S., Thiebaux, S.: Directed unfolding of petri nets. T. Petri Nets Other Model. Concurr. **1**, 172–198 (2008). https://doi.org/10.1007/978-3-540-89287-8_11

Transmitter Positioning of Distributed Large-Scale Metrology Within Line-Less Mobile Assembly Systems

Christoph Nicksch, Alexander K. Hüttner and Robert H. Schmitt

Abstract

In Line-less Mobile Assembly Systems (LMAS) the mobilization of assembly resources and products enables rapid physical system reconfigurations to increase flexibility and adaptability. The clean-floor approach discards fixed anchor points, so that assembly resources such as mobile robots and automated guided vehicles transporting products can adapt to new product requirements and form new assembly processes without specific layout restrictions. An associated challenge is spatial referencing between mobile resources and product tolerances. Due to the missing fixed points, there is a need for more positioning data to locate and navigate assembly resources. Distributed large-scale metrology systems offer the capability to cover a wide shop floor area and obtain positioning data from several resources simultaneously with uncertainties in the submillimeter range. The positioning of transmitter units of these systems becomes a demanding task taking visibility during dynamic processes and configuration-dependent measurement uncertainty into account. This paper presents a novel approach to optimize the position configuration of distributed large-scale metrology systems by minimizing the measurement uncertainty for dynamic assembly processes. For this purpose, a particle-swarm-optimization algorithm has been implemented. The results show that the algorithm is capable of determining suitable transmitter positions by finding global optima in the assembly station search space verified by applying brute-force method in simulation.

Keywords

Assembly · Large-scale metrology · Transmitter positioning

C. Nicksch (✉) · A.K. Hüttner · R.H. Schmitt
Chair of Production Metrology and Quality Management, RWTH Aachen University, 52074 Aachen, Germany
e-mail: c.nicksch@wzl.rwth-aachen.de

© The Author(s) 2022
T. Schüppstuhl et al. (eds.), *Annals of Scientific Society for Assembly, Handling and Industrial Robotics 2021*,
https://doi.org/10.1007/978-3-030-74032-0_2

1 Introduction

Fluctuating demands as well as increasing number of variants are today's main challenges for manufacturing companies in high-wage countries. To still remain competitive, they face the challenge of producing economically [1]. To transfer market dynamics into long-term competitive advantages, production must be flexible. Especially in the assembly of large-scale products, this is a challenge, since large-scale products are defined by large dimensions at narrow tolerances in the submillimeter range and require disproportionate effort for their manufacturing [2]. One example is the assembly of aircraft, which accounts for more than 50% of the workload of the entire aircraft manufacturing process [3].

The concept of Line-less Mobile Assembly Systems (LMAS) offers a solution for these challenges and can address new product variants by rapidly reconfiguring assembly stations by integrating mobile resources such as mobilized assembly robots and automated guided vehicles (AGV) for the transport of products [4]. To enable the implementation of LMAS, the mobilization of assembly resources significantly increases the necessary amount of positioning reference data needed on the shop floor. Thus, the need for a metrology infrastructure on the shop floor arises 5]. Distributed large-scale metrology (LSM) consists of multiple transmitter and receiver units and can be adapted to certain assembly processes covering the whole assembly station and providing positioning data with low measurement uncertainties [2]. Transmitter positioning influences the measurement uncertainty significantly and thus limits the maximum measurable tolerance to be checked [6]. As a result, optimal transmitter positions must be determined considering assembly system boundaries and measurement uncertainty.

This paper aims to develop a novel approach to optimize transmitter positions of distributed LSM-systems within LMAS. The approach focuses on an Indoor-GPS-System (iGPS) as an example for distributed LSM-systems. In the following, a brief state-of-the-art for transmitter positioning algorithms and derived algorithm requirements are presented. Subsequently, the new algorithm is described and verification based on a simulated assembly use case is shown.

2 Distributed Large-Scale Metrology Within Assembly Systems

2.1 Approaches for Transmitter Positioning

Franceschini et al. [7] defines *transmitter positioning* as a sensor configuration, characterizing each sensor by its spatial coordinates and orientation angles. The problem complexity is strongly related to the geometry of the working environment.

In recent years, there have been multiple publications on the configuration of distributed metrology. Quinders [6] developed a heuristic approach to determine iGPS transmitter positions in a 3D space. To reduce complexity, an expert has to manually select potential transmitter positions before an algorithm determines the final configuration in a discretized space. Wang et al. [8] proposed an algorithm for the positioning measurement of laser tracker stations for the error identification of heavy-duty machine tools. He simplifies working ranges using spheres and does not consider dynamics and obstacles (which could collide with the line-of-sight of the laser tracker). Wang and Quinders both use the measurement uncertainty (MU) as the target variable to be minimized. Ray and Mahajan [9] used a genetic algorithm to determine the positions of ultrasonic sensors to track robot positions. Still, he did not consider surrounding obstacles and only focuses on a feasible triangulation during the measurement and not on the MU. The approaches of Galetto and Pralio [10] and Laguna et al. [11] also concentrated on the positioning of ultrasonic sensors. Both approaches considered only a 2D solution space and simplify the solution space by using discretization.

Franceschini et al. [7] summarized that most works have addressed the 3D sensor positioning problem by considering reduced-size networks, reducing the problem to a 2D formulation, or referring to simple design goals. Table 1 compares the given approaches. Most of the approaches simplify the solution space and do not take spatial boundaries into account. Still, this is an important factor for metrology systems within LMAS because distributed LSM-systems require a line-of-sight (LOS) between transmitters and receivers. The MU should be considered in particular because it correlates with permitted minimum tolerance of the assembly process.

This work focuses on these deficits and will apply the developed algorithm to a Nikon iGPS measurement system. The iGPS is a suitable metrology system for LMAS scenarios due to its scalability and high accuracy [2, 12].

Table 1 Evaluation of approaches for transmitter positioning

Reference	Automated Configuration	Unrestricted Solution Space	Spatial Boundaries	Minimizing MU
Quinders	○	○	●	●
Wang	●	●	○	●
Ray	●	○	○	○
Galetto	●	○	○	○
Laguna	●	○	○	○

● criterion fulfilled   ○ criterion not fulfilled

2.2 Requirements for iGPS Transmitter Positioning Within LMAS

The requirements (R) for the iGPS transmitter positioning problem can be categorized as follows:

iGPS: R1: A technically feasible solution depends on the iGPS working range, which is defined by: a permitted distance between a transmitter and a receiver (*2 m–30 m*), a minimum distance between transmitters (*2 m*) and a permitted elevation angle between a transmitter and a receiver (± *30°*).

R2: To calculate a receiver position, a minimum number of three LOS is required. Figure 1 shows an iGPS configuration with three transmitters and one receiver defined by the elevation angle θ, the azimuth angle φ and the transmitter–receiver distance d (corresponds to LOS between transmitters and receivers) [6].

LMAS: R3: The algorithm has to consider all spatial restrictions based on a given LMAS assembly station and its process. This comprises all physical assembly resources involved in the process (e.g. AGVs, robots, products or fixtures) as well as its logical and time-related dependencies.

R4: To enable a simulation close to the real assembly process, complex data formats such as CAD-models need to be considered in terms of resolution and accuracy.

Deficits of the state-of-the-art: R5: Based on the deficits of state-of-the-art approaches, the algorithm should automatically determine a configuration for a distributed LSM system.

R6: The resulting MU is used as an optimization goal and without limiting the solution space and reducing the complexity of the real assembly and measurement environment.

To treat the MU as a target variable of the optimization, a MU-model is required. Quinders [6] developed an empirical MU-model for the iGPS which will be used within this work and explained in detail in Sect. 3.2.

3 Particle-Swarm-Optimization for Transmitter Positioning

3.1 Particle-Swarm-Optimization

The objective to find a transmitter configuration with the lowest MU for static and dynamic assembly scenarios with several obstacles (assembly resources or products) equals to a restricted and non-steady global optimization problem. For this kind of problems, particle-swarm-optimization has been applied successfully on various benchmark functions and is also used in this work [8, 13, 14].

A transmitter T is defined by its Cartesian coordinates x, y and z (see Eq. 1). The particle swarm consists of n_P individual particles where each particle s represents a possible solution. One particle comprises n_T transmitters (see Eq. 2). The particle is updated in every iteration k by the term v for k_{max} iterations (see Eq. 3). The update term v

in Eq. 4 is based on three scaling parameters α_1, α_2 and β. Two further variables generate the swarm intelligence: *pBest* saves the best solution of particle s (cognitive term) and *gBest* saves the best solution of all particles (social term). The term "best solution" equals the particle with lowest MU. Normally distributed random numbers $U(0, 1)$ invoke a stochastic exploration (exploring the whole search space, i.e. assembly stations) and exploitation (converging to local optima). The balance of those properties can be improved by adding a second social parameter α_3 and a variable *lBest* which provides information from neighboring particles [13].

$$T = \begin{bmatrix} x & y & z \end{bmatrix} \tag{1}$$

$$s_I = \begin{bmatrix} T_1 & \dots & T_{n_T} \end{bmatrix} \text{with} i \in [1, n_P] \tag{2}$$

$$s_i^{k+1} = s_i^k + v_i^{k+1} \text{with} k \in [1, k_{max}] \tag{3}$$

$$v_i^{k+1} = \beta \cdot v_i^K + \alpha_1 \cdot U(0, 1) \cdot \left(pBest_i^k - s_i^k \right) + \alpha_2 \cdot U(0, 1) \\ \cdot \left(gBest^k - s_i^k \right) + \alpha_3 \cdot U(0, 1) \cdot \left(lBest_i^k - s_i^k \right) \tag{4}$$

3.2 Fitness Function and Penalty Terms

To determine these best solutions, a fitness function F evaluates the particles at each iteration and for all time steps considering dynamic use cases. The fitness function in Eq. 5 contains a target function f and m penalty terms to consider technical constraints (see R1 and R3 in Sect. 2.2). A penalty term comprises a penalty parameter r and a measure of violation Ψ of the constraint. r is set to the amount of a theoretical loss of one LOS at the current solution s since a violation of technical constraints (e.g. measurements out of the working range of the iGPS or within an object) leads to a LOS collision in the worst case.

$$F\left(s^k\right) = f\left(s^k\right) + \sum_{i=1}^{m} r(s^k) \cdot \Psi_i^K(s^k) \tag{5}$$

If one transmitter is positioned in or above an assembly resource or handling space (prohibited space) at a certain time step, it is defined as a collision of the LOS for the corresponding transmitter–receiver pair. If collisions occur, Ψ is calculated as the number of collisions multiplied by the simulation time steps since a violation at one time step defines the considered configuration unacceptable for the whole process. If a transmitter is positioned too close or too far away from the target, too close to another transmitter or exceeds the maximum elevation angle, the Euclidean distance of the exceedances is normalized to the interval from 0 to 1. The respective Ψ are calculated as the sum of these normalized values over all transmitters. If the minimum number of LOS of the remaining valid transmitter positions is not reached, Ψ is calculated as the difference multiplied by the time steps. If a particle embodies a solution that exceeds the limit of the search space in one dimension, the particle's solution in this dimension is set to its respective limit (decoder approach). If all constraints were satisfied, Ψ for *gBest* would be null.

Quinders [6] determined empirically the MU of the iGPS depending on the positions of the transmitters (see Eq. 6): u_{MS} is the resulting MU of the iGPS, u_{RE} is the standard uncertainty of the resolution, u_{EV} is the standard uncertainty of the equipment variation, u_{Ref} is the standard uncertainty of the reference, u_{Bi} is the standard uncertainty of the systematic error (bias), u_{Cal} is the standard uncertainty of calibration, u_{Conf} is the standard uncertainty of the configuration and u_T is the standard uncertainty of the temperature.

$$u_{MS}^2 = \max\{u_{RE}^2, u_{EV}^2\} + u_{ref}^2 + u_{Bi}^2 + u_{Cal}^2 + u_{Conf}^2 + u_T^2 \tag{6}$$

Quinders [6] proved that only u_{Conf} and u_{EV} are affected by transmitter positions. Due to this fact, all remaining constant terms are neglected in Eq. 7 for the optimization problem. According to his experiments, only the average elevation angle $\bar{\theta}(s)$, the maximum azimuth angle $\varphi_{max}(s)$ and the number of LOS $n_{los}(s)$ have impact on the configuration-dependent MU f.

$$\begin{aligned} f(s) &= u_{Conf}^2(s) + u_{EV}^2(s) \\ &= (-11.7\ n_{los}(s) + 87.2)^2 + \\ &\quad (37.9 + 0.1\ \bar{\theta}(s) + 0.03\varphi_{max}(s) - 4,4\ n_{los}(s))^2 \end{aligned} \tag{7}$$

Based on the solution matrix in Eq. 2 and the fitness function in Eq. 5, the optimization problem for a dynamic assembly with a number of T_{max} discetized time steps T_s is defined by:

$$\min_{s} \frac{1}{T_{max}} \sum_{j=0}^{T_{max}} F_j(s) \tag{8}$$

3.3 Assembly Simulation for Particle Evaluation and Optimization Procedure

To calculate the evaluation function F, it is necessary to determine n_{los}, $\bar{\theta}$ and φ_{max} for all time steps of the assembly process. Therefore, the process is simulated based on CAD-models of the LMAS assembly station. For a first proof-of-concept, assembly resources and products are simplified as convex bodies (boxes or cylinders) to decrease the computation time. Positions and rotations of these bodies are updated for each time step using homogeneous transformation matrices. Due to this simplification, a LOS interruption between a receiver-transmitter pair can be detected by calculating an intersection between the connection line and considered physical objects within the assembly station (according to Chazelle and Dobkin [15]).

The integration of the simulation into the optimization problem consists of the following steps: First, the particles are initialized in a random circular distributed manner at middle height in the search space of the assembly station. Afterwards, the fitness function F evaluates all particles for T_{max} time steps using the assembly simulation. After determining the terms *pBest*, *gBest* and *lBest*, all particles are updated for the next iteration according to Eq. 3. This cycle repeats until the maximum number of iterations k_{max} is reached as a stop criterion for the algorithm. After the last iteration, *gBest* represents the final solution and, accordingly, the transmitter positions with the lowest MU within the particle swarm.

4   Verification on Assembly Use Case

4.1   Description of Simulated Use Case

The proposed approach is verified by a simulated, theoretical LMAS use case. To match a LMAS scenario, the assembly process contains all relevant types of LMAS resources. The use case is depicted in Fig. 2. The assembly station consists of stationary and mobile components. Three stationary bodies represent fixtures or tooling components. A plane on the floor represents a handling space, reserved for robot movements or logistics, on which it is prohibited to place transmitters. On the left side, an AGV moves along a straight line on the floor. In the center, a robot rotates along two axes simulating an assembly task on a product which is transported by another moving AGV along a circular path on the floor. All mobile resources move simultaneously during the process. An iGPS receiver is placed on the robot tool-center-point (see red dot in Fig. 2). The use case includes a predefined fix set of five transmitters, since more than five LOS do not have a significant impact on the MU [6]. The intention of the use case is to illustrate complex dynamics for a distributed LSM-system within LMAS. The complexity of the system can be further increased by the number and size of additional stationary and mobile resources as well as the simulation time.

4.2   Analysis of Simulation Results

To verify if the algorithm is able to detect a global minimum in terms of MU, the solution of the algorithm is compared to a brute-force solution. According to the brute-force method, the 3D space of the assembly station is discretized using a point-to-point distance

Fig. 1 iGPS configuration
with three transmitters and one
receiver according to Quinders
[6]

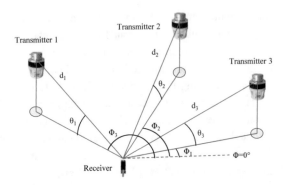

of 2 m. Each point represents a possible position for a transmitter. Based on this solution space, the MU for all possible combination for transmitter configurations are determined by Eq. 7. The transmitter configuration with the lowest MU corresponds to the global optimum as a reference value for the verification of the proposed algorithm.

Table 2 shows the algorithm parameters used within the verification scenario. The parameters α_1, α_2, α_3 and β decrease by increasing the number of iterations to invoke the exploitation behavior over the process time to find a local minimum through smaller steps (according to Xin et al. [14]). The simulation contains *five transmitters* using a sample time of *0.1* s. The simulated process takes *7 s*.

33 simulation runs were conducted to investigate the performance of the algorithm. Figure 3 shows the best solution after the initialization (red dots), the final best solution after 50 iterations (dark blue dots) out of the 33 simulation runs, the remaining 32 final solutions (light blue dots) and the brute-force solution (yellow dots).

Clearly, the solutions (including the brute-force solution) are arranged in a semicircle around the target trajectory. This can be explained by the fact that the other half is mainly shielded by the robot arm. The three green and two red lines represent the current LOS state (green = no collision; red = collision).

Figure 4 shows the resulting MU including the penalty terms over all iterations. After the initialization, the MU of the particle-swarm-optimization solutions lies between 0.3 mm and 0.5 mm. The amount of penalty terms for each constraint is depicted on the right side. After 16 iterations all constraints referring to the working range of the iGPS are satisfied. After 19 iterations there is no collision between a transmitter and any physical

Table 2 Parameters used within verification

n_p	k_{max}	α_1	α_2	α_3	β	r
40	50	$2 - \dfrac{1,5}{k_{max}} * k$	$1,5 - \dfrac{1}{k_{max}} * k$	$2 - \dfrac{1}{k_{max}} * k$	$0,9 - \dfrac{0,5}{k_{max}} * k$	$\dfrac{\partial f(s)}{\partial n_{LOS}}$

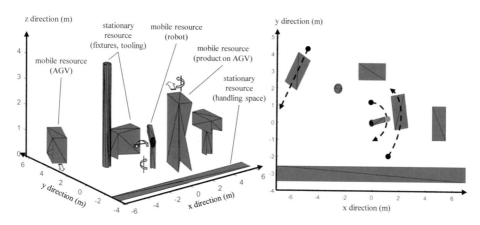

Fig. 2 Isometric view (left) and top view (right) of the LMAS use case

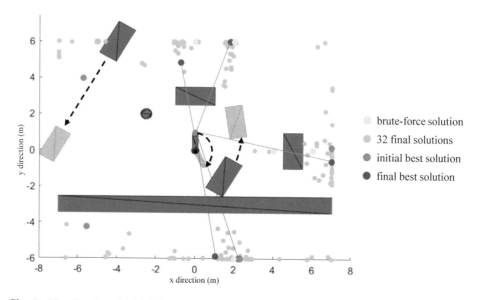

Fig. 3 Visualization of initial, final and brute-force solution with LOS (top view)

object within the assembly station for all 33 simulation runs. The most difficult constraint seems to be the LOS constraint (minimum number of three LOS). After 40 iterations all test runs provide a transmitter configuration with not less than three LOS over the whole process time.

The difference between the brute-force solution (0.1036 mm) and the median of all 33 particle-swarm-optimization solutions (0.1044 mm) is 0.0008 mm. The minimum and maximum of the particle-swarm-optimization solutions are 0.1034 mm and 0.1057 mm resulting in a span width of 0.0023 mm. It indicates that the algorithm finds even better

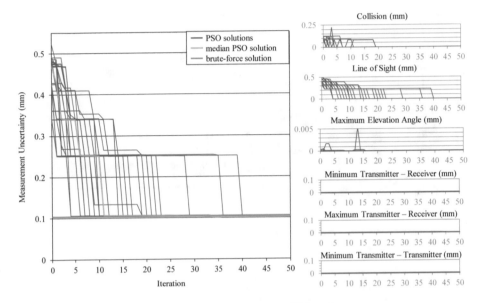

Fig. 4 Development of total costs and penalty terms over 50 Iterations

solutions than the brute-force method. This can be explained by the necessary discretization of 2 m used for brute-force method.

Referring to the requirements from 2.2, it has been demonstrated that that the proposed algorithm is capable of satisfying all iGPS constraints and thereby fulfils R1 and R2. R3 is satisfied by modeling all relevant LMAS resources (AGVs, robots, products etc.) with their time-dependent positions and orientations during the assembly process. These objects are modeled based on standard CAD-format (e.g. STEP-file) as a composition of convex bodies (R4 is partly fulfilled). The algorithm determines automatically all transmitter positions using the MU as a target variable fulfilling R5 and R6. In contrast to the evaluated research works from 2.1, the solution space (limitation by discretization) and the complexity (2D view) of the possible scenarios is not limited and can be increased as needed and therefore fulfilling R6.

5 Conclusion and Outlook

An algorithm for automated transmitter positioning of distributed large-scale metrology within Line-less Mobile Assembly Systems was hitherto unknown to the authors and has been presented in this paper. An iGPS system was used as an example for distributed large-scale metrology. The first implementation of the algorithm, which is based on a particle-swarm-optimization and minimizes the measurement uncertainty shows a good applicability even for a complex and dynamical assembly use case. Simulation results show that the developed algorithm satisfies system constraints (e.g. spatial constraints through assembly resources and working ranges of the metrology system). A verification by applying a brute-force method proves that the proposed algorithm converges to the global optimum with a deviation of 0.8 μm to the brute-force solution considering a discretized 3D space with a point-to-point distance of 2 m. This approach can also be transferred to other distributed large-scale metrology systems (e.g. laser trackers or laser scanners) by replacing the measurement uncertainty model of the iGPS.

To address the simplification in the simulation by using convex bodies, current research investigates how dynamic octree-collision-detection can be applied to the simulation for line-of-sight analysis in order to enable the import of non-convex bodies with higher resolution. The simulation will be validated through experiments on robot-based assembly stations comparing simulation and experimental results from the real world. Further work includes the enlargement of the iGPS measurement uncertainty model to consider the impact varying distances between transmitters and receivers.

Acknowledgements Funded by the German Federal Ministry for Economic Affairs and Energy (BMWi) within the research project iVeSPA.

References

1. Beheshti, Z., Shamsuddin, S.M., Sulaiman, S.: Fusion global-local-topology particle swarm optimization for global optimization problems. Mathematical Problems in Engineering (2014)
2. Chazelle, B., Dobkin, D.P.: Intersection of convex objects in two and three dimensions. J. Assoc. Comput. Mach. **34**, 1–27 (1987)
3. Franceschini, F., Galetto, M., Maisano, D., Mastrogiacomo, L., Pralio, B.: Distributed Large-Scale Dimensional Metrology. New Insights, p. 85 (2011)
4. Galetto, M., Pralio, B.: Optimal sensor positioning for large scale metrology applications. Precis. Eng. **34**(3), 563–577 (2010)
5. Hüttemann, G., Buckhorst, A.F., Schmitt, R.H.: Modelling and assessing line-less mobile assembly systems. Procedia CIRP **81**, 724–729 (2019)
6. Laguna, M., Roa, J.O., Jiménez, A.R., Seco, F.: Diversified local search for the optimal layout of beacons in an indoor positioning system. IIE Trans. **41**(3), 247–259 (2009)
7. Mei, Z., Maropoulos, P.G.: Review of the application of flexible, measurement-assisted assembly technology in aircraft manufacturing. Proc. Inst. Mech. Eng. Part B: J. Eng. Manuf. **228**, 1185 (2014)
8. Montavon, B., Peterek, M., Schmitt, R.: Communication architecture for multiple distributed large volume metrology systems. In: International Symposium on Systems Engineering, pp. 1–2 (2017)
9. Muelaner, J.E., Martin, O.C., Maropoulos, P.G.: Achieving low cost and high quality aero structure assembly through integrated digital metrology systems. Procedia CIRP **7**, 688–693 (2013)
10. Muelaner, J.E., Maropoulos, P.G.: Large volume metrology technologies for the light controlled factory. Procedia CIRP **25**, 169–176 (2014).
11. Quinders, S.: Virtueller Prototyp zur Optimierung und Absicherung der Konfiguration messtechnisch gestützter und roboterbasierter Montagesysteme. Dissertation, RWTH Aachen University 125–155 (2017)

12. Ray, P.K., Mahajan, A.: A genetic algorithm-based approach to calculate the optimal configuration of ultrasonic sensors in a 3D position estimation system. Robot. Auton. Syst. **41** (4), 165–177 (2002)
13. Schmitt, R.H., Peterek, M., Morse, E., Knapp W., Galetto, M.: Advances in large-scale metrology – review and future trends. CIRP Ann.-Manuf. Technol. **65**(2) 643–665 (2016)
14. Wang, Z., Forbes, A., Maropoulos, P.G. (eds.): Laser tracker position optimization (2014)
15. Xin, J., Chen, G., Hai, Y.: A particle swarm optimizer with multi-stage linearly-decreasing inertia weight. In: International Joint Conference on Computational Sciences and Optimization, p. 505 (2009)

Optimized High Precision Stacking of Fuel Cell Components for Medium to Large Production Volumes

Jens Schäfer and Jürgen Fleischer

Abstract

PEM fuel cells are well established in a number of niche markets. However, due to low production volume and manufacturer-specific designs, the assembly has been carried out manually most of the time. With new fields of application being exploited there is a rising demand for production systems. As there is no standardized design or material, production systems are often custom-made, thus being inflexible to design changes or different products. In combination with a volatile demand the need for flexible and scalable systems arises. In this paper special attention is paid onto pick and place operations of the catalyst coated membrane (CCM). Design criteria of a vacuum gripper are derived from the material properties. To meet the further requirements for a high position accuracy in an automated assembly the impact of process parameters onto the repeatability is investigated to identify optimization trends. The requirements and investigations lead to a conceptual assembly system that is able to cover several steps in fuel cell production.

Keywords

Fuel cell · Stacking · Manufacturing

J. Schäfer (✉) · J. Fleischer
Karlsruhe Institute of Technology, 76131 Karlsruhe, BW, Germany
e-mail: jens.schaefer@kit.edu

J. Fleischer
e-mail: juergen.fleischer@kit.edu

27

1   Introduction and Motivation

Proton exchange membrane fuel cells (PEM FC) can be characterized as a highly volatile market. One of the main challenges for a broad application of fuel cells in mobility is the available infrastructure. Initial fields of application will be developed when the necessary hydrogen infrastructure can be built up, which will be followed by a more widespread use. Consequently, the demand for fuel cells and production capacities will only gradually increase.

Manufacturing systems with fixed cycle time are most efficient at the designated production quantity. A demand higher than the capacity leads to opportunity costs for missed profit. A production volume lower than the planed yield results in low utilization rates and tied capital. In contrast, an agile manufacturing system is characterized by a piece number scalability, ensuring less tied capital and the avoidance of opportunity costs by scaling it up. By the complementary and substitutive use of similar systems the agility costs can be reduced, thus ensuring scalability by piece. For the application scenario of fuel cell production this means that a system must be able to cover a wide range of production steps with no loss of quality. The clocking process for fuel cell stack assembly is the individual part stacking, which shall be further investigated within this paper. The herein presented system is able to stack different fuel cell components with different designs. It is therefore necessary to determine the product and production interaction, with the focus of this paper being on single membrane handling and optimization of the parameter setting for this application. This includes the gripper design as well as the parameter settings.

2   Fuel Cell Stack Assembly

2.1   State of the Art

The base material for hydrogen fuel cells are proton exchange membranes (PEM), which are coated with catalysts to form a catalyst coated membrane (CCM). The CCM allows proton conductivity and gas tightness. To distribute the reactant gases to the membrane and drag electrons from the membrane a so-called gas diffusion layer (GDL) is at both sides of the CCM. To further improve the system durability the GDL is coated with a micro-porous layer on the sides facing towards the membrane. So-called bipolar plates (BPP) ensure the gas supply. The design to assemble the components as well as the sealing concept varies between manufacturers. CCM, GDL and a gasket can be assembled to a so-called membrane electrode assembly (MEA), for example by heat pressing. The seal can be applied onto the bipolar plate (seal-on BPP) or on the GDL (seal-on GDL) [1]. Figure 1 shows the important steps in fuel cell stack production with a seal-on GDL approach as also described in Stahl [1], Porstmann et al. [2].

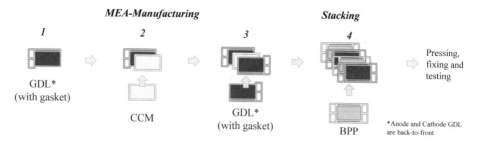

Fig. 1 Assembly of fuel cell stacks

The overall sealing concept has a huge impact on the production system, as the number of parts varies and the components can be more fragile or limp during handling. The above described single cell is then stacked with up to 400 individual cells. The assembly of fuel cell stacks has only come into focus in recent years and thus is yet not discussed in literature deeply. In Bobka et al. [3, 4] a fuel cell stacking system is presented which features various cameras mounted overhead for position measurement before and after handling. For the assembly itself preassembled MEA's have been used, the investigations are focused on an improvement of stacking accuracy. The gripper has been mentioned as a major source of inaccuracy but has not been investigated further. Another system is presented in Williams et al. [5], Laskowski and Derby [6]. Robots are used for the assembly and guidance pins are used for the alignment of the individual components. However, this system is limited in stack size, as the guidance pins are rather short and a stacking accuracy of only 0.51 mm was stated. In this concept preassembled MEA's have been used. There have been several research projects investigating fuel cell stack assembly with seal-on GDL such as MontaBS [7, 8]. and Fit-4-AMandA [2], however there is only little information on the assembly system itself and its core components. Numerous patents, such as Dreier [9], Munthe [10], HoKyun and Yoon [11] suggest that more attention is being paid to fuel cell stack assembly in an industrial environment. However, no information is available on the implementation of the registered property right. All the assembly systems mentioned above feature vacuum or low pressure grippers for handling the components in pick and place operations. Especially the gripping of the individual CCM and the porous GDL, that are not preassembled, have not been investigated yet, although being relevant for a seal-on GDL approach as mentioned in Porstmann et al. [2].

2.2 Concept for a Scalable and Flexible Production System

A comprehensive overview of requirements for an agile manufacturing system is given in Ramsauer and Rabitsch [12] and can be characterized by three key elements: proactive preparation, fast reaction time and optimized efficiency. On the machine level this requires

Fig. 2 Concept for a scalable
production system with
identical production machines

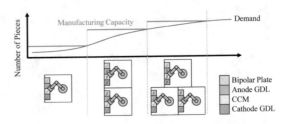

an understanding of the material-production interaction to cover future material developments. With pre-developed production modules for a successive expansion a wide
range of steps in the process chain can be covered. In Porstmann et al. [2] this is resolved
by a high degree of specialisation, so that each task is assigned to a single handling
system. The approach presented in Bobka et al. [4] is capable of doing several operations,
but can only grasp pre-assembled MEA. A rough concept for an agile manufacturing
system that is also capable of assembling MEA's is shown in Fig. 2.

Thus, there are two important optimization goals for agile systems:

(a) In the very first phase the system must be able to fulfil all handlings tasks.
(b) The system must easily be adapted to new products, which requires preliminary
     investigations and model building, which are conducted in this paper.

2.3   Materials in Fuel Cells

Table 1 shows the typical properties of materials for fuel cell components. Components
like the bipolar plates or the GDL show a great variation in their properties due to the lack
of standards. Furthermore, the components behave very different when being handled. The
bipolar plate is gas tight and rigid, whereas GDL and CCM are limp and prone to being
damaged during handling. Not only does this usually require a different set of grippers for
each component, it also makes it inevitable to adapt it to the desired specifications.

Table 1 Material properties relevant for stacking

	CCM	GDL	BPP
Material thickness	5–5 µm [13]	169–423 µm [14]	75 µm [15]
Structural thickness	–	–	0.5 mm [15]
Porosity	0%	39.2–89% [14]	0%

Table 2 Fuel cell membrane properties [16, 17]

Material	Thickness (μm)	E (MPa)	v	Basis weight (g/cm^2)
NC700	15	421	0.4	29.5
NR211	25.4	281	0.4	50

2.4 Constraints

For the assembly all the components need to be handled. Especially the membrane poses a great challenge for grippers, as it is not only very thin, but also has low mechanical properties. Common base materials are Nafion® NR211 and Nafion® NC700. Their respective properties are given in Table 2 [16, 17], the poisson value for Nafion® based materials is 0.4 [18]

Whilst gripping the pressure difference causes a bending of the membrane. The bending of a thin membrane can be described by Eq. (1) [13] with the materials young's modulus E and the poisson value v as well as the radius R of the suction openings of the gripper, the differential pressure Δp and the material thickness d.

$$w_0 = \frac{3}{16} * \frac{R^4}{d^3} * \frac{1 - v^2}{E} * \Delta p \tag{1}$$

Fuel cell membranes subjected to mechanical stress show no distinct transition between elastic and plastic behaviour. In general, plastic deformation should be avoided. As the membranes tend to become thinner for higher efficiency fuel cells this can be challenging for defect-free handling. Therefore, FEM-simulations were carried out with the membrane being subjected to a pressure difference at a circular opening, as it can be found in vacuum or low-pressure grippers. The membrane was modelled in Abaqus as a shell encastrated at the outer circumference with five integration points along the thickness and Simpson integration rule as well as Abaqus' standard static solver applied. The element size was set to 0.01 mm. The model as well as maximum Mises stress under variation of pressure difference and suction opening are shown in Fig. 3.

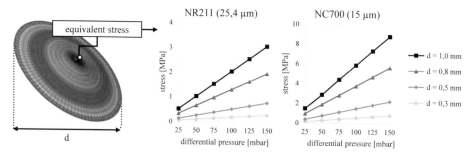

Fig. 3 Equivalent stress for different substrate materials, gripper diameters and differential pressure

Whilst NR211 membranes are uncritical for the investigated parameter setting, NC700 is prone to being damaged due to handling. In Kundu et al. [20] yield strength between 1.55 MPa and 2.5 MPa were obtained for Nafion based materials. Based on the FEM results a diameter of less than 0.5 mm is recommended for safe handling of membranes.

2.5 Investigation of Position Accuracy

For the assessment of an achievable repeatability, a test stand was designed. It features a z-axis, a pressure control and differential pressure measurement system and a camera mounted below the transparent table. The gripper is slightly bigger than the CCM, the openings have a diameter of 0.5 mm, Fig. 4 shows the corresponding experimental setup. The level settings were based on the upper and lower limits in preliminary investigations and are shown in Table 3, the differential air pressure results from the characteristics of the vacuum generator and set supply air pressure.

The central composite design has 16 different parameter settings for the full-factorial inner part (2^4-plan; - and + settings) and an additional nine parameter settings for centre and star points (+ + , 0 and - -). Each parameters setting has been carried out four times resulting in a total of 100 runs and position deviation measurements.

Figure 5 shows the analysis of means (ANOM) for the parameter settings in Table 3 with the average standard deviation represented in grey. To obtain an overall position deviation the mean value of x- and y-deviation on a single corner have been measured.

Fig. 4 Experimental setup for the design of experiment

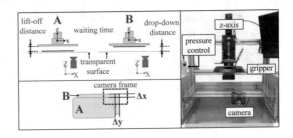

Table 3 Parameters for the design of experiments

Level	Supply air pressure (A) (bar)	Differential air pressure (mbar)	Lift-off distance (B) (mm)	Drop-down distance (C) (mm)	Waiting time (D) (s)
- -	1	21	0	0	0
-	2	45	1	2.5	0.5
0	3	91	2.5	5	1
+	4	112r	4	7.5	1.5
+ +	5	121r	5	10	2

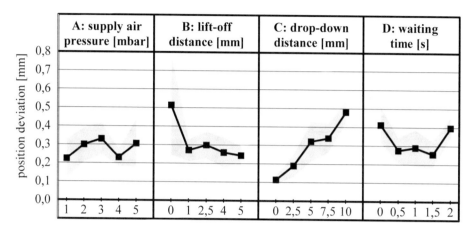

Fig. 5 Analysis of means for the complete central composite design of experiments

Table 4 ANOVA results for the design of experiments

	Supply air pressure (A)	Lift-off distance (B)	Drop-down distance (C)	Waiting time (D)	Error
(x + y)/2	4.1%	6.8%	23.5%	3.1%	62.5%

An analysis of variance (ANOVA) has been carried out with the results being shown in Table 4. After determination of the F-value the lift-off and drop-down distance are significant. The ANOVA shows, that a large fraction of the variation cannot be explained by the factor settings, thus leading to a relatively high error.

For the inner part of the central composite design first order interactions were investigated as well, showing no significant interactions. Hence the error cannot be explained by factor interactions alone. One of the reasons for a high variation might be the membrane rolling alongside its edges. The rolling is temperature and humidity induced. This curling leads to an unpredictable position after gripping. This also explains why a higher accuracy can be achieved for a higher pick-up distance: If the distance is bigger than the curling induced curvature, the CCM could be shifted before gripping. CCM flatness and curling have not been measured during experiments. Generally it is recommended to lay the CCM flat, as a higher drop-drown distance reduces accuracy.

2.6 Analysis of the Assembly Procedure and Tolerances

The assembly of fuel cell stacks involves several pick and place operations. The tolerance field for the final placement of the component itself is quite narrow, depending on the stack design from ± 0.1 mm to ± 0.4 mm [1]. Positioning itself cannot be carried out by a fixed stop as the components are too limp. For this reason optical position measurements have to be carried out. The total tolerance of the machine can be calculated as a

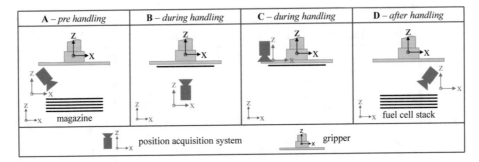

Fig. 6 Camera positioning in fuel cell stack assembly

Table 5 Tolerances in fuel cell pick and place-assembly

	A (mm)	B (mm)	C (mm)	D (mm)
Gripping	± 0.045	–	–	± 0.045
Placement	± 0.045	± 0.045	± 0.045	± 0.045
Camera	± 0.005	± 0.005	± 0.005	± 0.005
Measuring station	–	± 0.02	–	–
Sum	±0.095	± 0.07	± 0.05	± 0.095

combination between the handling device and the measurement device. The pixel density limits the accuracy of the camera, furthermore the handling device has a limited repeat accuracy. If the camera and gripper are not mounted to the same component and share a common origin these inaccuracies add up (camera position A, B and D in Fig. 6 with the camera coordinate system in blue, the gripper coordinate system in black and the world coordinate system in red).

Based on the results of Fig. 5 it is assumed that the gripper inaccuracy is ± 0.025 mm for gripping and placement. In contrast to the test setup there will also be an additional repeat inaccuracy of ± 0.02 mm by the robot itself, resulting in ± 0.045 mm for gripping and placement. The camera accuracy is set to ± 0.005 mm in all cases. The measuring station is assumed to have the same repeat accuracy as the robot. Concept D only allows for quality assurance, a perfect alignment of the component prior to gripping is assumed. The estimated inaccuracies are shown in Table 5 with concept C promising to have the lowest inaccuracy, which therefore will be pursued further in future.

3 Summary and Outlook

A flat surface gripper has been used to grasp and release a CCM for fuel cell production. Through FEM the critical suction hole opening size has been investigated and optimized for given materials. A direct link between material properties and gripper design was

Fig. 7 Target system for scalable fuel cell manufacturing

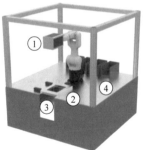

① multi-component gripper
② externally actuated position correction frame
③ lowerable stacking table
④ component magazines

made. This proactive preparation allows for fast reaction times if the CCM is changed to another material, therefore contributing to an agile manufacturing system, with the planned base unit shown in Fig. 7. The multi component gripper allows for high utilization rates in low volume production and can be easily extended by the integration of identical systems for higher volume production as outlined in Fig. 2.

Through the investigation of the parameter setting an optimization trend for higher pick and place accuracies was derived. Especially the distance between gripper and component during grasping and releasing plays a crucial role. It is concluded, that higher accuracies are achieved with slight distance during grasping and no distance during release. According to the results it still can be very challenging to reduce the inaccuracy below 0.1 mm. Possible countermeasures include an externally actuated position frame with integrated position detection, thus eliminating errors due to membrane curling prior to grasping. Cause-effect relationships need to be further investigated, for example air turbulences or electrostatic effects can be taken into account. With reference to the idea of an agile manufacturing system similarly designed investigations for GDL and bipolar plate will be carried out.

Acknowledgements The authors would like to thank the Federal Ministry of Transportation and Digital Infrastructure for funding the project EMSigBZ (Grant No. 03B11012C) in the National Innovation Programme Hydrogen and Fuel Cell Technology (NIP).

References

1. Stahl, P.: Edge effects on the single cell level of polymer electrolyte fuel cells. Doctoral thesis, Universität Stuttgart (2018)
2. Porstmann, S., Wannemacher, T., Richter, T.: Overcoming the challenges for a mass manufacturing machine for the assembly of PEMFC stacks. Machines **7**, 66 (2019)
3. Bobka, P., Gabriel, F., Römer, M., et al.: Fast pick and place stacking system for thin, limp and inhomogeneous fuel cell components. In: Wulfsberg, J.P., Hintze, W., Behrens, B.-A. (eds.) Production at the Leading Edge of Technology, pp. 389–399. Springer, Berlin (2019)
4. Bobka, P., Gabriel, F., Dröder, K.: Fast and precise pick and place stacking of limp fuel cell components supported by artificial neural networks. CIRP Ann. **69**, 1–4 (2020).

5. Williams, M., Tignor, K., sSigler, L., et al.: Robotic arm for automated assembly of proton exchange membrane fuel cell stacks. J. Fuel Cell Sci. Technol. **11**, 1 (2014)
6. Laskowski, C., Derby, S.: Fuel cell ASAP: two iterations of an automated stack assembly process and ramifications for fuel cell design-for-manufacture considerations. J. Fuel Cell Sci. Technol. **8**, 713–722 (2011)
7. Schmalz, J.: GmbH: MontaBS—Entwicklung von Montagetechnologie und Automatisierungskonzepten für die Fertigung von Brennstoffzellen: Schlussbericht (2018)
8. ElringKlinger: MontaBS - Entwicklung von Montagetechnologie und Automatisierungskonzepten für die Fertigung von Brennstoffzellen: Schlussbericht (2017)
9. Dreier, G.: Vorrichtung und Verfahren zum Herstellen eines Brennstoffzellen-Stacks (DE1 201 00 17 A1) (2017)
10. Munthe, S.: Manufacturing arrangement for a fuel cell stack and method for manufacturing a fuel cell stack (WO 2020/005137A1) (2019)
11. HoKyun, J., Yoon, J.: Vorrichtung zum automatischen Stapeln eines Brennstoffzellenstapels (DE102016214982A1) (2015)
12. Ramsauer, C., Rabitsch, C.: Agile Produktion - Ein Produktionskonzept für gesteigerten Unternehmenserfolg in volatilen Zeiten. In: Biedermann, H. (ed.) Industrial Engineering und Management, vol 88, pp 63–81. Springer Fachmedien Wiesbaden (2016)
13. Breitwieser, M.: Direct membrane deposition as novel fabrication technique for high performance fuel cells. Doctoral Thesis, Institut Für Mikrosystemtechnik; Department Of Microsystems Engineering; IMTEK (2017)
14. Rashapov, R.R., Unno, J., Gostick, J.T.: Characterization of PEMFC gas diffusion layer porosity. J. Electrochem. Soc. **162**, F603–F612 (2015)
15. Mohr, P.: Optimierung von Brennstoffzellen-Bipolarplatten für die automobile Andwendung. Doctoral Thesis, Universität Duisburg-Essen (2018)
16. Chemours: Nafion NR211 and NR212 Ion Exchange Materials. Solution Cast Membranes. https://www.chemours.com/en/-/media/files/nafion/nafion-nr211-nr212-p-11-productinfo.pdf (2020). Accessed 16 Dec 2020
17. Chemours: Nafion NC700 Reinforced PFSA Membrane https://www.chemours.de/-/media/files/nafion/nafion-nc700-p-23-product-info.pdf (2020). Accessed 16 Dec 2020
18. Solasi, R., Zou, Y., Huang, X., et al.: On mechanical behavior and in-plane modeling of constrained PEM fuel cell membranes subjected to hydration and temperature cycles. J. Power Sources **167**, 366–377 (2007)
19. Schomburg, W.K.: Introduction to Microsystem Design, 2nd edn., RWTHedition. Springer, Heidelberg (2015)
20. Kundu, S., Simon, L.C., Fowler, M., et al.: Mechanical properties of Nafion^{TM} electrolyte membranes under hydrated conditions. Polymer **46**, 11707–11715 (2005)

Mobile, Modular and Adaptive Assembly Jigs for Large-Scale Products

Sebastian Hogreve, Katharina Krist and Kirsten Tracht⬤

Abstract

The assembly of products is often supported by jigs. Especially for large dimensional products, jigs and fixtures are used to align the components and ensure the stability of the assembly until all parts are firmly mounted. This paper describes the development of mobile, modular and adaptive assembly jigs, which are designed to support ergonomic working in the production of high-lift systems for civil aircrafts. The jig supports the workers to adapt the position and orientation of the product to the current assembly operation. The fundamentals of the development are explained and the features of a concept, called assembly wheel, are presented. The assembly wheel consists of two or more robot arms on a circular seventh axis. The robot arms hold and position the components to be assembled so that all joining spots are freely accessible to the worker. The ergonomic benefits of the concept were examined in a study using a 3D model of the jig. A demonstrator on a scale of 1:2 was set up, with which real experiments with an adaptive jig can be conducted for evaluation.

Keywords

Assembly · Fixtures · Ergonomics · Aero-space industry

S. Hogreve (✉) · K. Krist · K. Tracht
University of Bremen, Bremen Institute for Mechanical Engineering, Badgasteiner Str. 1, 28359 Bremen, Germany
e-mail: hogreve@bime.de

K. Krist
e-mail: krist@bime.de

1   Introduction

To assemble a product, individual parts are put together. By means of different joining operations, complex goods are created. The functionality and quality of an assembled product often depends on the precision of the alignment of the parts. This is the reason why during assembling a lot of effort is spend on measuring, positioning and adjusting of parts. For the production of serial products, it is common to support assembly processes through jigs. The jigs help to align the parts to be assembled and keep them in position until the joining process is completed. This reduces the desired number of handling tasks, fastens the production process and reduces the risk of assembly errors. Besides these advantages, jigs may also be necessary to ensure the initial fixation of the loose parts. An assembly group is sometimes fragile and limp until all parts are added and fixed. The jig supports the assembly group and prevents it from collapsing. Especially during the production of large-scale products, like aircraft components or railcar bodies, jigs are necessary to bring the parts in position and to support the mechanical structure.

Assembly fixtures are usually designed for a single process. Technical restrictions of the jig influence the assembly sequence, operations to be performed and force a certain posture of the assembler due to accessibility. Even though jigs and fixtures are usually designed to support an ergonomic working procedure, they are mostly inflexible and immovable. The device can only be adapted to changes in the production process or the product with great effort. Some fixtures do have adjustment options so that the worker can bring the product into an ergonomic position and orientation. However, these usually have only one degree of freedom and a limited range of movement. With large-scale assemblies, however, there is the challenge of having several technicians with different physiques and different needs working simultaneously on one assembly. In order to provide all assemblers with ergonomic access to the product at the same time, a much more complex jig design is required.

This paper presents the concept and the investigation results for a mobile, modular and adaptive assembly device. In this context adaptive needs to be understood as the capability to adapt to the assembly situation and the workers physical needs. The jig has been developed according to an example from the aerospace industry and it has been investigated using its 3D CAD model. Furthermore, for evaluation of ergonomics and process capability a reduced scale demonstrator of a mobile, modular and adaptive jig has been designed and built. The application product used for these investigations is a high-lift system of a single aisle aircraft. The high lift system consists among other components of an outer landing flap, which is carried by two supports. The complete high-lift system is attached to the wing trailing edge as well as to the wing underside. In order to increase productivity in aircraft assembly, the high-lift system will be delivered pre-assembled to the final assembly line where it will be mounted to the wing following a plug and fly concept. Figure 1 shows a simplified model of the high lift system. It is derived on a scale of 1:2 from the real [1, 2] system and represents the outboard flap of a left wing.

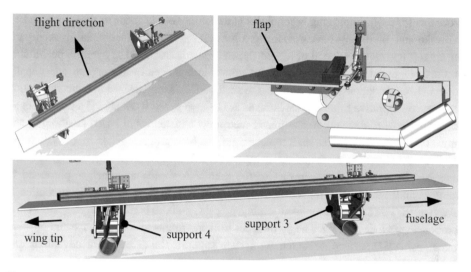

Fig. 1 Isometric, side and rear view of derivative product (simplified model of real product)

2 Jigs and Fixtures for Assembling Aircraft Components

Jigs and fixtures are used wherever components must be precisely aligned with each other. In the production of aircraft components, the greatest assembly effort is required when joining sheet metal parts and components made of carbon fiber reinforced plastics (CFRP). Therefore, most fixtures are used for this purpose. Conventionally, jigs and fixtures are inflexible and have no adaptability in case of product changes. Millar and Kihlman derive the need for reconfigurable fixtures from the example of wing box assembly [3]. Due to the size and the manufacturing effort, the introduction of a new jig can take more than 24 months, which contradicts a short time to market. They present a concept for a reconfigurable jig for the assembly of wing boxes. Therefore they propose the use of off-the-shelf components and the introduction of design tools [3]. Jigs can also consist of plug-in systems that ensure reconfigurability of the system. Zhang et al. present a modular system that allows a high reusability in case of product changes [4]. It consists of a reconfigurable frame system, relocatable clamps and additional parts. The flexible positioning of the clamps reduces the design effort and shortens the implementation time. An adaptation to the assembly situation is not considered. An ergonomic working method is therefore not promoted.

In many research projects, industrial robots are integrated into the jigs in order to create an adaptivity to the assembly situation. The goals are a faster adaptation to product changes, the compensation of shape and position errors and the full automation of joining processes [5, 6, 7]. Improving ergonomics is therefore not the focus of these approaches. Schwake and Wulfsberg have developed a handling system for manipulating shell parts for aircraft bodies [5]. Instead of a rigid jig, an industrial robot takes over the positioning of the components. Linear actuators integrated into the end effector can additionally correct shape deviations by pressing on the shell part. It is emphasized that the application requires collaborative robots, since humans are located in the work area [5]. Schmitt et al. pursue the goal to develop an automated metrology assisted robot based positioning and untwist process. The process shall replace rigid jigs by a programmable robot system [6]. They are developing a model of the components deformation behavior to determine the necessary compensatory movements of the robots. A similar approach is presented in Ramirez and Wollnack [7]. Here a flexible assembly system for CFRP structures is investigated. For a fully automated joining process, a six-axis robot carrying the tool is combined with a flexible jig. The jig includes hexapod robots to correct shape and position deviations. A 3D surface measurement system supports the detection of deviations and enables the robots to make corrections.

In other approaches, an attempt is made to work without a special jig. Mozillo et al. present an assembly process without a jig [8]. The authors describe that jigs and tools are monolithic and have to be designed rigidly due to the high tolerance requirements. The use of laser trackers to determine the exact position and alignment of the components makes rigid jigs unnecessary. This approach also improves the ergonomics of the work process [8]. All presented work and approaches focus on the assembly of structural components for wing box or fuselage. Adaptable jigs supporting the ergonomic assembly of systems with moveable parts were not found.

3 Development of Adaptive Assembly Jig

In the development of the adaptive device the concept was designed according to the construction guideline VDI 2221 [9]. The challenges in the assembly process of landing flaps, as will be described in the section below, such as the ergonomic risks of overhead work, were taken into account. Therefore, four general aims are essential for the definition of detailed requirements. The first goal is to promote an ergonomic and intuitive working method. Second, the device should ensure accessibility to the assembly spots and third, the device must support the simultaneous assembly activity of several assembly workers at the same product. Last but not least the assembly jig must support the plug & fly concept of the high lift system. In a detailed analysis of the requirements, the following major criteria were identified.

3.1 Requirements Analysis

In terms of the basic function of the assembly jig, it must be capable of receiving and holding the assembly and all components required for it at all stages of the assembly process. Loose parts must be fixed at least until they are fully attached to the assembly. In order to ensure a high utilization of the assembly jigs, the jigs must be able to accommodate both left and right wing high lift systems as well as inboard and outboard systems. The worker should be able to freely adjust the working height to support an ergonomic working procedure. There should also be a possibility to rotate the product so that it can be brought into different orientations depending on the assembly spot. The assembly workshops previously conducted with the derivative product have shown that employees would like to swivel and rotate the product around two axes in order to reach all assembly spots [10, 11]. The motion axes and the motion range must be dimensioned in such a way that no overhead work and no work in kneeling or bending occurs. Ideally, the degrees of freedom of the jig are selected so that the assembly can be brought into a position and orientation that allows workers of different heights to work on the assembly simultaneously and ergonomically. The jig must be designed in such a way that its elements do not impede or obscure the accessibility of the assembly spots at any time.

There can be significant dimensional variations during manufacturing and assembly of the wing box [12]. To support the plug & fly concept the actual dimensions of the attachment points generated and measured during the production of the wing box must be exactly reproduced by the assembly jig. Therefore, appropriate adjustment possibilities must be provided inside the jig. The completely assembled high-lift system should also be adjusted in the assembly jig and a function test should be carried out, so that the adjustment and calibration work in the final assembly line can be reduced to a minimum. In order to support the entire assembly, the jig must have a total load-bearing capacity of at least 600 kg. Even during the functional test, where the center of gravity is shifted by the extended flap, the load capacity and stability of the jig must be ensured. To protect the employees, all moving components must be secured against the danger of squeezing. An automatic force and power limitation during contact between man and machine would be desirable.

It is expected that the sample product will be manufactured in a cycle line to increase productivity. Since the assembly cannot be transferred from one assembly jig to the next due to its instability, the assembly jig itself must be mobile to transport the product from station to station. Since only the supports are assembled during the first assembly steps, and thus the dimensions of the assembly are still very limited in relation to the finished product, a modular design of the jig is desirable. It allows the jig to adapt its size and functions according to the progress of the production process.

3.2 Basic Concept for an Adaptive Jig

As described in the requirements, the assembly jig must adapt the pose of the assembly to both the physical conditions of the workers and the respective assembly step. For this purpose, several systems were drafted with creativity techniques and assessed afterwards. For lack of space, only the final concept, called assembly wheel, is presented in this article.

The principle sketch in Fig. 2 represents the components of the basic concept by oversimplified elements in order to essentialize the idea of the assembly wheel. The significant components of the system, like the robot arms and the ring shaped structure are greatly simplified to facilitate the understanding of the system. The concept suggests the collaboration of two or more robot arms. The robot arms lift and hold the components and position them both in relation to each other and in the desired pose relative to the worker. The special feature of the assembly wheel is the arrangement of the robot arms on a ring shaped seventh axis. This creates motion redundancy for the robots. They can thus change their arrangement without moving the product and create situation-specific work space for the worker. This is necessary to ensure accessibility to all assembly spots. The two or more robot arms must hold the assembly together. The task allocation could be such that one robot arm carries the main load of the assembly and the other arms hold and position the components to be assembled. The worker then carries out the actual joining. This would not only reduce the physical strain on the worker, but also relieve him of cognitive work as pre-programmed robotic motions guide him through assembly operations. Another major advantage is the free positioning and orientation of the workpiece corresponding to the demands of the workers. The improved accessibility through individual height adjustment promotes ergonomic assembly work, even when several technicians work together on the assembly object. Furthermore, the jig can easily be adapted to

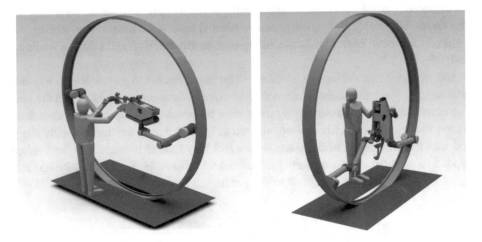

Fig. 2 Principal sketch of assembly wheel concept (not to scale)

changes in the production process or to new variants of the assembly product. Since all assembly steps can be executed with the same jig, there is no need to transfer the product between different specialized jigs during the assembly process. The presented concept leads to a very complex system which is more expensive than a conventional jig. But the posted advantages of the assembly wheel surpass this drawback. A patent has been applied for the principle of the assembly wheel.

3.3 Draft Design of Assembly Wheel

From the basic concept shown in Fig. 2, a draft design with real dimensions was developed, taking the example product into account. Figure 3 shows a 3D model of the draft. The assembly wheel is made of a circular bent steel beam. Profiled rail guides at both sides of the wheel provide precise guidance for the carriages on which the robots are installed. Even with a very stiff design of the wheel there will be some deformation of the wheel while the robots travel along the circumference. This will have negative influence on the position accuracy of the robots. The implementation of an indoor GPS can help to decrease the relative positioning error between the cooperating robots [13]. Norman et al. show that the achievable accuracy can be at least within 0.3 mm [13]. How a workpiece can be handled by two cooperating industrial robots is discussed in Spiller and Verl [14] for example. Since conventional collaborative robots do not provide sufficient payload, the usage of standard industrial robots with high payload is proposed. To avoid injuries to the worker, the robots need to be enabled for human–robot-collaboration. In Behrens [15] different safeguarding techniques are presented. The authors successfully developed and tested devices for hand-guiding, power and force limiting as well as for speed and separation monitoring [15]. For the assembly wheel a combined system is most suitable. A hand-guiding function gives the workers the ability to manually set the position of the components while a power and force limiting function protects the workers during autonomous movements of the robots.

 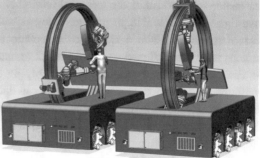

Fig. 3 Advanced 3D model of assembly wheel concept in real scale

The platform is designed as an automated guided vehicle. The Mecanum wheels allow omnidirectional driving maneuvers and thus facilitate the positioning in a workstation. High performance batteries inside the platform provide the necessary power for the traction motors and the robot systems. Charging can be done in the assembly stations while the platform stands still.

To study the feasibility of the assembly process with an adaptive jig, human modelling in Siemens NX has been used. Therefore the assembly process has been divided into thirteen assembly steps. For each assembly step, the robots of the assembly wheel are positioned so that they do not obstruct accessibility for the respective assembly spots. Human models for the 95th and 5th percentiles of both the male and female population were loaded and placed at the jig. The position and orientation of the assembly was then adapted to the human models so that their working posture is largely ergonomic. To estimate the degree of physical stress of the workers, the key indicator method (developed by Federal Institute for Occupational Safety and Health [16]) was used. It has been observed that the adaptive jig sufficiently supports ergonomic working for every human model in every assembly step. None of the working steps causes increased or even high physical stress. However, the study with CAD models cannot answer any questions regarding the workers' personal feeling of safety. It is assumed that the collaboration with the powerful robots and the high rising assembly wheel can lead to fears among the employees.

4 Derivation of a Physical Demonstrator

For the practical evaluation of the idea of adaptive jigs for the assembly of large-scale components a physical demonstrator is required. In the first step, the basic influences of an adaptive jig on the assembly process, the support of an ergonomic working method and the personal feelings of the workers when working with an adaptive jig are to be investigated. A simplified jig without industrial robots is sufficient for the experiments required for this purpose. Only in the second step, the construction of a complex experimental fixture according to Fig. 3 with a circular track and industrial robots is reasonable to demonstrate the technical feasibility of the assembly wheel concept and to investigate the achievable position accuracy as well as methods for path planning.

A demonstrator with a simplified kinematic concept was designed and built. The demonstrator was designed on a scale of 1:2 to match the derivative product from Fig. 1. As can be seen in Fig. 4 (left), the assembly jig consists of three platforms. The outer platforms are supported on air cushions and can thus be easily moved in all directions. On these platforms the supports are pre-assembled. The center platform is stationary and hosts the landing-flap pre-assembling. The right side of the picture shows how the outer platforms with the pre-assembled supports are docked to the center platform for finalizing the assembly of the derivative product.

Fig. 4 The jig demonstrator with two mobile jigs and one stationary platform

All platforms have a lifting column for adjusting the working height. The lifting columns can be moved individually or synchronously by means of a hand switch or pedal. In the mobile platforms a rotary module is placed on the lifting columns. This allows the support to be rotated about the transverse axis during pre-assembly. As a reference to the assembly wheel concept an octagonal frame made of aluminum profile is added to the mobile platforms. This structure gives the workers the feeling of working inside the wheel and provides interfering contours similar to the circular track. Integrated lamps provide adequate lighting for the assembly areas.

In Fig. 5 the pre-assembly of a support and the landing flap is shown. The mobile jig docks to another stationary platform for pre-assembly. On this platform, the material and tools are provided in small rack trolleys. The pre-assembly platform also increases the movement area for the worker. The basic bodies of the supports are provided on small carriages. The worker attaches the holder to the support and then lifts it out of the cart with

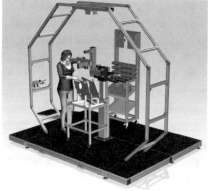

Fig. 5 Mobile jig stopped at preassembly platform (left). Preassembly of landing flap at the base platform (right)

Fig. 6 Preparation for mounting support #3 to jig on mobile platform (left). Mounting of landing flap to the jig on the center platform (right)

the lifting column. The photos in Fig. 6 show the demonstrator prepared for an assembly test with the derivative product. The design of the demonstrator allows to use it with and without the adaptive functions. This way the performance of an adaptive jig can be benchmarked against a conventional jig.

5   Conclusion and Outlook

The presented results show that assembly jigs for future high-lift systems mandate special requirements that are not fulfilled by traditional jig designs. To support the assembly of plug and fly components, jigs are required that can position components dynamically in relation to each other, that are mobile and can adjust the orientation of the component to be assembled according to the assembly situation. The assembly wheel was developed as a concept to meet these requirements. The components are carried by several industrial robots, which allow a free positioning in space. To increase the freedom of movement, the robots are mounted on a circular path. This allows them to move in different positions to clear the workspace for the worker. In a study with a 3D model it was shown that the concept of the assembly wheel allows an ergonomic assembly of the components and that especially people of different heights can work comfortably on the product. The automatic generation of the robot target positions and the determination of the required path movements have not been considered so far and represent a great challenge. Therefore, further research projects have to be carried out in order to enable the full potential of the assembly wheel concept. For the practical evaluation of adaptive jigs, a simplified demonstrator on a scale of 1:2 was built. It will be used in a practical test series to examine in particular the process capability and the physical and psychological effects of adaptive jigs.

Acknowledgements The results presented in this paper were developed in the research project "Next.Move – Next Generation of Moveables" funded under the program Luftfahrt-forschungsprogramm LuFoV-2, FKZ 20W1512G by Federal Ministry for Economic Affairs and Energy (BMWi).

References

1. Bader, A., Gebert, K., Hogreve, S., Tracht, K.: Derivative products supporting product development and design for assembly. Procedia Manuf. **19**, 143–147 (2018)
2. Gebert, K., Bader, A., Tracht, K.: Decision tool for designing derivative products for supporting assembly planning of large-volume assembly groups. Procedia CIRP **76**, 31–35 (2018)
3. Millar, A., Kihlman, H.: Reconfigurable flexible tooling for aerospace wing assembly. SAE Technical Paper 2009–01–3243 (2009)
4. Zhang, H., Zheng, L., Chen, X., Huang, H.: A novel reconfigurable assembly jig based on stable agile joints and adaptive positioning-clamping bolts. Procedia CIRP **44**, 316–321 (2016)
5. Schwake, K., Wulfsberg, J.: Robot-based system for handling aircraft shell parts. Procedia CIRP **23**, 104–109 (2014)
6. Schmitt, R., Witte, A., Janßen, M., Bertelsmeier, F.: Metrology assisted assembly of airplane structure elements. Procedia CIRP **23**, 116–121 (2014)
7. Ramirez, J., Wollnack, J.: Flexible automated assembly systems for large CFRP-structures. Procedia Technology **15**, 447–455 (2014)
8. Mozillo, R. Iaccarino, P., Vitolo, F., Franciosa, P.: Design and development of jigless assembly process: the case of complex aeronautical systems. In: 2019 II Workshop on Metrology for Industry 4.0 and IoT, pp. 132–136. IEEE (2019)
9. Jänsch, J., Birkhofer, H.: The development of the guideline VDI 2221 – the change of the direction. In: Marjanović, D. (ed.) Proceedings of the DESIGN 2006 / 9th International Design Conference, vol. 1, pp. 45–52. Faculty of Mechanical Engineering and Naval Architecture, University of Zagreb; The Design Society, Glasgow (2006)
10. Gebert, K., Onken, A.-K., Tracht, K.: Assembly workshops for acquiring and integrating expert knowledge into assembly process planning using rapid prototyping model. In: Schüppstuhl, T., Tracht, K., Franke, J. (eds.) Tagungsband des 3. Kongresses Montage Handhabung Industrieroboter, pp. 13–21. Springer Vieweg, Berlin, Heidelberg (2018)

11. Krist, K., Sievers, T., Onken, A.-K., Kodjo, Y., Tracht, K.: Application of derivative products for integration expert knowledge into assembly process planning. Procedia CIRP **88**, 88–93 (2020)
12. Saadat, M., Cretin, C.: Dimensional variations during Airbus wing assembly. Assem. Autom. **22**(3), 270–276 (2002)
13. Norman, A.R., Schönberg, A., Gorlach, I.A., Schmitt, R.: Validation of iGPS as an external measurement system for cooperative robot positioning. Int. J. Adv. Manuf. Technol. **64**, 427–446 (2013)
14. Spiller, A., Verl, A.: Force controlled handling with cooperating industrial robots. In: ROBOTIK 2012 – 7th German Conference on Robotics, pp. 496–501. VDE Verlag, Berlin (2012)
15. Behrens, R., Saenz, J., Vogel, C., Elkmann, N.: Upcoming technologies and fundamentals for safeguarding all forms of human-robot collaboration. In: Proceedings of 8th International Conference Safety of Industrial Automated Systems – SIAS 2015, pp 18–23. Deutsche Gesetzliche Unfallversicherung (DGUV), Berlin (2015)
16. Gefährdungsbeurteilung bei physischer Belastung – die neuen Leitmerkmalmethoden (LMM) – Kurzfassung. 3rd edn. Bundesanstalt für Arbeitsschutz und Arbeitsmedizin (BAuA), Dortmund, Berlin, Dresden (2019) (in German)

Design of an Automated Assembly Station for Process Development of All-Solid-State Battery Cell Assembly

Arian Fröhlich, Steffen Masuch and Klaus Dröder

Abstract

Today, lithium-ion batteries are a promising technology in the evolution of electro mobility, but still have potential for improvement in terms of performance, safety and cost. In order to exploit this potential, one promising approach is the replacement of liquid electrolyte with solid-state electrolyte and the use of lithium metal electrode as an anode instead of graphite based anodes. Solid-state electrolytes and the lithium metal anode have favorable electrochemical properties and therefore enable significantly increased energy densities with inherent safety. However, these materials are both, mechanically and chemically sensitive. Therefore, material-adapted processes are essential to ensure quality-assured manufacturing of all-solid-state lithium-ion battery cells. This paper presents the development of a scaled and flexible automated assembly station adapted to the challenging properties of the new all-solid-state battery materials. In the station various handling and gripping techniques are evaluated and qualified for assembly of all-solid-state battery cells. To qualify the techniques, image processing is

A. Fröhlich (✉) · S. Masuch · K. Dröder
Institute of Machine Tools and Production Technology, Technische Universität Braunschweig, Langer Kamp 19b, 38106 Braunschweig, Germany
e-mail: a.froehlich@tu-braunschweig.de

A. Fröhlich · S. Masuch · K. Dröder
Battery LabFactory Braunschweig, Technische Universität Braunschweig, Langer Kamp 8, 38106 Braunschweig, Germany

T. Schüppstuhl et al. (eds.), *Annals of Scientific Society for Assembly, Handling and Industrial Robotics 2021*,
https://doi.org/10.1007/978-3-030-74032-0_5

set up as a quality measurement technology. The paper also discusses the challenges of enclosing the entire assembly station in inert gas atmosphere to avoid side reactions and contamination of the chemically reactive materials.

Keywords

All-Solid-State Battery · Lithium-Ion Battery · Battery Cell Assembly · Lithium Metal Anode

1 Introduction

Lithium-ion battery cells are one of the key technologies to promote the global and sustainable energy revolution. However, their electrochemical performance, safety and cost-effectiveness over the entire life cycle need to be further improved [1]. For this reason, research focuses on future battery generations with different materials and designs. These include all-solid-state battery cells, whose structure differs from conventional battery cells due to the utilization of e.g. pure lithium metal anode and a solid-state electrolyte. The use of a pure lithium metal anode increases the energy and power density at cell level and improves the fast charging capability, as there is no host structure with limited intercalation processes. By exchanging the liquid, reactive electrolyte with the solid-state electrolyte, the risk of leakage and thus the danger of exothermic side reactions is reduced [1, 2].

However, the aforementioned positive properties of the new materials are in conflict with their high mechanical and chemical sensitivity, which has a significant negative effect on the production processes of cell manufacturing. The multitude of handling operations during battery cell manufacturing need to be adjusted to the new material properties with consideration of the acting loads, since they have a major impact on the quality of the battery cells. For this reason, a design of an automated assembly station for all-solid-state battery cell manufacturing is proposed, that considers the material and process specific requirements. The station performs the handling operations with the new materials, which are so far limited to laboratory scale, using industrially established production technology, so that the findings can be transferred directly to industrial production. Due to a flexible system design, various handling processes can be executed and the resulting interactions between process parameters such as low deposition accuracy of electrodes and electrolyte can be investigated. A low deposition accuracy has a strong negative impact on the performance of conventional battery cells and has therefore to be analyzed for the new battery generation as well [3].

2 Handling-Affecting Material Properties of Lithium Metal Anode and All-Solid-State Electrolyte

Lithium metal anodes are widely used in high-energy all-solid-state battery cells because of their advantageous electrochemical properties (e.g. pure lithium has a theoretical specific capacity of 3860 Ah/kg and the lowest standard potential of all metals). Because of these excellent electrochemical properties, an electrode thickness of 20 μm is sufficient for high-energy cells [4, 5, 6]. However, the low density and thickness of the material result in very light and limp electrodes, which are sensitive to mechanical stress due to low mechanical strength (e.g. Young's modulus of lithium is 4.91 GPa). Even small loads can cause surface damage that easily leads to deformation and surface breakouts due to the strong adhesion tendency of lithium. Material damage is the source of loss of direct contact between the solid-state electrolyte and the lithium metal electrode, which results in high electrical resistances in battery cells operation [7, 8]. In addition to the mechanical sensitivity, lithium has a strong reactivity with various elements of the ambient atmosphere, especially with water. The reaction products formed have poor ionic conductivity, which is why these reactions have to be avoided by means of dry room or inert gas atmosphere during the production processes [5, 6].

As with lithium metal electrode, solid-state electrolytes offer great potential for improvement in battery cells while at the same time increasing the challenges in material processing. On the one hand, a group of materials based on organic polymers and on the other hand, groups of inorganic sulphidic or ceramic materials are high ionically conductive with low interfacial resistance [9]. A multitude of different materials and designs exist, which are examined in cells on laboratory scale with regard to electrochemical properties, but are rarely characterized with regard to processing relevant mechanical properties [7, 8, 9, 10]. For this reason, only a general tendency of the mechanical behavior can be derived from established base-materials. Polyethylene oxide is a widely used material in polymer-based electrolytes and has a Young's modulus of 330 MPa, which indicates low stiffness [10, 12]. In contrast, ceramic or sulphidic electrolytes have a high stiffness (10–200 GPa) and tend to brittle behavior [10, 11]. Regardless of the material group, the electrolytes react strongly with the elements of the ambient atmosphere, especially water, requiring a dry room or inert gas environment during cell manufacturing [13].

The brief review of the material properties shows the close link between high electrochemical performance with challenging mechanical and chemical properties of the new materials. Due to the strong dependence of electrochemical performance on the mechanical integrity of the electrodes and the electrolyte, damage-free handling is essential in the manufacturing processes of all-solid-state batteries.

3 Handling in Battery Cell Production

3.1 Battery Cell Assembly Process

In lithium-ion battery production, the assembly of the battery cells is subsequent to the electrode manufacturing process and is carried out in several interlinked process steps. Electrodes are handled in many of the process steps (e.g. drying, cutting, stacking), but the most crucial one is the stacking step. During stacking, the electrodes and the separator or solid-state-electrolyte are successively built up to a compound. For conventional intercalation electrodes, different methods of building the compound are established. Using the winding method, the webs are rotated together around a winding core and wound into round or prismatic electrode-separator composites. If the mechanical properties of the electrodes prevent winding, they are cut from the electrodes web and stacked between web-guided and folded separator by handling systems (z-folding method, stacking) [1]. In z-folding and stacking, the electrodes are withdrawn from a magazine and deposited on an adjustment table. Image processing is used to measure the position and orientation of the electrodes on the adjustment table. A second handling system then grasps the electrodes from the alignment table and uses a correction vector to deposit the electrode on the target position in the stack. The measurement and automatic correction of the position and orientation of the electrodes is essential, since smallest deviations of the surface overlap of the electrodes cause significant losses in the electrochemical performance [3].

Presumably, a stack of single electrodes will be used for lithium metal electrodes and solid-state electrolyte, since the mechanical properties of the solid-state electrolyte do not allow bending around tight winding radii. In addition, both the high and uniform compression required to ensure that the electrodes are in contact at cell level and the high volume changes during charge and discharge cycles are contrary to a wound cell design [11].

3.2 Handling Principles in Battery Cell Assembly

A handling system consists of connected modules, e.g. an automated movement system, a gripper, and peripherals. The following section discusses the gripping principles, since the gripper interacts with the electrodes in particular.

Generally, the electrodes are gripped with pneumatic grippers, which generate the gripping force via an airflow [14]. The established pneumatic grippers can be categorized according to vacuum or overpressure actuation. A vacuum actuation principle with homogeneous distribution of the lifting force is given by the area vacuum gripper. Although the vacuum causes mechanical contact between the electrode and the gripper surface, the surface loads during mechanical contact are reduced by distributing the vacuum through numerous small openings. The mechanical contact nevertheless may result in material adhesion and surface damage to the electrodes, making precise and undamaged electrode removal difficult. These disadvantages are the reason for using overpressure-operated grippers for handling. The overpressure-actuated Bernoulli gripper generates a fast airflow in the gap between the gripper and the electrode, so that the pressure in the gap decreases. Consequently, the ambient pressure pushes the electrode towards the gripper. Since the lifting force generated is mainly dependent on the airflow in the gap, there is always a gap between the gripper and the electrode. In summary, the Bernoulli gripper offers quasi-contactless handling, but can thus only poorly compensate for relative lateral movements between the gripper and electrode.

The impact of the gripping process on the materials to be handled is not only dependent on the gripping principle. The set lifting force, the distance between the gripper and the gripping object for pick-up and deposition as well as the velocity and acceleration profile are further important handling parameters. In addition to these control variables, disturbance variables, which can only be influenced to a limited extent, have an effect on the process, e.g. the static charge resulting from the absence of humidity in inert gas atmosphere for discharging surfaces. Consequently, the evaluation of gripping principles for material-adapted handling has to consider all variables influencing the process, which is why a research set-up is required which allows the relevant parameters to be precisely adjusted.

4 Design of an Automated Assembly Station

Material-adapted handling processes are an important prerequisite for a quality-assured industrial production of all-solid-state batteries. The characteristics of the new materials are also challenging in the usually manual production of small-scaled laboratory cells, but these cells address material development rather than the development of cost-efficient industrial high-throughput processes. The design of the automated assembly station proposed in this paper therefore extends the target values of damage-free and reproducible handling in existing approaches to laboratory automation by the target values of flexibility and scalability as well as cost efficiency and throughput [15]. In the following sections, the single modules of the automated assembly station and their communication and interaction are described in detail.

The process development requires a constant and reproducible process environment, which minimizes the influence of environmental disturbances on material and process properties. The high reactivity of the electrodes and the electrolyte necessitates specific conditioning of the ambient process atmosphere to reduce the amount of potential reactants (e.g. water and oxygen). For this reason, the entire assembly process is set up in a gas-tight glove box whose internal atmosphere can be precisely adjusted (e.g. $H_2O < 0.1$ ppm, $O_2 < 0.1$ ppm) by filling with inert gas or defined gas mixtures. Pneumatics based on compressed air, which are established in industrial automation technology, may only be implemented if the inserted pneumatic components are sufficiently gas-tight and the consumed compressed air (e.g. return stroke of cylinders, venting of components) is not blown off into the glovebox but returned. An alternative operation with inert gas as pressurized medium reduces the requirements for components gas tightness and gas recirculation, but demands continuous active pressure regulation in the gas-tight glovebox due to the supplied or discharged gas volume. Generating the pressurized medium inside the glovebox is only reasonable for small-scaled actuators because of the very limited and cost-intensive installation space.

To avoid the above-mentioned disadvantages of pneumatics, an electrically operated handling system is installed in the assembly station. The handling system combines the Festo EXCM-30 planar surface gantry with Festo EGSC-32 lifting axis and Festo ERMO-12 rotary axis to form a cartesian kinematic with a degree of freedom of 4. The H-parallel kinematic drive concept of the planar surface gantry and the consistent application of compact stepper motors in all axes allow a high degree of space utilization while maintaining high dynamics (x–y velocity 0.5 m/s, x–y-z acceleration 10 m/s^2) and repeatability ($\pm$0.05 mm). An operating voltage of 24 V meets the low dielectric strength and poor heat conduction of inert gases, which is challenging for electrical components. Figure 1 depicts the handling system integrated into the glovebox, supplemented by a gripper, a turntable with different surface materials for electrode supply and deposition, and an image acquisition system.

For the detection and measurement of the position and orientation of the electrodes in the handling process an image processing system is developed. By means of the image processing, the position and orientation of the electrodes to be grasped can be communicated to the handling system and thus a corrected grasping is achieved. In addition, the precision of deposition (absolute accuracy and repeatability) is assessed after the handling process. For image acquisition, a 20 MP industrial camera and a ring light as incident

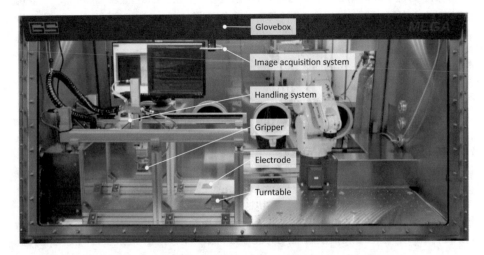

Fig. 1 Automated assembly station in glovebox, with the front glass open

light, enhanced by a linear polarizing filter to reduce reflections of the metallic surfaces of lithium electrodes, are mounted above the turntable. This setup is chosen so that overlapping electrodes can be detected during the stacking process. The image is imported into a program developed in PyCharm and processed by successive algorithms of the OpenCV library. The image processing is based on a monochrome image of the electrode, which is distortion-free by calibration. The processing begins with a Gaussian filter, which reduces noise and smoothes edges. Then, using the Canny algorithm based on the Sobel operator, the edges in the image are extracted and contours resulting from linked edges are fitted using polygon approximation. The approximated contours are used afterwards to define a reference in the first handling step and the comparing contour in subsequent handling steps. For each defined contour, the centralized and normalized image moments are calculated to provide a scaling invariant comparison of contours, even if they are different in position and orientation in the image area. By subtracting the image moments and an approximation of ellipse main axes to the contours, both position and orientation deviations between the captured electrodes can be determined. Figure 2 shows the measurement of position and orientation deviations of an electrode from the electrode reference contour.

In the PyCharm program, in addition to the image processing, the master control of all components involved in the automated assembly station is implemented (see Fig. 2). All subordinate control components are industrially established technologies (e.g. programmable logic controller Festo CECC-D) or are based on open platforms close to industrial applications (e.g. programmable logic controller Controllino Mega) and apply standardized protocols for communication (TCP, serial interface USB and UART). Summarizing all the modules, the automated assembly station is similar to the production technology of industrial stacking processes and at the same time flexibly adaptable and expandable.

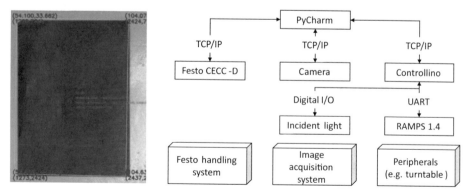

Fig. 2 Result of the image processing for measurement of position and orientation deviation of a lithium metal electrode (red contour) from the electrode reference contour (green contour) (left), block diagram of linked modules and communication protocols (right)

5 Experimental Evaluation of Handling Processes in the Automated Assembly Station

In order to evaluate different process set-ups and to develop a material-adapted handling process, lithium electrodes composed of a 20 μm lithium layer on a 10 μm copper substrate with overall dimensions of 50×70 mm^2 are used. The electrodes are supplied on one half of the turntable on a polypropylene substrate since polypropylene is established in processing as a separating layer between lithium surfaces. As a surface for deposition, a polymer-based electrolyte made of polyethylene oxide is attached to the other half of the turntable. By rotating the turntable, the corresponding surface can be provided for pick-up and deposition. As variation parameters, different grippers are mounted to the handling system and the set lifting force as well as the distance between gripper and electrode during pick-up and deposition are varied. The variation steps of the lifting force are based on the weight of the electrode and start with the minimum force for lifting the electrode that was identified in preliminary tests. The variation steps of all other parameters are similar to the industrial stacking of conventional electrodes. Table 1 gives an overview of the variation parameter settings.

The target value of the experimental evaluation is the repeatability of the position and orientation of the deposited electrode. This repeatability is measured according to the ISO 9283, which contains performance criteria and related test methods of manipulating industrial robots. Following this standard, a motion sequence close to the real stacking process is performed and repeated 30 times for each variation. The motion sequence consists of pick-up, vertical and horizontal movement with simultaneous rotation, and deposition. All movements are executed with the maximum acceleration and velocity of the handling system (see Sect. 4). Prior to the motion sequence, the image processing assesses the supplied electrodes position in x, y and orientation a around the z-axis of the image coordinate system. These values are recorded as reference contour and serve as comparison to the position and orientation of the electrode captured after the handling process (see Fig. 2). By comparing the values of x, y, and a measured prior to and after handling, the repeatability is calculated with the following formulas of ISO 9283 [16].

$$l_j = \sqrt[2]{\left(x_{j,\text{prior}} - x_{j,\text{after}}\right)^2 + \left(y_{j,\text{prior}} - y_{j,\text{after}}\right)^2} \tag{1}$$

Table 1 Variation steps of the parameters

Gripper	Lifting force (mN)	Pick-up distance (mm)	Deposition distance (mm)
Bernoulli	30	0.5	0.5
Area vacuum	60	1.5	1.5
	120		

$$\tilde{l} = \frac{1}{30} \sum_{j=1}^{30} l_j \tag{2}$$

$$S_1 = \sqrt[2]{\frac{\sum_{j=1}^{30} \left(l_j - \tilde{l}\right)^2}{29}} \tag{3}$$

$$RP_{x,y} = \tilde{l} + 3S_1 \tag{4}$$

$$RP_a = \pm 3 \sqrt[2]{\frac{\sum_{j=1}^{30} \left(a_{j,\text{prior}} - a_{j,\text{after}}\right)^2}{29}} \tag{5}$$

The repeatability achievable in the process comprises the linked repeatabilities of the handling system, the camera, and the gripper. Therefore, the combined repeatability of the handling system and the camera is initially measured by moving and measuring a reference contour inflexibly attached to the end effector, with a mass corresponding to that of the grippers to be evaluated, 30 times with the specified motion sequence. The resulting values $RP_{x,y} = 0.088$ mm and $RP_a = 0.5°$ only indicate the repeatability of the combined systems for this motion sequence and process load, since, different to ISO 9283, no multiple points and trajectories are measured in the workspace. The repeatabilities shown in Fig. 3 are consequently composed of the sum of the repeatability of the handling system, the camera and the gripper. However, it is obvious that the combined repeatability of the handling system and the camera is considerably smaller than the repeatability of the grippers.

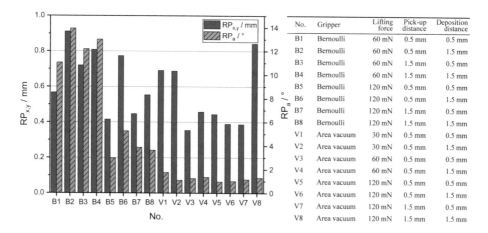

No.	Gripper	Lifting force	Pick-up distance	Deposition distance
B1	Bernoulli	60 mN	0.5 mm	0.5 mm
B2	Bernoulli	60 mN	0.5 mm	1.5 mm
B3	Bernoulli	60 mN	1.5 mm	0.5 mm
B4	Bernoulli	60 mN	1.5 mm	1.5 mm
B5	Bernoulli	120 mN	0.5 mm	0.5 mm
B6	Bernoulli	120 mN	0.5 mm	1.5 mm
B7	Bernoulli	120 mN	1.5 mm	0.5 mm
B8	Bernoulli	120 mN	1.5 mm	1.5 mm
V1	Area vacuum	30 mN	0.5 mm	0.5 mm
V2	Area vacuum	30 mN	0.5 mm	1.5 mm
V3	Area vacuum	60 mN	0.5 mm	0.5 mm
V4	Area vacuum	60 mN	0.5 mm	1.5 mm
V5	Area vacuum	120 mN	0.5 mm	0.5 mm
V6	Area vacuum	120 mN	0.5 mm	1.5 mm
V7	Area vacuum	120 mN	1.5 mm	0.5 mm
V8	Area vacuum	120 mN	1.5 mm	1.5 mm

Fig. 3 Repeatability of the handling processes with varied process parameters

The Bernoulli gripper is unable to grip the electrode from the turntable at a set gripping force of 30 mN even at a small pick-up distance. This is because the gripper does not cover the electrode completely and therefore the non-fixed, limply down hanging electrodes areas generate a lateral force, which the gripper cannot compensate. With increased lifting force a permanent fixation of the electrode is achieved, which repeatability is dependent on the distance during pick-up and deposition. During the free flight phase when lifting from the turntable towards the gripper, the electrode is strongly influenced by the airflow of the Bernoulli gripper, which leads to high position and orientation deviations with increasing pick-up distance and respectively longer free flight phase. These deviations are further increased during deposition by the static charge of the electrode, which results from the continuous friction of the flowing air on the surface of the electrode, since the resulting force of the static charge lead to an inhomogeneous attraction and consequently displacement of the electrode during deposition.

In contrast, the area vacuum gripper has significantly better orientation repeatability for all variations. In addition, increased lifting force results in a position repeatability superior to that of the Bernoulli gripper. The repeatability achieved is nearly independent of the pick-up and deposition distances, because the electrode in contact with the gripper closes the vacuum openings during gripping and thus neither electrostatically charging airflow nor lateral displacement occurs. However, according to Leithoff et al. [3] already minor inaccuracies have a strong negative effect on the electrochemical cell performance, which is why the achieved repeatability of the position in particular is insufficient for quality-assured stack assembly. The results identify the gripper and the set gripping parameters as decisive influence on the repeatability. Consequently, the handling of lithium metal electrodes and solid-state electrolytes requires gripping principles that are adapted to the material properties (e.g. light, limp, mechanically sensitive) and the ambient conditions (inert gas atmosphere, static charge).

6 Conclusion

The battery materials of the future, such as lithium metal electrodes and solid-state electrolytes, have, in contrast to favored electrochemical properties, a high mechanical and chemical sensitivity. For this reason, the production processes in cell manufacturing require new technical solutions. Based on industrially established process technology, a scaled and flexible automated assembly station was designed for the development and evaluation of material-adapted processes. This station consists of a handling system, image acquisition system and peripherals, all enclosed in a glovebox. By means of this interconnected equipment, handling processes with lithium metal electrodes were carried out varying the process parameters and evaluated based on the achievable repeatability. The experiment results indicate that the gripping principle applied is the major influence on the repeatability and that the established grippers do not achieve a sufficient repeatability with any parameter set tested. Consequently, further development of the gripping principles and a precise determination of suitable process parameters are necessary.

Acknowledgements The authors thank the German Federal Ministry of Education and Research for supporting the project ProLiMA (03XP0182F). The authors are grateful to Festo SE & Co. KG for supplying and supporting the operation of the handling system within the project ProLiMA.

References

1. Bockwinkel, K., Nowak, C., Thiede, B., Nöske, M., Dietrich, F., Thiede, S., Haselrieder, W., Dröder, K., Kwade, A., Herrmann, C.: Enhanced Processing and Testing Concepts for New Active Materials for Lithium-Ion Batteries. Energy Technol. **8**(2), 1900133 (2019)
2. Dirican, M., Yan, C., Zhu, P., Zhang, X.: Composite solid electrolytes for all-solid-state lithium batteries. Mater. Sci. Eng.: R: Rep. **136**, 27–46 (2019)
3. Fleischer, J., Ruprecht, E., Baumeister, M., Haag, S.: Automated Handling of Limp Foils in Lithium-Ion-Cell Manufacturing. In: Dornfeld, D.A., Linke, B.S. (eds.) Leveraging Technology for a Sustainable World, pp. 353–356. Springer, Berlin Heidelberg, Berlin, Heidelberg (2012)
4. Hao, F., Han, F., Liang, Y., Wang, C., Yao, Y.: Architectural design and fabrication approaches for solid-state batteries. MRS Bull. **43**(10), 775–781 (2018)
5. ISO 9283:1998: Manipulating industrial robots—Performance criteria and related test methods (1998)
6. Janek, J., Zeier, W.G.: A solid future for battery development. Nat. Energy **1**(9), 1167 (2016)
7. Jee, A.-Y., Lee, H., Lee, Y., Lee, M.: Determination of the elastic modulus of poly(ethylene oxide) using a photoisomerizing dye. Chem. Phys. **422**, 246–250 (2013)
8. Kwade, A., Haselrieder, W., Leithoff, R., Modlinger, A., Dietrich, F., Droeder, K.: Current status and challenges for automotive battery production technologies. Nat. Energy **3**(4), 290–300 (2018)
9. Leithoff, R., Fröhlich, A., Dröder, K.: Investigation of the influence of deposition accuracy of electrodes on the electrochemical properties of lithium-ion batteries. Energy Technol. 1900129 (2019)
10. Lin, D., Liu, Y., Cui, Y.: Reviving the lithium metal anode for high-energy batteries. Nat. Nanotechnol. **12**(3), 194–206 (2017)
11. Mauger, A., Armand, M., Julien, C.M., Zaghib, K.: Challenges and issues facing lithium metal for solid-state rechargeable batteries. J. Power Sour. **353**, 333–342 (2017)
12. Ma, J., Chen, B., Wang, L., Cui, G.: Progress and prospect on failure mechanisms of solid-state lithium batteries. J. Power Sour. **392**, 94–115 (2018)
13. Schnell, J., Günther, T., Knoche, T., Vieider, C., Köhler, L., Just, A., Keller, M., Passerini, S., Reinhart, G.: All-solid-state lithium-ion and lithium metal batteries – paving the way to large-scale production. J. Power Sour. **382**, 160–175 (2018)
14. Sun, C., Liu, J., Gong, Y., Wilkinson, D.P., Zhang, J.: Recent advances in all-solid-state rechargeable lithium batteries. Nano Energy **33**, 363–386 (2017)
15. Takada, K.: Progress in solid electrolytes toward realizing solid-state lithium batteries. J. Power Sour. **394**, 74–85 (2018)
16. Zheng, F., Kotobuki, M., Song, S., Lai, M.O., Lu, L.: Review on solid electrolytes for all-solid-state lithium-ion batteries. J. Power Sour. **389**, 198–213 (2018)

Grasping

Combined Structural and Dimensional Synthesis of a Parallel Robot for Cryogenic Handling Tasks

6

Moritz Schappler, Philipp Jahn, Annika Raatz and Tobias Ortmaier

Abstract

The combined structural and dimensional synthesis is a tool for finding the robot structure that is suited best for a given task by means of global optimization. The handling task in cryogenic environments gives strong constraints on the robot synthesis, which are translated by an engineering design step into the combined synthesis algorithm. This allows to reduce the effort of the combined synthesis, which provides concepts for alternative robot designs and indications on how to modify the existing design prototype, a linear Delta robot with flexure hinges. Promising design candidates are the 3$\underline{P}$RRU and 3$\underline{P}$RUR, which outperform the linear Delta (3$\underline{P}$UU) regarding necessary actuator force.

Key words:

Combined structural and dimensional synthesis • Cryogenic work environment • Flexure hinge • Joint range constraint • Parallel robot

M. Schappler (✉) · T. Ortmaier
Leibniz Universität Hannover, Hannover, Germany
E-mail: moritz.schappler@imes.uni-hannover.de

Institut für Mechatronische Systeme, An der Universität 1, 30823 Garbsen, Germany

P. Jahn · A. Raatz
Institut für Montagetechnik, An der Universität 2, 30823 Garbsen, Germany

T. Schüppstuhl et al. (eds.), *Annals of Scientific Society for Assembly,*
Handling and Industrial Robotics 2021,
https://doi.org/10.1007/978-3-030-74032-0_6

1   Introduction and State of the Art

The automation of handling processes is an omnipresent factor in industry and research institutions. A robot-supported automation solution is also desirable for extremely niche areas such as the cryogenic storage of biological materials [1]. The development of task-adapted structures for such exceptional cases poses significant challenges for designers: From an almost infinite variety of design possibilities, the optimal design for the task and the underlying geometric and situational constraints must be found. Even if the application's basic parameters are entirely known, it is impossible to manually design and evaluate all possible variations of the robot structure. One approach to realize all these variations is the computer-aided structural analysis using optimization algorithms. In this paper's context, such an optimization strategy is investigated using the example of a parallel robot for use in a cryogenic working environment, and the results are compared with the structure aimed at so far.

It is well established that the performance of parallel robots is highly subject to their kinematic parameters which can be determined for a given structure in a *dimensional synthesis* [2]. The selection of the specific structure, i.e. the *structural synthesis*, is usually performed manually with the help of design and construction principles [3]. As the systematic structural synthesis of parallel robots by means of screw theory [4] or evolutionary morphology [5] provides a high number of suitable structures, the selection of the optimal solution is an exhaustive task. The concept of *combined structural and dimensional synthesis*, introduced in [6] for parallel robots, assumes that the optimal solution can be found by independently optimizing all possible structures and selecting the best one. This requires a high number of simulations of the robots kinematics and dynamics and is only applicable with a general, yet efficient model and its implementation and a suitable optimization algorithm.

The engineering solution to the considered handling problem is the linear Delta robot. It was already subject to parameter optimizations regarding workspace-related objectives [7] or objectives related to kinematics and dynamics [8]. A dimensional synthesis for both the classical Delta robot and the linear Delta was performed in [6] and used for a systematic comparison of the two.

The comparison of multiple parallel robots (whether two different structures or two sets of parameters for one structure) has to be performed using multiple criteria [2], representing all requirements to the robot. Often genetic algorithms are employed such as the Strength Pareto Evolutionary Algorithm [6, 8] or Nondominated Sorting Genetic Algorithm [9]. Particle swarm optimization (PSO) is reported to have better convergence than genetic algorithms for constraint nonlinear optimization problems. One reason is that not only the parameters of the current iteration carry information but also past iterations are taken into account to generate a new set of parameters [10]. Constraint handling [11] is central for the validity of the robot synthesis and the convergence of the PSO.

This paper presents results for the combination of the engineering solution and the combined synthesis presented above by taking the most restricting constraints of the task into

account for the structural synthesis and thereby vastly reducing the amount of possible structures, for which a dimensional synthesis has to be performed. The contributions of the paper
are

- transferring the specific constraints of cryogenic handling tasks in a suitable form for
  parameter optimization,
- proving the applicability of multi-objective PSO on the dimensional synthesis of parallel
  robots as opposed to genetic algorithms in literature,
- presenting design alternatives of the linear Delta for cryogenic handling.

The remainder of the paper is structured as follows: Sect. 2 gives an overview of the
constraints of the cryogenic handling task. The engineering approach to the robot synthesis is
presented in Sect. 3, followed by the combined synthesis in Sect. 4. The results are discussed
in Sect. 5.

2   Task Definition and Requirements

The freezing and storage of biological material in biobanks at temperatures below $-130\,°C$
is commonly referred to as cryopreservation. Manual handling of biological or toxic samples
is still the norm in research institutions. In such systems, the samples are often transferred
in, out, or moved by hand using bulky protective clothing. This poses considerable risk of
injury to the worker through cold burns as well as a threat to the sample integrity.

To overcome these problems, a parallel robot for the realization of full automation is
being developed.

2.1   Requirements for the Parallel Robot

The possibility of placing the drives in the warm area outside the storage container makes
the parallel structure interesting for use in cryogenic environments, as it allows the drive
technology to be decoupled from the cryogenic handling area. The drive movement is then
transferred to the end effector platform via passive joints.

The robot to be developed is subject to a number of geometric constraints. The installation
space dimensions correspond to the internal dimensions of the Cryotherm BIOSAFE cryogenic
storage container, which is to be used for the demonstrator (Fig. 1, left). The usable interior
space (height 680 mm and inner diameter 600 mm) is highlighted by the red dashed line in
the middle of Fig. 1.

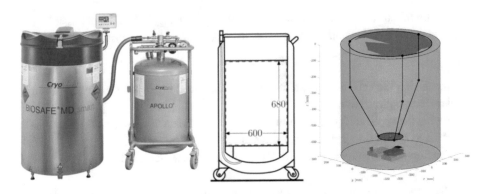

Abb. 1 Left: Cryotherm BIOSAFE cryogenic storage container, middle: dimensions of the installation space, right: robot handling scenario in a MATLAB simulation

For the storage of sample tubes in the cryogenic storage container, racks of type Micronic 96-3 are to be used, between which the tubes are to be transferred by the manipulator (see Fig. 1, right). The rack's height, including the sample tubes, is 45.2 mm, and the sample tubes height is 44 mm. To avoid collisions between the sample tubes to be transferred and the sample tubes stored in the racks during the pick-and-place process, the height of the necessary working space is set to 110 mm, to ensure a safety distance of approx. 20 mm. To keep the working space area as small as possible, the racks are placed lengthwise next to each other. The space next to it is used for a scanner, which will be used to identify the sample tubes. The resulting square area of the working space is 200 mm wide. To ensure a good thermal insulation of the cold area, the moving parts of the parallel robot have to cover a constant area on the cap of the container, favoring a vertical arrangement of linear drives.

2.2   Requirements for the Solid-State Joints

Extremely high demands are placed on the robot's passive joints: The extreme temperatures of below −130 °C do not allow the use of classic rigid body systems such as ball joints due to freezing of lubricants or jamming of components through cold shrinking. To avoid these disadvantages, flexure hinges in the form of cohesive swivel joints are used. Due to their monolithic structure, there are no parts that move against each other. Clamping is not possible and, in addition, the use of lubricants is not necessary. A major disadvantage of flexure hinges, however, is their low range of angles compared to conventional joints. Therefore, a parallel robot based on flexure hinges – depending on the required rotation angle limitation – can have a significantly reduced workspace compared to an otherwise identical parallel robot with conventional joints [12]. Furthermore, the negative influence of

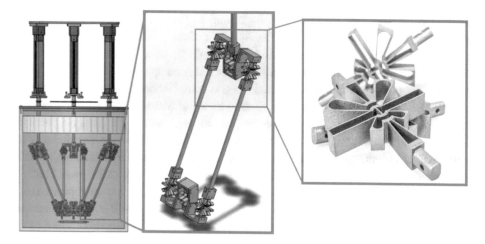

Abb. 2 Left: CAD rendering of a possible parallel robot structure (from [15]), middle: detail on the leg chain, right: flexure hinge photograph

cryogenic environmental conditions on flexure hinges' deformation behavior has not been investigated in detail so far.

A cascading flexure hinge, depicted in the right of Fig. 2, was developed based on the work of Fowler and Henein [13, 14]. The cohesive hinge is made of the titanium alloy TiAl6V4 by laser sintering due to this material's superior properties under cryogenic conditions. In preliminary work [15], it could be shown that a rotation angle of up to 30° (in one direction from the neutral position) and therefore a joint range of up to 60° can be realized with the developed flexure hinges. This *joint range* presents a major *constraint* regarding the robot's kinematics.

3 Engineering Approach and First Prototype

In a first approach the selection of the parallel structure was limited to one variant: Each kinematic leg chain consists of a vertically aligned linear drive and two passive universal joints, representing the common *linear Delta robot* [3, 8], see Fig. 2, left. Since both the inner and outer axes of the two universal joints are parallel to each other, a change in orientation is prevented, cf. [2–5], see Fig. 2, middle.

The system only has three translational degrees of freedom, required for handling the sample tubes during cryopreservation. In preceding works, a MATLAB tool was developed for a dimensional synthesis of this specific structure in the confined space. The main goal was to determine the parameter set from the set of possible combinations of the geometric parameters, in which the required joint angle ranges of the passive, solid-state joints are minimal. In addition, the optimal installation angles of the passive joints were calculated, at which the deflections from the rest position are minimal. Also, a workspace analysis was carried out for the determined optimum parameter set. A comprehensive description of the developed MATLAB tool based on a particle swarm optimization is omitted for the sake of brevity. The analysis showed that the optimized structure with a maximum angle range of the passive joints of only 46° experiences the least stress in the passive joints but poses the danger of singularities of the first type. Singularities of this type lie on the boundaries of the workspace and result, for example, from the stretching positions of individual link chains. It was assumed, that presetting the inclination in the universal joints to 26° would make it possible to avoid any singularities of the first type. Furthermore it was anticipated, that a larger inclination would reduce the necessary drive forces and thus result in smaller and more cost-effective drives. Due to the nature of the kinematic chains, the working space of the developed parallel robot structure can be represented as an overlap of three cylinders, in the sectional area of which the square area to be covered is located, which contains the bearing racks and the scanner. The minimum achievable application range of the actuator platform is, therefore a circle with the radius 141.42 mm. Based on the workspace restrictions in Sect. 2.1, the resulting bar length was calculated to 334.6 mm. As an illustration, a possible configuration of the resulting parallel robot is shown in Fig. 2. However, the construction shown here is only one of many possible configurations. With the experience gained from the reasoning of the manual design phase, the following systematic synthesis is performed in order to explore all possible solutions for the task and validate the preliminary design.

4   Combined Structural and Dimensional Synthesis

The parallel handling robot can – theoretically – be built up of a vast amount of possible leg chains [4, 5]. With a *structural synthesis* similar to [5], 51 unique leg chains consisting of revolute (R), prismatic (P) and universal (U) joints were identified for the cryogenic handling task described above. In the following, only serial kinematic leg chains without the parallelogram elements of Fig. 2 are selected. In a possible design step after the synthesis, joints with parallel axes can be kinematically replaced by parallelograms [3]. The alignment of base and platform coupling joint is not considered explicitly in textbooks on structural synthesis [4, 5]. However, to make use of the structural synthesis in combination with the dimensional synthesis, this aspect plays a crucial part. A general set of four possible alignments of the base coupling joint (radial, tangential, vertical or conically inclined to the base circle) and three alignments of the platform coupling joint (vertical, tangential and radial to the platform circle) are selected for evaluation. A brute-force approach by performing the dimensional synthesis for all $51 \times 4 \times 3$ combinations without task constraints has proven to be feasible, allowing automation and avoiding symbolic calculations, e.g. of screw vectors

[4]. Not all combinations provide a feasible parallel robot with full mobility and only 328 remaining valid structures are stored in a database. As minimizing motion in the area of thermal insulation is a hard requirement, only the vertical and conically inclined alignment of actuated prismatic base coupling joints is taken into consideration.

This leaves 33 specifiable parallel robots for the following *dimensional synthesis*, where the kinematic parameters of these structures are optimized. The 5–10 optimization parameters (depending on the structure) include the base and platform size and the inclination of conical base joints. Kinematic lengths are expressed with the Denavid-Hartenberg (DH) parameters in the notation of Khalil. An additional offset length between the prismatic joint and the next revolute joint is added to separate joints in cold and warm areas.

The robot is modeled to be of an aluminum alloy with thin struts as hollow cylinders ($\varnothing 53$ mm, strength 3 mm) and a thin circular platform plate (strength 10 mm). An additional payload of 3 kg at the platform takes the gripper into account. The robot structure is modeled with rigid body dynamics by neglecting link elasticity [16]. The flexure hinges, i.e. all passive joints, are assumed to have a linear joint elasticity. The stiffness of 1.4 N m/27.5° is obtained using the finite element method within ANSYS of the joint depicted in Fig. 2, [15]. A reference trajectory for a pick-and-place application between the two racks as described in Sect. 2 is simulated for 37 positions. The inertial forces are simulated, but only play a minor part compared to forces from gravity and joint elasticities.

The *overall procedure of the dimensional synthesis* of a single robot structure was extended w.r.t. the authors previous work [16] and is sketched in Fig. 3. The *first major step* of the fitness function for a particle is the calculation of the inverse kinematics (IK) in all

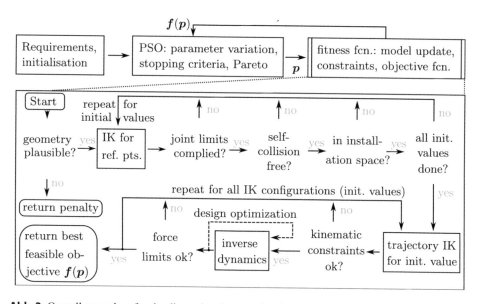

Abb. 3 Overall procedure for the dimensional synthesis of a robot

configurations for the 37 reference points. The IK configurations ("elbow up/down") are found by setting random initial values for the gradient-based IK algorithm and significantly change the outcome of the fitness evaluation, as constraints in this task are mostly only complied in one configuration. The violation of a constraint immediately leads to the abortion of the current configuration with the corresponding penalty term. After the computation of the IK, the prismatic joint offset is determined using a trust-region optimization and geometric considerations. This slightly influences the inertial forces due to the offset's mass. As a *second step*, the trajectory inverse kinematics and dynamics is calculated for all valid configurations, using a general methodology [2, 5, 16]. All constraints are again checked for the trajectory.

As the torques of the joint elasticities (considered as torsion springs) and therefore the actuator forces strongly depend on the flexure hinge rest positions, these present additional design parameters. A choice of the rest positions in the middle of the joint range for the trajectory minimizes the spring torque, but not the actuator forces, which present the design objective.

Therefore, an additional *design optimization* is performed for the rest positions of the flexure hinges as four parameters, assuming a symmetric robot. A pretensioning of the flexure hinges within the 55° angle range was allowed, producing a partial compensation of gravity by the spring torque. The design optimization loop is performed using a single-objective PSO minimizing the maximal actuator force.

The *constraints* are checked in the order of graveness of their violation and the computational effort to determine them. This presents a variation of the PSO "static penalty" approach [11] (static w.r.t. iteration count), termed "hierarchical constraints" in this work [16]. Examples of checked constraints in this order are

- geometric plausibility (leg length matching base/platform),
- success of the inverse kinematics (using a gradient-based solution),
- range of joint angles ($< 55°$ for the flexure hinges),
- self-collisions (using capsules as elementary geometry and axis-aligned bounding boxes as a first check),
- installation space (joint positions have to be inside the cylinder of Fig. 1),
- everything aforementioned for the trajectory IK,
- condition number of the manipulator Jacobian (< 200),
- actuator force in a reasonable range ($< 100\,\text{N}$),
- material stress (within a 50% safety distance of the material's limits).

The violation of an earlier check leads to a higher *penalty term* for the fitness value, where each constraint has a reserved range of values and all constraints are continuous depending on the degree of their violation (corresponding to inequality constraints). Each constraint violation leads to an immediate abortion of the current iteration to reduce the computational effort.

By this means, the time for one fitness evaluation ranges from nearly 0 s (quick check for invalid geometry or failure in reference point IK) over 0.5 s (kinematics constraints after trajectory IK), to 1.8 s (full objective function without additional design optimization) and 13 s (including design optimization of spring rest positions with 120 evaluations of the inverse dynamics). The computation was performed on a state-of-the-art Intel Xeon computing cluster system, using a MATLAB implementation. If all constraints are met, the maximum position error and the maximum actuator force are taken as two *objective functions*, using the multi-objective PSO algorithm from [10]. The position error is obtained by standard methods from [2] (with absolute values of the manipulator Jacobian), assuming 10 μm encoder accuracy of the linear drives. The physical values are normalized and saturated to a value smaller than the constraint penalties [16].

If a constraint is violated, both fitness values are equally set to the penalty. The feasible results and good convergence of the optimization show that the advantages of this approach (no constraint handling parameters, computationally efficient, MATLAB implementation available [10]) prevail the disadvantages (loss of diversity in the particle swarm [11]) for the optimization problem at hand.

5 Discussion of the Results of the Synthesis

Using the presented framework with 9 repetitions of the dimensional synthesis, 100 generations and 100 particles each, results with qualitatively good convergence were obtained for robots 5–12 displayed in Figs. 4 and 6, with around 500 valid results out of the 10,000 evaluations of the fitness function. Every optimization in this setting only takes about one to three hours (on the computing cluster), depending on the success rate and the IK convergence rate. Robots 1–4 of Fig. 4 needed more evaluations, which were provided by running 50 generations with 400 particles and an initial population consisting of the best results of the previous runs. The causing complexity of the kinematics can be deduced from the figures and the number of parameters n, ranging from 5 to 10. The Pareto fronts of all repetitions are combined into one and are shown in Fig. 5. A position accuracy of 40 μm was selected as a reference for the following detailed comparison in Table 1. Of the 33 different structures discussed above, 18 remain which fulfill the constraints and are regrouped for the sake of simplification (by neglecting their difference in platform coupling) to 12 remaining robots.

The structures 1–4 are clearly dominant over all others due to their actuation force lower than 40 N. However, the kinematic structure of numbers 3 and 4, visible in Fig. 4, is significantly more complicated than the engineering solution (structure 9 in Fig. 6), making their realization less likely. Structures 1 and 2 have moderate complexity and are even able to reach full isotropy (cond(J) = 1 in the whole workspace) for some particles on the Pareto front, which generally is very favorable [5]. A detailed analysis shows that a low actuator force in general is mainly enabled by a compensation of the effects of gravity and joint elasticity, together with a good force transmission from actuators to platform. From other runs of the optimization it is known that passive joint angle ranges of only 22° are possible for some structures, but the optimal results are over 40°. Therefore minimizing the angle range or the elastic joint torque does not directly benefit this objective, supporting the results from Sect. 3.

The 3PRUR-structures (numbers 5 and 6) present the second best alternative, as an actuator force of 56 N can be achieved with a conical alignment of the prismatic joints. The engineering solution of Sect. 3 corresponds to the 3PUU-kinematics (number 9), which has a similar performance as structures number 7 to 12. The engineering solution is evaluated in the last row of the table and also lies on the Pareto front, validating the different tools. All these structures (number 7 to 12) have a similar parallel and vertical alignment of joint axes, noted by ṘṘŔŔ. The main difference between numbers 7/8, 9/10 and 11/12 is the replacement of ṘŔ-pairs of revolute joints by universal joints, which sets the intermediate DH parameters to zero, but does not change the kinematic structure. For these structures a conical joint alignment is not sufficiently beneficial.

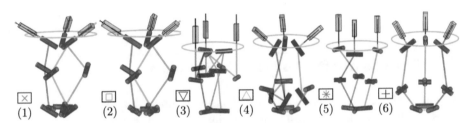

Abb. 4 Visualization of selected robot kinematics from Table 1 with markers from Fig. 5. Leg chains are printed in different colors and the circle marks the tank's upper edge

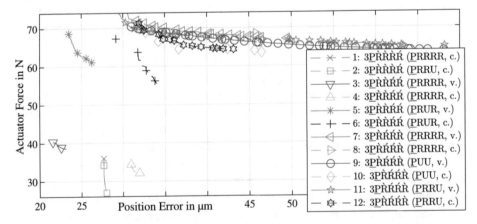

Abb. 5 Pareto fronts for all robot structures. The parallel robot notation is taken from [2] and the kinematic chain notation is taken from [4], where all Ṙ and Ŕ are parallel to each other, respectively. Base alignment noted with "v" (vertical) and "c" (conical)

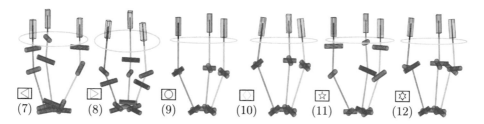

Abb. 6 Second half of the robot kinematics visualizations from Table 1

Tab. 1 Summary of one typical particle for each robot from the Pareto front. Abbreviations: "Cond." (condition number of Jacobian), range (of passive joint angles), mass (articulated, legs and platform without payload), n (number of optimization variables), r_B (base radius), φ_B (prismatic joint inclination), r_P (platform radius), q_{1off} (prismatic joint offset), a_i,d_i (DH parameters). Row "Eng.": engineering solution from Fig. 2

		Performance					n	Kinematic parameters								
		Err.	Force	Cond.	Range	Mass		r_B	φ_B	r_P	q_{1off}	a_3	d_3	a_4	d_4	a_5
		μm	N		deg	kg		mm	mm	deg	mm	mm	mm	mm	mm	mm
1	×	28	28	1.2	51.0	4.3	10	374	59	81	208	263	135	332	144	24
2	□	28	27	1.1	49.3	4.2	9	359	57	83	200	271	164	309	127	—
3	▽	23	39	5.3	53.7	6.1	8	223	0	80	408	330	106	151	91	390
4	△	32	32	3.5	46.8	5.1	9	297	50	80	212	314	316	151	118	233
5	∗	26	61	2.5	46.7	4.2	6	206	0	80	235	282	150	—	—	307
6	+	34	56	2.5	53.0	3.7	7	252	30	80	164	275	229	—	—	222
7	◁	40	69	3.8	40.8	3.6	8	225	0	80	158	165	52	321	143	34
8	▷	44	81	3.8	40.3	3.8	9	177	174	81	367	109	28	258	47	20
9	○	40	67	3.4	36.9	3.6	5	225	0	80	369	—	—	347	41	—
10	◇	42	64	3.9	33.5	3.6	6	191	172	80	306	—	—	399	104	—
11	☆	40	68	3.8	40.5	3.6	7	225	0	80	153	204	97	323	94	—
12	✪	40	65	3.5	34.4	3.7	8	207	174	80	327	22	38	374	69	—
Eng.		37	67	3.1	39.3	3.6	6	230	0	80	395	—	—	335	0	—

6 Summary and Outlook

Enhancing the assumptions in the combined structural and dimensional robot synthesis with knowledge from the engineering approach allows to vastly reduce the complexity of the optimization problem, without limiting the combined synthesis in the highly constrained cryogenic handling task. The comparison already proves the feasibility of the chosen design relative to other possible structures. The theoretical improvement of a design change is quantified to reduce the already low actuator force about 60%. This would require using two single revolute joints instead of one universal joint and may reduce the structural stiffness. Further investigations on replacing consecutive parallel joints by parallelogram subchains have to be performed before considering the design change. The findings on compensating gravity with elastic joint moments may be used in a pretensioning of the flexure hinges and in the control of the robot.

Danksagung The authors acknowledge the support by the Deutsche Forschungsgemeinschaft (DFG) under project numbers 341489206 (combined synthesis) and 349906175 (cryogenic handling). MATLAB code to reproduce the results is available at GitHub under https://github.com/SchapplM/robsynth-paper_mhi2021.

Literatur

1. Borchert, G., Löchte, C., Brumme, S., Carbone, G., Ceccarelli, M., Raatz, A.: Design methodology for a compliant binary actuated parallel mechanism with flexure hinges. In: F. Viadero, M. Ceccarelli (eds.) New Trends in Mechanism and Machine Science, pp. 171–179. Springer, Netherlands, Dordrecht (2013). https://doi.org/10.1007/978-94-007-4902-3_18
2. Merlet, J.P.: Parallel robots. Solid Mechanics and Its Applications, vol. 128, 2nd edn. Springer Science & Business Media (2006). https://doi.org/10.1007/1-4020-4133-0
3. Frindt, M., Krefft, M., Hesselbach, J.: Structure and type synthesis of parallel manipulators. In: Robotic Systems for Handling and Assembly, pp. 17–37. Springer (2010). https://doi.org/10.1007/978-3-642-16785-0_2
4. Kong, X., Gosselin, C.M.: Type Synthesis of Parallel Mechanisms. Springer, Berlin (2007). https://doi.org/10.1007/978-3-540-71990-8
5. Gogu, G.: Structural synthesis of parallel robots, part 1: methodology. Solid Mechanics and Its Applications, vol. 866. Springer, Netherlands (2008). https://doi.org/10.1007/978-1-4020-5710-6
6. Krefft, M.: Aufgabenangepasste Optimierung von Parallelstrukturen für Maschinen in der Produktionstechnik. PhD thesis, Technische Universität Braunschweig (2006)
7. Stock, M., Miller, K.: Optimal kinematic design of spatial parallel manipulators: application to linear delta robot. J. Mech. Des. **125**(2), 292–301 (2003). https://doi.org/10.1115/1.1563632
8. Kelaiaia, R., Company, O., Zaatri, A.: Multiobjective optimization of a linear delta parallel robot. Mech. Mach. Theory **50**, 159–178 (2012). https://doi.org/10.1016/j.mechmachtheory.2011.11.004
9. Jamwal, P.K., Hussain, S., Xie, S.Q.: Three-stage design analysis and multicriteria optimization of a parallel ankle rehabilitation robot using genetic algorithm. IEEE Trans. Autom. Sci. Eng. **12**(4), 1433–1446 (2015). https://doi.org/10.1109/TASE.2014.2331241
10. Coello, C.A.C., Pulido, G.T., Lechuga, M.S.: Handling multiple objectives with particle swarm optimization. IEEE Trans. Evol. Comput. **8**(3), 256–279 (2004). https://doi.org/10.1109/TEVC.2004.826067. Code from V. Martínez-Cagigal
11. Mezura-Montes, E., Coello, C.A.C.: Constraint-handling in nature-inspired numerical optimization: past, present and future. Swarm Evol. Comput. **1**(4), 173–194 (2011). https://doi.org/10.1016/j.swevo.2011.10.001
12. Hesselbach, J., Raatz, A., Kunzmann, H.: Performance of pseudo-elastic flexure hinges in parallel robots for micro-assembly tasks. CIRP Ann. **53**(1), 329–332 (2004). https://doi.org/10.1016/S0007-8506(07)60709-4

13. Fowler, R., Maselli, A., Pluimers, P., Magleby, S., Howell, L.L.: Flex-16: a large-displacement monolithic compliant rotational hinge. Mech. Mach. Theory **82**, 203–217 (2014). https://doi.org/10.1016/j.mechmachtheory.2014.08.008

14. Henein, S., Spanoudakis, P., Droz, S., Myklebust, L.I., Onillon, E.: Flexure pivot for aerospace mechanisms. In: 10th European Space Mechanisms and Tribology Symposium, San Sebastian, Spain, pp. 285–288 (2003)

15. Jahn, P., Raatz, A.: Numerical simulation and statistical analysis of a cascaded flexure hinge for use in a cryogenic working environment. In: Annals of Scientific Society for Assembly, Handling and Industrial Robotics, pp. 81–94. Springer, Berlin (2020). https://doi.org/10.1007/978-3-662-61755-7_8

16. Schappler, M., Ortmaier, T.: Dimensional synthesis of parallel robots: unified kinematics and dynamics using full kinematic constraints. In: 6. IFToMM D-A-CH Konferenz. Lienz, Österreich (2020). https://doi.org/10.17185/duepublico/71211

Secure Clamping of Parts for Disassembly for Remanufacturing

Simon Rieß, Jonas Wiedemann, Sven Coutandin and Jürgen Fleischer

Abstract

Robot based remanufacturing of valuable products is commonly perceived as promising field in future in terms of an efficient and globally competitive economy. Additionally, it plays an important role with regard to resource-efficient manufacturing. The associated processes however, require a reliable non-destructive disassembly. For these disassembly processes, there is special robot periphery essential to enable the tasks physically. Unlike manufacturing, within remanufacturing there are End-of-Life (EoL) products utilized. The specifications and conditions are often uncertain and varying. Consequently the robot system and especially the periphery needs to adapt to the used product, based on an initial examination and classification of the part. State of the art approaches provide limited flexibility and adaptability to the disassembly of electric motors used in automotive industry. Especially the geometrical shape is a limiting factor for using state of the art periphery for remanufacturing. Within this contribution a new kind of flexible clamping device for the disassembly of EoL electrical motors is presented. The robot periphery is systematically developed

S. Rieß (✉) · J. Wiedemann · S. Coutandin · J. Fleischer
Karlsruhe Institute of Technology, 76131 Karlsruhe, BW, Germany
e-mail: simon.riess@kit.edu

J. Wiedemann
e-mail: jonas.wiedemann@student.kit.edu

S. Coutandin
e-mail: sven.coutandin@kit.edu

J. Fleischer
e-mail: juergen.fleischer@kit.edu

T. Schüppstuhl et al. (eds.), *Annals of Scientific Society for Assembly, Handling and Industrial Robotics 2021*,
https://doi.org/10.1007/978-3-030-74032-0_7

regarding the requirements stemming from the remanufacturing approach. It consists of three clamping units with moveable pins. Utilizing two linear axes, a two dimensional working space is realized for clamping the parts depending on their conditions and shape.

Keywords
Clamping System · Disassembly · Remanufacturing

1 Introduction

Circular economy and remanufacturing as part of it has dragged a lot of attention within the last years. It is expected, that the numerous advantages will lead to an increased number of realized remanufacturing solutions in future Tolio et al. [1]. Among the advantages, there are fewer investments for manufacturers in producing and selling a product [2], less energy consumption during production and a smaller environmental impact [3]. However, there are some major challenges to overcome for the realization and operating of remanufacturing production lines. After assembly and during its life cycle, products naturally underlie changes and physical effects. Therefore the products appear with alterations after their lives at the remanufacturing line. Products are referred to be of *unknown specifications* upon feeding into the remanufacturing production system. Consequently, today's remanufacturing lines include human labor to handle the uncertainties with human intuition [4]. To be economically competitive on the other hand, industry has an interest to automate the remanufacturing process. Automating processes with a high number of product variations demands a high flexibility from the production system [5]. Handling devices and utilized machine equipment need to fit the processed products. This contribution presents a clamping system for fixation of parts for disassembly operations. The novelty of the system is the application of reconfigurable support modules in the domain of disassembly for remanufacturing of electric motors. Due to the adaptability of the clamping system, a high flexibility regarding product variations is realized. The clamping system is presented in a CAD model and a functional prototype, which proves the functionality of the approach.

2 Current State of the Art of Clamping Devices in Disassembly for Remanufacturing Applications

Many of today's products are not designed for remanufacturing or disassembly, but for their usage within the life-cycles [6]. Consequently, disassembly for remanufacturing is rather complicated to realize. For example, hardly reachable connections, certain fastening and assembly principles, surface coatings and untracked influences during the life-cycle on the connections complicate the dismantling. At the same time, the disassembly step is considered as one of the key challenges to realize for its direct influence on the subsequent re-assembly of the product. The relevant connectors need to be dismantled in such a way, that following process steps need not rework the connection leading to additional expenses and thus a minimization of the economic benefit [7]. During the dismantling of connectors, there are physical process forces occurring which need to be taken into account. Therefore a fixation of the part is mandatory. In industry, different kinds of grippers and clamping systems are used. In academia, there is research being conducted to develop adequate gripping and clamping systems. The following paragraph gives a summarizing overview of related work.

The current state of the art of industrial clamping devices for remanufacturing applications has a twofold character on the level of flexibility. On the one end of the spectrum, there are automated disassembly systems which are typically capable of processing a small spectrum of products and provide limited adaptability towards different products. Typical examples can be found in the disassembly of mobile phones or single use cameras [8, 9]. Utilized clamping devices are thus specially designed tools covering a small number of product derivatives and product variations. On the other end of the spectrum, there are disassembly systems with a rather large degree of flexibility based on manual labor. Examples are to be found in the disassembly for remanufacturing in gearbox remanufacturing [10]. Utilized clamping devices for disassembly are typically manually actuated clamping vices with the human deciding on the best clamping position, orientation and forces. Because of the mandatory flexibility, clamping devices have been subject to intensive research efforts. In an early work, until then existing optimizations methods for the selection of support positions in fixture designs are summarized by Menassa and DeVries [11]. It is found, that kinematic analysis from CAD simulations and finite element analysis can solve the problem appropriately, given the processing forces. In a more recent contribution, Kaya optimizes the layout of a fixture system with the aid of genetic algorithms [12]. It is thus possible to solve fixture layout problems with regard to previously defined process forces from a simulation. The adaption of clamping systems to the physical shape of the product is a related question also. The state of the research tackles this question by reconfigurable systems. There are systems which can configure towards different geometrical shapes. The contribution of Jonsson et al. serves as example [13]. In this contribution, there are multiple fixture elements being placed on a rig depending on the operations. After one part has been processed, the clamping system may be reconfigured for different parts. Another work by Brost and Peters studies the automated design of 3-D fixtures [14]. With the developed pin board tool, arbitrarily shaped parts can be clamped on a planar surface and the system may be reconfigured afterwards. Modern trends in the design of clamping layouts can be summarized to active fixturing [15]. Clamping devices of this kind are equipped with elements, which modify the applied forces on the part with respect to variable inputs. A representative contribution is the work of Valisek et al. Within this contribution, a pneumatic clamping device equipped with multiple sensors is developed [16]. Summarizing the state of the art one can say that many solutions for the clamping of parts in different applications exist and related questions were answered. Clamping systems in disassembly for remanufacturing applications in special have not been subject to research however. Thus existing solutions for the configuration towards different parts which are subject to external influences during their lifecycles has not been regarded. Especially for the use case of automated disassembly of electric motors, a clamping system is required.

3 Concept of the Clamping System for Disassembly for Remanufacturing

3.1 Extracting Requirements

Functional requirements are extracted from the analysis of the state of the art and explained in detail below. The given list is may not be complete for all disassembly tasks. It sums up the requirements for manipulation tasks for the remanufacturing of electrical motors. Different use cases may have special requirements.

1. Processing of multiple part derivatives on same remanufacturing production system

   As discussed, economically operated remanufacturing lines need to adapt physically to multiple product derivatives. This is because of the assumption, that multiple products are being processed on one solely remanufacturing line. Correspondingly, the clamping devices need to be of such kind to enable the clamping despite the possible physical variations of derivatives. One possible solution is the configuration of clamping elements to different products by adaptive clamping elements.

2. Reliable adaption of clamping unit to uncertain classification of parts

   Even in case of an ambiguous, vague or an erroneous classification of parts, the fixation of the part has to be reliable. If such a fixation cannot be realized, the clamping system needs to report this information to the higher control device. In the best case, the clamping device is capable of sensing critical data, to improve the classification.

3. Withstand process forces without irreversible distortion and appropriate clamping forces to enable dismantling process without damaging the core at the same time

   The clamping device has to withstand the applied process forces. Therefore, the mechanical structure needs to be of a solid construction. Applied clamping forces need to be large enough to enable the dismantling of the product. Yet at the same time, they cannot exceed the physical limits of the part, leaving irreversible damages.

4. Exposition of connectors to handling device for dismantling

In order to expose the areas of interest to other handling devices, the positioning and orientation of the part within the clamping device needs to be planned. Therefore, robust and adaptive planning algorithms are mandatory.

The mainly addressed products to be disassembled within the regarded use-case are two kinds of electric motors for automotive applications. One is a starter motor for combustion engines and the other one is a motor which is used for different auxiliary tasks such as window lifting or the actuation of wipers. In Fig. 1 some of the motor derivate are shown. Due to the utilization in different applications within the car, the smaller motor has different kinds of flanges (3) which are varying in shape and size. The disassembly of the motors includes the removal of the screws (4) and the dismantling of housing components (2).

During the disassembly there are process forces occurring. Especially while unscrewing a certain normal force is mandatory to avoid contact loss of the screw driver and the screws. For the development of the clamping system, the disassembly forces and the possible clamping positions have been analyzed. The regard revealed that the mandatory screws are positioned in an offset location from the feasible clamping positions on the housing at the small motor. Therefore, the applied force during unscrewing generates a momentum on the clamped part. The generated torque exceeds the possibilities from a clamping application in a two position grip and a support position closer to the force vector stemming from the unscrewing process is mandatory. Therefore, the special novelty in this development is the utilization of a supportive, third form clamping system which can be moved to an adequate position in order to cope the process forces of the disassembled part. Since there are different motor types being processed on the clamping system, it needs to adapt individually to the motor type because of the processed part derivatives of the smaller motor. This requires a modification of the support system which is realized by a gantry system.

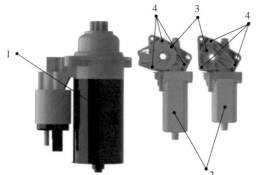

Comp.	Description
1	Starter Motor for combustion Engines
2	Motor housing
3	Flange
4	Screws

Fig. 1 Different kinds of addressed electric motors

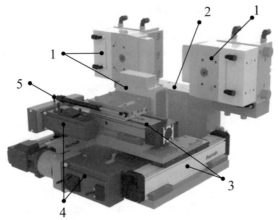

Comp.	Description
1	Matrix form clamping system
2	Pneumatic piston
3	XY Gantry system
4	Motors
5	Linear sensor

Fig. 2 CAD model of the clamping device for disassembly in remanufacturing applications

3.2 Concept Proposal

Figure 2 shows the CAD model of the system. It consists of three matrix pin clamping elements (1), which can be pneumatically locked. Those are the adaptive clamping units and they consists of several metal pins on top of a spring each. By pushing an arbitrary shape onto the pin surface, the negative contour is formed. After pneumatic actuation, the pins are held in place and are not pushed back by the spring. Thus, it is possible to generate form closure independent from the shape and condition of the part. The clamping force is generated by a pneumatic piston (2), actuating one of the three clamping elements against a second, fixed one. The third clamping unit is located on a XY Gantry system (3) in such a form, that it is possible to reconfigure the clamping system to different parts by moving the clamping elements. The motors for the actuation of the gantry system (4) are arranged in such a way that disturbing volume in the clamping area is avoided. Using linear sensors (5), the position of the clamping devices can be detected. The selection of an appropriate clamping orientation and position of the parts is part of the research project, but not of this contribution.

4 Prototype Clamping System

For the verification of the approach, a functional prototype is built and described in the following. The prototype uses the same pin clamping elements and the generation of the clamping force is likewise in a pneumatic fashion. At the same time, the gantry system is not used, but the system can be manually configured towards the motor derivate by placing the vertically oriented pin clamping element on bolts. Figure 3 shows an image of the prototype.

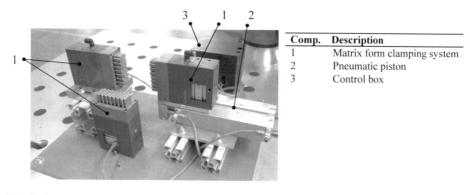

Comp.	Description
1	Matrix form clamping system
2	Pneumatic piston
3	Control box

Fig. 3 Prototype system of a clamping device for disassembly in remanufacturing applications

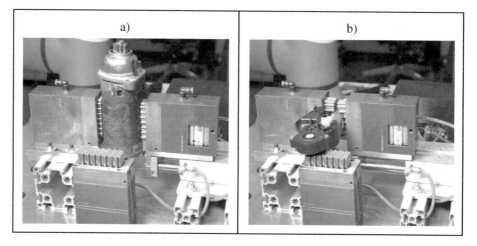

Fig. 4 Depiction of clamping results

The clamping device is faced with the task of clamping both different motor types and therefore setting a corresponding configuration. Afterwards there are screws being removed and the utilized components thereby proven for the capability of withstanding the process forces. Figure 4. shows the clamping of the different motors in comparison. In both cases, the motor is being clamped securely and the components withstand the process forces during disassembly.

5 Conclusion and Future Work

In this contribution, the results of the development of a clamping device for applications in disassembly for remanufacturing has been presented. The requirements emerging from the remanufacturing application have been summarized and respected in the development. The prototypical realization of a comparable module has been shown. The conducted tests verify the approach and the conceptual result. The special novelty of the clamping system is the adaptation of the support position with regard to the disassembly for remanufacturing operations of electric motors at their after-life stage. For future examinations, the presented clamping system needs to be built up to the final expansion stage. Additionally, there are optimization methodologies required for the configuration of the equipment. With adapted methodologies it is possible to identify ideal clamping positions for subsequent processes such as unscrewing or handling tasks.

Acknowledgements The authors would like to express their appreciation to all industry and research partners for supporting the project "AgiProbot". The project is funded by the Carl Zeiss Foundation.

References

1. Bakker, O., Papastathis, T., Popov, A., et al.: Active fixturing: literature review and future research directions. Int. J. Prod. Res. **51**, 3171–3190 (2013)
2. Brost, R., Peters, R.: Automatic design of 3-D fixtures and assembly pallets. Int. J. Robot. Res. **17**, 1243–1281 (1998)
3. D'Adamo, I., Rosa, P.: Remanufacturing in industry: advices from the field. Int. J. Adv. Manuf. Technol. **86**, 2575–2584 (2016)
4. Graham, I., Goodall, P., Peng, Y., et al.: Performance measurement and KPIs for remanufacturing. J. Remanufacturing 5(10), 2210–2227 (2015)
5. Heinrich, B., Linke, P., Glöckler, M.: Grundlagen Automatisierung. Springer Fachmedien Wiesbaden, Wiesbaden (2020)
6. Jonsson, M., Kihlman, H., Ossbahr, G.: Coordinate controlled fixturing for affordable reconfigurable tooling. In: Proceedings of the 2nd CIRP Conference on Assembly Technologies and Systems, vol.1, pp. 1–11 Elsevier B.V, Amsterdam (2008)
7. Kaya, N.: Machining fixture locating and clamping position optimization using genetic algorithms. Comput. Ind. **57**, 112–120 (2006)
8. Liam—An Innovation Story. https://www.apple.com/environment/pdf/Liam_white_paper_Sept2016.pdf. Last accessed 17 Dec 2020
9. Menassa, R., DeVries, W.: Optimization methods applied to selecting support positions in fixture design. J. Eng. Ind. **113**, 412–418 (1991)
10. Nave, M.: Beitrag zur automatisierten Demontage durch Optimierung des Trennprozesses von Schraubenverbindungen. Doctoral Thesis, Universität Dortmund (2003)
11. Peeters, J., Vanegas, P., Mouton, C., et al.: Tool design for electronic product dismantling. In: The 23rd CIRP Conference on Life Cycle Engineering. Procedia CIRP, vol. 48, pp. 466–471. Elsevier B.V, Amsterdam (2016)

12. Remanufacturing Market Study, For Horizon. https://www.remanufacturing.eu/assets/pdfs/remanufacturing-market-study.pdf (2020). Last accessed 17 Dec 2020
13. Tolio, T., Bernard, A., Colledani, M., et al.: Design, management and control of demanufacturing and remanufacturing systems. CIRP Ann. **66**(2), 585–609 (2017)
14. Velíšek, K., Košt'ál, P., Zvolenský, R.: Clamping Fixtures for Intelligent Cell Manufacturing. Springer, Berlin (2008)
15. Vongbunyong, S., Chen, W.: Disassembly Automation: Automated Systems with Cognitive Abilities. Springer Cham, Heidelberg (2015)
16. Wegener, K., Chen, W., Dietrich, F., et al.: Robot assisted disassembly for the recycling of electric vehicle batteries. In: The 22nd CIRP conference on Life Cycle Engineering. Procedia CIRP, vol. 29, pp. 716–721. Elsevier B.V, Amsterdam (2015)

Aerial Grasping and Transport Using an Unmanned Aircraft (UA) Equipped with an Industrial Suction Gripper

Markus Lieret, Benedikt Kreis, Christian Hofmann, Maximilian Zwingel and Jörg Franke

Abstract

Due to the availability of highly efficient unmanned aircraft (UA) and the advancement of the necessary technologies, the use of UA for object manipulation and cargo transport is becoming a more and more relevant research area. A reliable identification and localization of cargo and interaction objects as well as maintaining the required flight precision are essential to guarantee a successful object handling. Within this paper we demonstrate the successful application of an autonomous UA equipped with a lightweight suction gripper for object interaction. We discuss the approach used for precise localization as well as the identification and pose estimation of individual gripping objects. Concluding, the overall system performance is evaluated within an industrial-oriented use case.

M. Lieret (✉) · B. Kreis · C. Hofmann · M. Zwingel · J. Franke
Institute for Factory Automation and Production Systems (FAPS), Friedrich-Alexander-Universität Erlangen-Nürnberg (FAU), Erlangen, Germany
E-mail: markus.lieret@faps.fau.de

B. Kreis
E-mail: benedikt.kreis@fau.de

C. Hofmann
E-mail: christian.hofmann@faps.fau.de

M. Zwingel
E-mail: maximilian.zwingel@faps.fau.de

J. Franke
E-mail: joerg.franke@faps.fau.de

© The Author(s) 2022
T. Schüppstuhl et al. (eds.), *Annals of Scientific Society for Assembly, Handling and Industrial Robotics 2021*,
https://doi.org/10.1007/978-3-030-74032-0_8

Keywords

Autonomous unmanned aircrafts • Flying robots • Aerial grasping • Object detection • Localization

1   Introduction

Over the last years the application of UA for cargo transport has gained importance due to rapid technological advancements and increasing efficiency, payload and flight time. UA are already used to deliver units of stored blood, emergency medication, parcels, food and other goods within urban and rural areas. In terms of industrial usage, UA are increasingly being considered for transport as they allow direct routes, reduced delivery times and can provide a cost-efficient complement to existing logistics solutions. The applications considered in this paper include not only the transport of urgently needed spare parts and tools, but also the cyclical delivery of small and lightweight parts and components [1].

Besides the periodic cargo transport UA are also convenient for the interaction with objects and their manipulation, for example in search and rescue scenarios, construction or maintenance. Focusing on industrial applications, UA equipped with suitable grippers can provide efficient solutions to the extraction of individual parts from various load carriers. Within the load carriers, the components can be provided parcelled and ordered (e.g. cardboard boxes), parcelled and unordered (e.g. plastic bags) as well as loose and unordered.

Key requirements for those applications are the precise localization of the UA in relation to the interaction object as well as a gripping device suitable for the specific task. Based on those requirements, we developed a solution for the precise localization and grasping of different objects. Focusing on object interaction and manipulation the proposed method ensures a precise localization between the UA and the interaction object and is suitable for applications with grippers or different tools.

To validate the proposed methodology, we use an industrial use-case whereby the UA picks up components parcelled in cardboard boxes from a small load carrier (SLC). The evaluation consists of the automated localization and pick-up of the cargo objects, the subsequent transport as well as the defined drop-off of the goods.

The paper is structured as follows: After presenting related work and selecting a suitable gripping technology for the targeted use-case we present the system architecture and provide details on the used approaches for object localization and positioning. The overall system is then evaluated within the aforementioned use case and possible improvements and future work are deduced.

2   Related Work

The working principles of grippers used for ground based robots do not differ from those designed for aerial grasping. Nevertheless, a successful aerial grasping operation is more challenging than handling objects with industrial robots. UA are constantly moving even when hovering at a fixed position. That is why it is crucial to compensate these movements with an intelligent gripper design. Many different approaches have been presented to allow the aerial gripping of cargo objects such as impactive, ingressive, astrictive and contiguitive grippers. The most relevant of these approaches are discussed below [2, 3].

Extensive research has already been conducted in the field of aerial gripping. Many of the approaches focus on the application of mechanical grippers. Those grippers show promising results and allow the grasping of various objects but are limited in their ability to deal with position errors of an UA [4, 5]. Other highly flexible manipulators with several joints such as the one designed by Zhang et al. need a complex control system in order to be usable [6].

Furthermore, magnetic grippers as used by Gawel et al. [7] or Bähnemann et al. [8] require additional constructions to be mechanically compliant and their use is restricted to ferrous objects. Nevertheless, they have a significant advantage over mechanical grippers: Only one object surface needs to be accessible, which is an important requirement in many industrial picking scenarios using SLCs.

Another possibility is to use adhesive grippers that offer similar advantages as magnetic grippers but can handle non-ferrous objects. An exemplary use case applied to UA was studied by a research group of the University of Pennsylvania [9]. They have evaluated a self-constructed adhesive gripper regarding its ability of perching a UA on inclined and vertical surfaces. Adhesion may be used as well for object grasping considering certain prerequisites. It is especially important that the surface of the object is clean so that a grasping force can be applied. In an industrial environment this is hard to ensure except a clean room is available as it is required for the production of electronics.

Besides the mentioned grippers, suction grippers are a valid choice for UA as they can handle large and broad objects and require less positioning accuracy than jaw grippers. On the other hand, the transport objects are limited by their surface structure and to obtain the required holding forces, heavy and energy-consuming vacuum pumps are necessary, leading to lower payloads and reduced flight time. Kessens et al. [10] has presented a system featuring an autonomous quadrotor, an on-board vacuum pump and a gripper with four individual self-sealing suction cups. The gripper provides a maximum holding force of 6 N and is capable of gripping multiple objects with irregularly structured surfaces. Another gripper designed by Kessens et al. [11] is even able to hold forces up to 150 N and is suitable for lateral grasping.

In accordance with the aforementioned forms of provision, suction grippers are especially suitable to grasp objects out of a SLC, as they enable a frictional connection along the surface normal, thus do not require an enclosure of the part, and can apply forces to objects having different surface structures typically appearing in storage environments such as cardboard boxes and plastic bags.

3   Methodology and System Design

In this section the individual approaches for precisely localizing the UA and the detection and localization of the interaction objects will be presented. As indicated before, we focus on an industrial picking process and the corresponding general steps required for aerial object interaction and grasping as depicted in Fig. 1. After the take-off phase (not depicted) the UA flies to an initial position, roughly over the interaction object (1). When hovering above the interaction object, it is first identified (2) and then located in order to calculate a suitable grasping point using the object's estimated pose (3). As soon as the grasping point is reached, the suction gripper is activated and the object is extracted from the SLC (4). After the transport phase (5) the object is dropped off at the destination (6).

3.1   Object Identification and Localization

Approaches available for object identification and localization range from tags and optical markers attached to the objects to feature matching approaches and machine learning solutions. Each method has its own advantages and disadvantages which can be found in the relevant literature.

Within our procedure we apply an object matching algorithm based on features extracted from gray-scale images. Even though independent optical markers provide more accurate

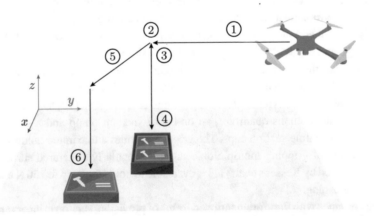

Fig. 1 Steps of the presented aerial grasping and transport process. The movement to an initial position (1) roughly above the interaction object is followed by the detection (2), localization (3), grasping (4), transport (5) and drop-off at the destination (6)

results and convolutional neural networks allow faster identification and localization of objects with images, the feature-based approach is still more suitable. It does not require the attachment of additional markers and does not need the extensive acquisition and annotation of training data. The feature-based solution can therefore easily be extended to different objects, which is especially suitable for warehouse applications where a large number of products is stored.

To be able to identify and locate the required objects the procedure is implemented as follows. A gray-scale image of every object side is taken leading to six images in the case of a cubic cardboard box. To uniquely identify the individual objects, we calculate a set of Speeded Up Robust Features (SURF) of every image as described by Bay et al. [12] and store them in a database to quicken the following matching process. During the flight phase the camera image is analysed and the calculated SURF descriptors are compared to the ones stored in the database. The matching is thereby done between the six training images of the grasping object and the camera image using a k-nearest neighbors algorithm in combination with Lowe's ratio test [13].

Based on the matching features, the perspective transformation between the camera image and the corresponding database image is calculated using the Random Sample Consensus algorithm. For further processing only the detected contour of the grasping object, which is described by its four corner points in the image plane, is required. To increase the robustness of the object detection, the length-width ratio of the rectangular contour is compared with the real object dimensions stored in the object database.

The spacial position of the interaction object in relation to the camera coordinate system is described as a Perspective-n-Point problem using the detected corner points, the known object dimensions and the intrinsic camera parameters. The problem is solved with the direct linear transformation followed by the Levenberg-Marquardt algorithm [14].

To transform the object's location from the camera coordinate system into the world frame, the known pose of the object and the camera mounting position are used. The overall transformation is described in Eq. 1 and its illustration and the denotation of the indices is shown in Fig. 2.

$$\substack{O\\W}T = \substack{U\\W}T \cdot \substack{C\\U}T \cdot \substack{O\\C}T \tag{1}$$

3.2   UA Localization

As stated before, we focus on a system architecture suitable for industrial applications. Therefore, we do not use a Motion Capturing or satellite-based localization system for localizing the UA but a Kinexon ultra-wideband (UWB) tracking system instead, which provides a tracking accuracy of less than ten centimetres and is widely used to track autonomous guided vehicles or cargo objects in industrial facilities.

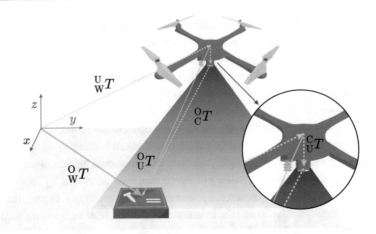

Fig. 2 Coordinate transformations used to determine the interaction object's global position. The indices have the following meaning: world (W), unmanned aircraft (U), camera (C), interaction object (O)

While the accuracy of the UWB system is adequate for basic flight operations within a production plant it is not sufficient for gripping the desired objects. Therefore AprilTags, optical black-and-white markers, are used to achieve the required position accuracy. They are placed on the ceiling above the object's storage place and face downwards so that they can be located using a camera mounted to the upper side of the UA.

The methodology used to fuse UWB tracking data and optical positioning information, an evaluation of the achieved in-flight positioning accuracy as well as the general system architecture used for object interaction with UA is presented in [15]. Thereby it has been shown, that the positioning accuracy based on UWB and additional optical tracking information is sufficient for grasping objects with lateral dimensions of 10 cm x 10 cm or more.

Since the used suction cup can only compensate vertical position deviations of up to 2 cm, an additional distance sensor is used to measure the vertical distance between the suction cup and the interaction object with millimeter precision. The sensor data is then used to detect a contact between the suction cup and the object as well as to determine whether the object was successfully gripped.

4   Evaluation

To evaluate the presented system several tests using the developed UA as shown in Fig. 3 were conducted and the results will be discussed below. The evaluation includes performance tests of the suction gripper as well as the detection and grasping of objects.

The UA is based on the DJI F450 airframe with suitable motors and electronic speed controllers, a Pixracer flight control unit (FCU) running the 1.9.2 stable release of the PX4

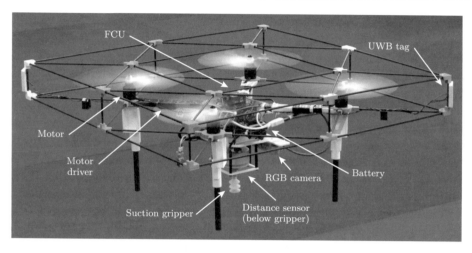

Fig. 3 Hardware setup of the UA developed for object grasping

autopilot and an Aaeon UP Board as companion computer. To generate the required retention force we use an electric vacuum pump provided by Schmalz [16]. To detect and locate the interaction objects two additional sensors are mounted to the UA: an Intel RealSense D435 camera and a STM VL6180X time-of-flight distance sensor.

4.1   Retention Force of the Suction Gripper

To evaluate the retention force of the suction gripper, a specimen with adjustable weight is grasped. The specimen's weight is thereby incrementally increased until the generated retention force is not enough to hold the specimen anymore. After the static test, the retention force is evaluated in-flight. In order to do this, the specimen is gripped and a take-off with maximum velocity of $1.0\,\text{m/s}$ and with maximum acceleration of $2.0\,\text{m/s}^2$ to a flight altitude of $1.0\,\text{m}$ is conducted, followed by a square shaped flight trajectory with a maximum velocity of $1.5\,\text{m/s}$ and a maximum acceleration of $2.0\,\text{m/s}^2$. The weight is increased again until the specimen falls down during the flight maneuvers.

The static test shows that the suction gripper can produce a holding force of $10.81\,\text{N}$ whereas during the flight phase objects with a weight force of up to $10.01\,\text{N}$ can be retained safely.

4.2   Object Detection and Localization

To evaluate the performance of the object detection, the detection rate and the corresponding computation time are measured. The evaluation is divided again into a static test where the UA is placed on an elevated platform and a dynamic in-flight test where the UA hovers above the grasping objects to analyse the influence of flight movements.

In each evaluation 100 measurements are taken in different distances to the interaction object to determine the influence of the object size in the image plane. The images are taken at a fixed resolution of 640×480 pixels and the evaluation is performed on a desktop computer equipped with an i7-8700 CPU, 32 GB RAM and a GTX 1080 GPU.

The results of the detection rate are shown in Fig. 4. The size of the green circles represents the percentage of successful detections and its center the distance to the object. In both cases the detection rate decreases as the object distance increases, but the effect is more significant when the UA is hovering. This is due to unavoidable in-flight movements of the UA which result in a reduced image quality and thus less features can be matched. With the used image resolution the detection rate drops significantly as soon as the distance between camera and interaction object exceeds a value of 60 cm. Therefore, the UA should hover around 30 cm to 50 cm above the interaction object to ensure a reliable detection and localization of the object.

Besides the detection rate, the time required to locate the interaction object is evaluated as described before. The measured execution time for the in-flight tests are shown in the right diagram of Fig. 4. It shows that with increasing distance to the object the execution time decreases as the number of extracted object features decreases and thus the feature matching requires less time. The worst observed execution time during flight is 242.20 ms while the median execution time is 100 ms at a distance of 36 cm and 80 ms at a distance of 76 cm. As the execution time directly depend on the number of detected features, during a

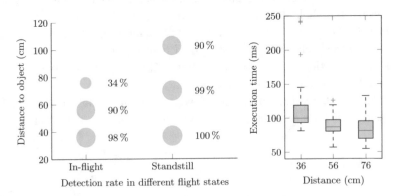

Fig. 4 Detection rate dependent on the distance between camera and grasping object (left) and time required for object detection and localization while hovering above the object (right)

static test under same conditions, the time required to locate can be larger than in-flight as more features need to be compared.

Based on the results of the examined detection rate, the UA should hover about 30 cm to 50 cm above the interaction object during the following evaluation of the grasping success rate in order to guarantee a reliable detection and localization of the object.

4.3   Grasping Success Rate

To evaluate the grasping success rate, the UA performs the procedure as described in Sect. 3. The storage position of the SLC containing the cardboard boxes is roughly determined by placing a UWB tag inside the SLC. The initial position setpoint sent to the UA is around 40 cm above the SLC to allow a precise localization of the cardboard box. After grasping, the UA ascends 50 cm and drops the object off at a random position 1.5 m away from SLC. The whole process is conducted autonomously and repeated ten times. The attempt is only considered successful when the object is precisely located, grasped and extracted from the SLC without being dropped. Based on this metric the object was grasped successfully nine out of ten times, resulting in a grasping success rate of 90 %. A successful grasp of a cardboard box is shown in Fig. 5.

Fig. 5 In-flight grasping of a cardboard box out of a small load carrier. The box tilts to the side, as it is not grasped exactly in its center point

The failed grasping attempt was caused by a tilting of the cardboard box so that it was caught on the rim of the SLC and the suction gripper could not provide enough retention force to hold the object.

5   Summary and Conclusion

Within this paper, we have presented an approach for the object interaction and aerial grasping of lightweight cargo objects using an UA equipped with an industrial grade suction gripper. While the overall system works well and the gripping was successful in 90 % of the attempts, there are still limitations that need to be addressed in future research.

As shown within the evaluation, using a single suction cup leads to objects leaning to the side and reduces the load restraint during the transport. Within future research we will therefore use additional suction cups in a linear or triangle arrangement.

This leads to other issues, such as improving the positioning accuracy to guarantee that all suction cups fit on the object's surface. To improve the accuracy of the position and eliminate the currently required optical markers at the same time, we plan to replace the marker based localization by an alternative tracking approach such as optical flow and visual odometry.

Acknowledgements The project receives funding from the EU ECSEL Joint Undertaking under grant agreement n737459 (project Productive4.0) and the German federal ministry of education and research under grant agreement 16SE0198.

References

1. Maghazei, O., Netland, T.: Drones in manufacturing: exploring opportunities for research and practic. J. Manuf. Technol. Manag. vol. Forthcoming (2019)
2. Mohiuddin, A., Tarek, T., Zweiri, Y., Gan, D.: A survey of single and multi-UAV aerial manipulation. Unmanned Syst. **08**(02), 119–147 (2020)
3. Wolf, A., Schunk, H.: Grippers in motion: the fascination of automated handling tasks. Hanser, Hanser eLibrary, München (2018)
4. Thomas, J., Loianno, G., Polin, J., Sreenath, K., Kumar, V.: Toward autonomous avian-inspired grasping for micro aerial vehicles. Bioinspiration Biomim. **9**(2) (2014)
5. Pounds, P.E.I., Bersak, D.R., Dollar, A.M.: Practical aerial grasping of unstructured objects. In: 2011 IEEE Conference on Technologies for Practical Robot Applications, pp. 99–104, IEEE; Mo 11.04.2011 - Di 12.04.2011
6. Zhang, G., He, Y., Dai, B., Gu, F., Yang, L., Han, J., Liu, G., Qi, J.: Grasp a moving target from the air: system & control of an aerial manipulator. In: 2018 IEEE International Conference on Robotics and Automation (ICRA), pp. 1681–1687, IEEE; Mo 21.05.2018 - Fr 25.05.2018
7. Gawel, A., Kamel, T., Novkovic, M., Widauer, J., Schindler, D., von Altishofen, B.P. , Siegwart, R., Nieto, J.: Aerial picking and delivery of magnetic objects with MAVs. In: 2017 IEEE International Conference on Robotics and Automation (ICRA), pp. 5746–5752 (2017)

8. Bähnemann, R., Pantic, M., Popović, M., Schindler, D., Tranzatto, M., Kamel, M., Grimm, M., Widauer, J., Siegwart, R., Nieto, J.: The ETH-MAV team in the MBZ international robotics challenge. J. Field Robot. **36**(1), 78–103 (2019)
9. Thomas, J., Loianno, G., Pope, M., Hawkes, E.W., Estrada, M.A. , Jiang, H., Cutkosky, M.R., Kumar, V.: Planning and control of aggressive maneuvers for perching on inclined and vertical surfaces. In: Volume 5C: 39th Mechanisms and Robotics Conference, American Society of Mechanical Engineers, 08022015
10. Kessens, C.C., Thomas, J., Desai, J.P., Kumar, V.: Versatile aerial grasping using self-sealing suction. In: 2016 IEEE International Conference on Robotics and Automation (ICRA), pp. 3249–3254 (2016)
11. Kessens, C.C., Horowitz, M., Liu, C., Dotterweich, J., Yim, M., Edge, H.L.: Toward lateral aerial grasping manipulation using scalable suction. In: 2019 International Conference on Robotics and Automation (ICRA), pp. 4181–4186 (2019)
12. Bay, H., Tuytelaars, T., van Gool, L.: Speeded-up robust features (surf). Comput. Vis. Image Underst. **110**(3), 346–359 (2008)
13. Lowe, D.G.: Distinctive image features from scale-invariant keypoints. Int. J. Comput. Vis. **60**(2), 91–110 (2004)
14. Moré, J.: The levenberg-marquardt algorithm: Implementation and theory. In: G. Watson, (ed.), Numerical Analysis, volume 630 of Lecture Notes in Mathematics, pp. 105–116. Springer, Berlin (1978)
15. Lieret, M., Lallinger, M., Tauscher, M., Franke, J.: Localization and grasping of small load carriers with autonomous unmanned aerial vehicles. In: Annals of Scientific Society for Assembly, Handling and Industrial Robotics, pp. 241–250. Springer, Berlin (2020)
16. Vacuum generator ECBPM - operating instructions. https://pimmedia.schmalz.com/Dokumente/Bedienungsanleitung/10/1003/100301/10030100556/BAL_10.03.01.00556_en-EN_00.pdf (2020). Accessed 22 Sept 2020

Computing Gripping Points in 2D Parallel Surfaces Via Polygon Clipping

Ludwig Vogt, Yannick Zimmermann and Johannes Schilp

Abstract

To generate suitable grasping positions between tessellated handling objects and specific planar grippers, we propose a 2D analytical approach which uses a polygon clipping algorithm to generate detailed information about the intersection between both objects. With the generated knowledge about the intersection we check whether its shape fits to the set criteria of the operator and represents a valid grasping position. Before the polygon clipping algorithm is applied, a preprocessing step is performed, where appropriate surfaces from the handling object and the gripper are extracted. After rotating all surfaces into a common plane, potential clipping positions are detected and the clipping is performed to get an accurate intersection detection. The validation shows comparable running times to a OBBTree algorithm (0.1 ms per grasping position) while increasing the stability of the results from 30 to 100% for the evaluated test objects.

Keywords

Handling · Polygon clipping · Gripping point determination

L. Vogt (✉) · Y. Zimmermann · J. Schilp
Chair of Digital Manufacturing, Faculty of Applied Computer Science, Augsburg University,
Eichleitnerstr. 30, 86159 Augsburg, Germany
e-mail: Ludwig.Vogt@informatik.uni-augsburg.de

J. Schilp
Fraunhofer Research Institution for Casting, Composite and Processing Technology–IGCV,
Provinostr. 52, 86153 Augsburg, Germany

T. Schüppstuhl et al. (eds.), *Annals of Scientific Society for Assembly,*
Handling and Industrial Robotics 2021,
https://doi.org/10.1007/978-3-030-74032-0_9

1   Introduction

With a rising degree in automated process chains and a continuously demand of smaller lot sizes the requirements for automated product handling are increasing. Additionally, additive manufacturing enables the production of topology optimized parts which can have a very complex and delicate shape or structure and are therefore difficult to handle [1].

The automated handling process and the determination of gripping points has been a popular research field in recent times [2, 3]. For the determination of gripping points two main approaches exist: an analytical approach, taking mechanical and physical properties of the object into account and an empirical approach, trying to replicate the movement of the human hand [2]. While the empirical grasping approach uses neural networks [4] or fuzzy logic [5], the analytical grasping approach includes mathematical models to determine the contact between the gripper and the object [6]. A central part of both strategies is to find suitable grasping positions on the object. To form stable grasps, it is aspired to generate a maximum overlap between the gripping surface and the object. For some handling tasks, only partial overlapping is realizable without changing the grippers. To control this, our main target is to generate a complete grasp set which contains all planar grasps with information about the overlapping between gripper and handling object. Therefore, it is necessary to accurately determine the shape of the contact area. While the empirical approaches show a fast and computational efficient solution they need extensive and high quality training data to generate accurate models for each gripper [7]. Trained models are also gripper specific and difficult to use for other gripper shapes. In analytical methods a collision detection can be used to check for a correct gripper alignment and intersection area, but often contacts are calculated via bounding boxes with primitive shapes [8] and the contact shape is approximated or neglected.

To generate information about the intersection between the object and the gripper a 3D collision detection based on a Oriented Bounding Box Tree (OBBTree) from the vtk library was first used [9]. Although the collision detection was robust and performing well for boundary contacts, the intersection determination was not suitable for our application. Tests showed a good approximation for basic intersection shapes e.g. rectangle, triangle and circle but unsatisfying or no output at all for concave shapes containing holes and crooked intersection planes (cf. Fig. 1).

For this reason, a 2D analytical approach where we use a polygon clipping algorithm to determine area gripping points is proposed.

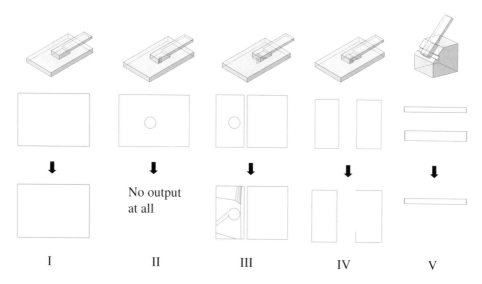

Fig. 1 Results from the intersection detection (red) with an OBBTree algorithm. Second row shows the desired intersection shape, last row shows the computed output. Intersecting cases I-IV lay in the XY plane and case V in a crooked plane

2 State of the Art

Before giving an overview about the various methods to determination gripping points for planar grippers and their intersections, we give a short introduction about the basic concept of polygon clipping and detail the most used algorithms.

2.1 Polygon Clipping

Originally polygon clipping algorithms were developed for basic operations in creating graphic output [10]. Polygon clipping is the calculation of the intersection of two given polygons: a subject polygon and a clipping polygon (cf. Fig. 2] (Rappoport 11]. Because the intersection output from the 3D collision is unsuitable for our application (cf. Chap. 1) and Triangle-Triangle intersection algorithms are either case dependent [12] or perform poorly for bigger applications [13], we develop an alternative algorithm. The usage of polygon clipping algorithms [14] enables a stable and accurate intersection detection between two polygons. In the Literature several algorithms have been proposed to solve this problem, but in the following three of the more prominent ones are described.

Sutherland and Hodgman [15] propose an algorithm which is able to clip a convex clipping polygon and a concave subject polygon against each other but not two concave polygons. The Weiler-Atherton algorithm [14] is able to clip two non-self-intersecting

Fig. 2 Initial situation for a polygon clipping algorithm (**a**) and the resulting solution (**b**)

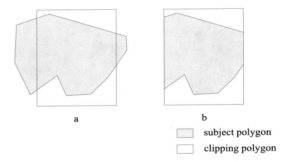

a   b

subject polygon

clipping polygon

concave polygons against each other. They are even allowed to contain holes. The algorithm is based on classifying each intersection between both polygons and creating the total intersection from it with a runtime complexity of $O(n2)$. The Vatti algorithm [16] is also able to clip two concave polygons against each other. After the determination of the left and right bound of each polygon the intersection is computed via the use of scanbeams with a runtime complexity better than $O(n2)$.

2.2 Gripping Point Determination and Intersection Detection

Because the developed approach is restricted to force closure with area contacts, this paragraph focusses on the relevant literature in this subject area. Some Methods 17, 18] focus on identifying contact points which fulfill the wrench equilibrium without deriving shape information for the intersection. Li et al. [19] calculated a grasp synthesis by shape matching the human hand with the form of the handling object based on a comparison of self-created feature sets. While this approach is suitable to find positions for a full overlapping, it is not applicable for partial overlappings. A triangular clustering is proposed in Harada et al. [20] to determine the contact points of a two finger parallel gripper with soft fingertips. Therefore, neighbor triangles of a surface are clustered via a comparison of their normal vectors and saved to a new rectangular grasping plane. In the next step the gripper plane is matched with the grasping plane on the object but no intersection between the two shapes is derived. Bonilla et al. [21] used bounding boxes to decompose the handling object and find suitable grasps with the use of basic geometries. Their testing showed a robust and flexible method with success rates of at least 77,61% but no shape information for the contact is created and used. Lin et al. [22] decompose objects, represented via a an RGB image, into several ellipses and build a grasping rectangle for each decomposition. The missing depth data prevents a calculation of the intersection area. Spenrath and Pott [23] use a heuristic search to select grasping positions in bin picking applications with the use of predefined contact regions. Trained neural networks [3, 4] show a good performance with high success rates (up to 95%) but need a specific training set for each gripper and do not determine specific shapes for the intersections between the handling object and gripper. Therefore, our main goal is to get a stable and computational efficient intersection detection for grippers and handling objects even for complex intersection shapes.

3 Gripping Point Determination with an Intersection Detection

The algorithm is built in a Python 3 environment and can be divided in 6 sequences (cf. Fig. 3). To find a stable, planar and accurate matching between the gripper surface and the handling object, the algorithm takes CAD data from both objects as an input and calculates valid gripping points. This means that the intersection surface satisfies the set criteria (cf. Chap. 4.2). Note that the results of the algorithm ignore the reachability and the 3D collision between the gripper and the object. These steps will be implemented in future work.

3.1 Preprocessing

First, the CAD-data of the handling object is imported in a.stl-format and all surface triangles are stored in a $(n \times 9)$ matrix, where n denotes the number of triangles in the mesh. Afterwards a similar approach as in Harada et al. [20] is used to cluster the triangles in 2D surfaces, with the difference that all triangles lay in the spanned surface and don't cut the newly generated surface. From the surfaces in this data set, pairs of parallel surfaces are calculated and stored via a comparison of their plane normal vectors. To take rounding errors into account, or if it is desired to allow a small angle between the surfaces, it is also possible to set a specific threshold. Additional restriction for this step is the geometry of the gripper (max. & min. opening of the gripper). Finally, the contour of the gripper surface is derived from the gripper model (.stl-format). The identification of the gripper surface is done manually at the moment.

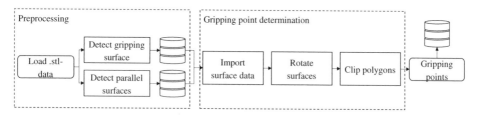

Fig. 3 Workflow of the program to determine gripping points with a polygon clipping algorithm

3.2 Gripping Point Determination Via Polygon Clipping

These two generated data sets are handed over to the gripping point determination algorithm. First, the surfaces are imported and afterwards their contours are extracted. For this, the 3 connection vectors of all triangles n are calculated and saved. For each surface, an empty set is created and afterwards an iteration over the created vector tuples is done. If the vector is already existent in the data set, it is deleted from the set, otherwise it is included. After this step the remaining vectors in the contour are strung together and represent the contour of the surface. Next, is a rotation of all surfaces into the XY-plane. For the following steps the algorithm uses *default* variables which can be set manually by the operator and are further explained in Chap. 4.2. After the rotation, an iteration over the contour area is done and then the clipping is performed (cf. Fig. 4).

To generate different grasping positions, distributed on the object surface, the gripper surface is shifted along the grid from Fig. 4. At every (x, y) combination a rotation of the gripper surface with a *default_roation* value is done. Afterwards the intersection between both polygons is determined with a polygon clipping algorithm and if the following two criteria are satisfied, a valid gripping point is detected:

- The overlapping area is greater or equal than the set *default_roation*.
- At the second object surface, a matching position which also satisfies the *default*_overlapping is found.

The implemented extension for the Weiler-Atherton algorithm enables the clipping of polygons against each other with an arbitrary shape. As long as they are not self-intersecting, the Vatti algorithm can be used directly from a clipper library. In the last step the *gripping_point* data set is rotated back to its original orientation. The whole algorithm is summarized in the following:

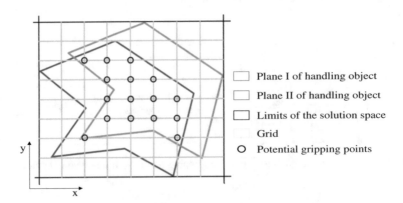

Fig. 4 Determination of potential gripping points (circles) from the two surfaces on the handling object (red, blue) with a superimposed grid in the XY-plane

Gripping point determination via a polygon clipping algorithm

```
## Contour detection
for i in 1:n do
append {p[i][1]-p[i][0], p[i][2]-p[i][1], p[i][2]-p[i][0]} to vectors
contour = []
for i in 1:n do
  for j in 1:n do
    if vectors[i[j]] in contour then
      delete vectors[i[j]] in contour
    else then
      append vectors[i[j]] to contour
sort contour so that vectors are strung together
## Surface rotation
angle = calculate angle between plane normal vector &
XY-plane
rotate contour around origin with angle
## Polygon clipping
max_x, max_y = maximum x & y values in contour
min_x, min_y = minimum x & y values in contour
for i in max_x:min_x with step size default_distance do
  for j in max_y:min_y with step size default_distance do
    for k in 1:360 with step size default_rotation do
      set position of gripping_surface to x = i, y = j
      rotate gripping_surface with k°
      if polygon clipping of contour and gripping_surface ≥
      default_overlapping & counterpart at second object surface
      also, then
        append [x = i, y = j, rotation = k] to gripping_points
rotate back gripping_points around the origin with −angle
end
```

p denotes the points of all triangles in a surface, *vectors* all the calculated connection vectors and *contour* all the derived vectors of the surface outline. Also, *gripping_surface* denotes all the contour vectors of the gripping surface and *gripping_points* denotes the set of calculated grasping positions which satisfy the set criteria.

---

4 Evaluation

4.1 Test Objects and Evaluation Criteria

To evaluate the algorithm, three self-created handling objects (cf. Fig. 5) are combined with a rectangular gripping surface. The surface of the gripper is 20 mm × 30 mm, while

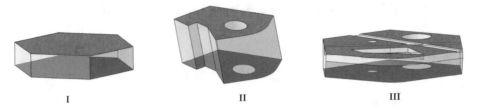

I   II   III

Fig. 5 Representation of the three (I, II, III) handling objects for the test scenario and the derived parallel planes (red) for the gripping point determination

the size of the handling objects is noticeably bigger to enable suitable grasps. The boundary of the 2D surfaces on the handling objects have varying complexity and contain different shapes and holes. To enable a comparison between our approach and existing strategies, we implemented the Vatti algorithm and a custom extension of the Weiler-Atherton algorithm and compared their performance against the Triangle-Triangle intersection detection [13] and the 3D intersection detection from the OBBTree vtk method. As mentioned in Chap. 1, the intersection detection from OBBTree is giving unsatisfying results but is included to act as a reference for computational efficiency as it represents a state-of-the-art 3D collision detection.

We computed the runtime of all algorithms for each test case 30 times and calculated the average computation time per position. The runtime computation is restricted to the core process because it is assumed that the preprocessing for every algorithm is more or less the same. To measure the stability of the methods we check all intersections if they are correct or not, so a score of 50% equals 5 correct outputs for 10 examples.

4.2 Restrictions

First tests of the algorithm without restrictions showed a solution set with an infinite number of grasping positions. Therefore, it is possible to set the following constraints via variables to reduce computational resources and the size of the solution set:

- Distance between two gripping points [mm]: In a radius with this value around a detected gripping point no other gripping point is allowed. Otherwise the algorithm would generate gripping points with an offset which is close to zero.
- Number of rotations at one gripping point [−]: Because a generated gripping point can be the center of many grips if the gripper is rotated around that point, the number of rotations are restricted for each position.
- Minimal overlapping of the gripper surface [%]: A successful grip is also possible if less than 100% of the gripper surface are in contact with the handling object. To realize that, a variable was introduced to set the minimal overlapping between the two surfaces on each gripping side.
- Maximum number of gripping points [−]: For each set of surface pairs a threshold for the maximum number of gripping points can be set.

Table 1 Set default_ parameters for the gripping point determination in the test cases

	Default_distance [mm]	Default_rotations [−]	Default_overlapping [%]	Default_#points [−]
Value	5	6 every 60°	100	None

Table 1 shows the *default_*parameters used in the test cases.

5 Results

The results of the proposed algorithm are visualized in Fig. 6 for the three handling objects. Table 2 contains the runtime evaluation for the four algorithms. For sake of a clear visualization, only a fraction of the full solution set of handling object I, containing 2576 grasping positions with the given variables, is visualized.

Comparing the four algorithms, the Vatti- and OBBTree algorithm delivered the best performance followed by the Triangle-Triangle and the customized Weiler-Atherton algorithm. Considering the average clipping time per position, the results show a dependency to the intersection complexity, more specifically the number of intersecting

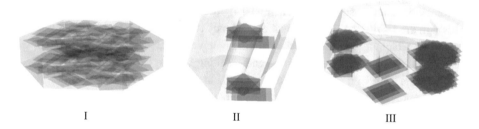

I　　　　　　　　　　II　　　　　　　　　　III

Fig. 6 Resulting grasping positions with the Vatti Algorithm on the three handling objects with the marked intersections (red) between the gripper and handling object

Table 2 Performance of the intersection detection with the Weiler-Atherton, Vatti, OBBTree and Triangle-Triangle algorithm. The proposed methods are marked bold

Handling object	Potential grasping positions [−]	Suitable grasping positions [−]	Avg. clipping time per position [ms] (stability)			
			Weiler-Atherton	Vatti	OBBTree	TriangleTriangle
Object I	4068	2576	1.02 (100%)	0.023 (100%)	0.0354 (30%)	0.742 (100%)
Object II	786	12	11.26 (100%)	0.063 (100%)	0.0364 (30%)	1.2961 (100%)
Object III	2484	130	30.83 (100%)	0.11 (100%)	0.0362 (31%)	2.78 (100%)

triangles for the two clipping algorithms and the Triangle-Triangle algorithm. The OBBTree algorithm does not indicate that dependency. These results affirm the suspected performances from Chap. 2. Due to the self-extension of the Weiler-Atherton algorithm, it's runtime complexity is bigger than (On4), in square the original runtime complexity and in the Triangle-Triangle algorithm 6 inequalities have to be solved, which can be a time-consuming process. Even for object I where a proportionately small number of triangles intersect, the Vatti algorithm and OBBTree algorithm show by far the best performance with computing times smaller than 4/100 ms while the other algorithms took approximately 1 ms to compute one position. While the Vatti algorithm and OBBTree algorithm show comparable running times, their stability shows a clear difference because the OBBTree algorithm correctly detected just 30% of the intersections. The other 70% of the classifications were false and looked comparable to the output in Fig. 1.

6   Conclusion

As shown, our algorithm represents an alternative way to accurately detect area grasping positions and their intersection for tessellated based handling objects and planar grippers. With the graphical representation of the grasping positions for three test cases we showed the validity and robustness of our approach even for complex intersection shapes. A performance comparison between existing approaches from the literature showed the suitability of our algorithm concerning computational complexity as it showed a better performance than a standard Triangle-Triangle intersection determination algorithm. Although the results showed a runtime dependency to the number of intersecting triangles for polygon clipping algorithms, the runtime complexity of the Vatti clipping was still comparable to the OBBTree algorithm and resulting in a higher stability score.

While the algorithm was successfully tested, a few parts were identified for further improvement. The algorithm uses settable *default_* variables which are not optimized because the solution size was sufficient for all test cases. To optimize these parameters, further test cases have to be evaluated. Another point to expand the flexibility of the gripping point determination is to extend the algorithm for grippers with different gripping surfaces at each finger and implement parallel computing. At last, as part of a greater gripping point determination project the algorithm will be implemented into a thorough gripping point determination routine. There, an analysis of the reachability and security of the grips will be done.

References

1. Bonilla, M., Resasco, D., Gabiccini, M., et al.: Grasp planning with soft hands using Bounding Box object decomposition. In: 2015 IEEE/RSJ International Conference on Intelligent Robots and Systems (IROS), pp. 518–523. IEEE (2015)
2. Gottschalk, S., Lin, M.C., Manocha, D.: OBBTree. In: Fujii, J. (ed.) Proceedings of the 23rd annual conference on Computer graphics and interactive techniques—SIGGRAPH '96, pp. 171–180. ACM Press, New York, USA (1996)

3. Guigue, P., Devillers, O.: Fast and robust Triangle-Triangle overlap test using orientation predicates. J. Graph. Tools **8**, pp 25–32 (2003)
4. Harada, K., Tsuji, T., Nagata, K., et al.: Grasp planning for parallel grippers with flexibility on its grasping surface. In: 2011 IEEE International Conference on Robotics and Biomimetics, pp. 1540–1546. IEEE (2011)
5. Hsiao, K., Chitta, S., Ciocarlie, M., et al.: Contact-reactive grasping of objects with partial shape information. In: 2010 IEEE/RSJ International Conference on Intelligent Robots and Systems, pp. 1228–1235. IEEE (2010)
6. Jiang, S., Zhao, X., Cai, Z., et al.: Single-grasp detection based on rotational region CNN. In: Ju, Z., Yang, L., Yang, C., et al. (eds.) Advances in Computational Intelligence Systems, vol. 1043, pp. 131–141. Springer International Publishing, Cham (2020)
7. Kim, I., Inooka, H.: Determination of grasp forces for robot hands based on human capabilities. Control Eng. Pract. **2**, 415–420 (1994). https://doi.org/10.1016/0967-0661(94)90778-1
8. Lachmayer, R., Lippert, R.B., Kaierle, S.: Additive Serienfertigung. Springer, Berlin (2018)
9. Li, Y., Fu, J.L., Pollard, N.S.: Data-driven grasp synthesis using shape matching and task-based pruning. IEEE Trans. Vis. Comput. Graph. **13**, 732–747 (2007). https://doi.org/10.1109/TVCG.2007.1033
10. Liang, Y.-D., Barsky, B.A.: An analysis and algorithm for polygon clipping. Communications of the ACM. Commun. ACM **26**, 868–877 (1983). https://doi.org/10.1145/182.358439
11. Lin, H., Zhang, T., Chen, Z., et al.: Adaptive fuzzy gaussian mixture models for shape approximation in robot grasping. Int. J. Fuzzy Syst. **21**, 1026–1037 (2019). https://doi.org/10.1007/s40815-018-00604-8
12. Mahler, J., Liang, J., Niyaz, S., et al.: Dex-Net 2.0: Deep Learning to Plan Robust Grasps with Synthetic Point Clouds and Analytic Grasp Metrics (2017)
13. Nguyen, V.-D.: Constructing force-closure grasps. In: Proceedings, 1986 IEEE International Conference on Robotics and Automation. Institute of Electrical and Electronics Engineers, pp. 1368–1373 (1986)
14. Rakesh, V., Sharma, U., Murugan, S., et al.: Optimizing force closure grasps on 3D objects using a modified genetic algorithm. Soft Comput. **22**, 759–772 (2018). https://doi.org/10.1007/s00500-016-2377-6
15. Rappoport, A.: An efficient algorithm for line and polygon clipping. Vis. Comput. **7**, 19–28 (1991). https://doi.org/10.1007/BF01994114
16. Roa, M.A., Suárez, R.: Grasp quality measures: review and performance. Auton Robot. **38**, 65–88 (2015). https://doi.org/10.1007/s10514-014-9402-3
17. Sabharwal C.L., Leopold, J.L.: A Trianlge-Triangle intersection algorithm. In: Computer Science & Information Technology (CS & IT). Academy & Industry Research Collaboration Center (AIRCC), pp 27–35 (2015)
18. Sahbani, A., El-Khoury, S., Bidaud, P.: An overview of 3D object grasp synthesis algorithms. Robot. Auton. Syst. **60**, 326–336 (2012). https://doi.org/10.1016/j.robot.2011.07.016
19. Spenrath, F., Pott, A.: Gripping point determination for bin picking using heuristic search. Procedia CIRP **62**, 606–611 (2017). https://doi.org/10.1016/j.procir.2016.06.015
20. Su, K.-H., Huang, S.-J., Yang, C.-Y.: Development of robotic grasping gripper based on smart fuzzy controller. Int. J. Fuzzy Syst. **17**, 595–608 (2015). https://doi.org/10.1007/s40815-015-0042-3
21. Sutherland, I.E., Hodgman, G.W.: Reentrant polygon clipping. Commun. ACM **17**, 32–42 (1974). https://doi.org/10.1145/360767.360802
22. Vatti, B.R.: A generic solution to polygon clipping. Commun. ACM **35**, 56–63 (1992). https://doi.org/10.1145/129902.129906
23. Weiler, K., Atherton, P.: Hidden surface removal using polygon area sorting. In: Unknown (ed.) Proceedings of the 4th annual conference on Computer graphics and interactive techniques—SIGGRAPH '77, pp. 214–222. ACM Press, New York, USA (1977)

Concept for Robot-Based Cable Assembly Regarding Industrial Production

Daniel Gebauer, Jonas Dirr and Gunther Reinhart

Abstract

The assembly of cables in industrial production is still a largely manually performed task. Therefore, automatic cable assembly offers much potential in terms of efficiency. The major challenge of automating this task lies in the formlessness of the cables, which entails unknown and inconstant states of the assembly objects. In this paper, a process chain and a concept are presented for the automated cable assembly in an industrial context. The process chain consists of five process steps, which are used to structure existing approaches and system configurations for automated cable assembly from a production technology perspective. The emphasis is on the coverage of the process steps and the system technology. The presented concept represents an approach for robotic cable assembly focusing on the flexibility to process multiple product variants. Basis for the ability to handle a variety of variants is the avoidance of a forced shape on the cables. For this approach, system technology as well as challenges and possible solutions are presented.

Keywords

Assembly · Cable · Deformable objects · Gripper · Machine vision

D. Gebauer (✉) · J. Dirr (✉) · G. Reinhart
Institute for Machine Tools and Industrial Management, Technical University of Munich, Boltzmannstr. 15, 85748 Garching, Germany
e-mail: Daniel.Gebauer@iwb.tum.de

J. Dirr
e-mail: Jonas.Dirr@iwb.tum.de

1 Introduction

The productivity and competitiveness of companies can be increased by automation and the use of industrial robots [1]. However, there are production processes that are difficult to automate due to technical reasons. One example is the production of control cabinets. The particular challenge is caused by the formlessness of the cables, which entails complex material behavior. In the production of control cabinets, wiring takes by far the most time [2]. Thus, automation of this process step offers great potential for increasing efficiency. Although a few robot-based approaches for cable assembly exist, there is currently no approach that takes into account all process steps ranging from the supply of cables to quality inspection. Therefore, we examine the robot-based cable assembly as a process chain from the point of view of production technology. Following this introduction, in Sect. 2 we present existing approaches in detail and discuss their coverage of the individual steps of the introduced process chain. Subsequently, we introduce our holistic concept for robot-based cable assembly in Sect. 3. Challenges and possible solutions are discussed, structured by the process steps. Section 4 summarizes this work and gives an outlook on further investigations.

2 State of the Art

This section aims to present the state of the art in the automated cable assembly using robot systems. First, commercial solutions for industrial applications are presented. Second, an existing structure for assembly functions is adapted to the robot-based cable assembly resulting in a five-step process chain. Present research approaches are described and classified concerning the process steps. Lastly, the state of the art is reflected critically and evaluated.

2.1 Industrial Solutions

So far, few commercial solutions exist for the automated assembly of cables. The wiring center Averex from Rittal GmbH & Co. KG is a portal machine, which covers the control cabinet assembly from the preprocessing of the cables to the mounting in the cabinet terminal and the routing in the cable duct [3]. The cables are supplied within tubes to the end effector. Therefore, the system can only handle a limited range of cable diameters. Despite its comprehensive coverage of the cable assembly process, the system is limited to few specific use cases.

System Robot Automazione Srl offers the Universal Robotized Wiring System SYNDY [4]. It is an assembly system for ceiling lights based on a six-axis industrial robot and an end effector specifically designed for the following use case. The robotic system feeds the cables in the end effector to the gripper, where it can mount cables in two different types of terminals from both vertical and horizontal directions. The system seems to be limited due to its narrow use case in the lighting sector and the restriction to specific cable diameters.

Besides the commercial solutions mentioned above, cooperations of academia, research institutes, and industry exist. The Chair of Production Systems at the Ruhr University Bochum collaborates in the research project, RoboSchalt with SCHUBS GmbH [5]. In another project, Fraunhofer IPA and Rittal GmbH & Co. KG have implemented the *Automatic control cabinet cabling* [6]. Both concepts address the automated cable assembly in control cabinets with two industrial robots, but only provide little information on details.

2.2 Research Approaches

In Lotter and Wiendahl [7], four assembly functions are presented: *supply, handling, joining,* and *inspection*. Due to the uncertain deformation of cables during *handling*, we propose to subdivide this step into *grasping* and *manipulation*, as cables are particularly complex here. Based on Lotter and Wiendahl [7], we propose a five-step process chain for the robot-based cable assembly as shown in Fig. 1.

The supply step represents the deployment at the very beginning of the assembly process. For an undefined supply pose, the initial object recognition and classification is part of the supply step. During *grasping*, *manipulation*, and *joining*, the cable is subjected to various external forces, e.g. applied by the gripper. In addition to gravity, these forces can influence the shape of a cable and thus make it more difficult to carry out the assembly. The *inspection* can be performed as a downstream process step or simultaneously to joining. In the following, the existing approaches are described in general terms and evaluated on which steps of the process chain the research is focused.

Table 1 summarizes this classification and shows the system technology that is used. The system technology is listed by three further subcategories: sensors, modeling, and gripping technology.

In Shirakawa et al. [8], an enhanced method for string shape recognition from 3D point clouds is presented. The method ranges from data acquisition and string shape recognition to the derivation of a point chain model. A detailed description of the procedure for the string shape recognition is provided, which demonstrates the extraction of a centerline point cloud from the 3D raw data. Subsequently, the point chain model is reconstructed from the point cloud. The described method is independent of the system technology, but

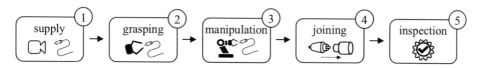

Fig. 1 Five-step process chain for robot-based cable assembly

Table 1 Classification of existing approaches

References	Title	Year	Process chain*	Sensors			Modeling			Gripping technology
				Optical	Tactile	Force/torque	Geometrical	Physical	Model-free	
Shirakawa et al. (2015)	String Shape Recognition Using Enhanced Matching Method From 3D Point Cloud Data	2015	●○○○○	●	○	○	●	○	○	○
Wnuk et al. (2017)	Concept for a Simulation-based Approach Towards Automated Handling of Deformable Objects – A Bin Picking Scenario	2017	○○◐○○	◐	○	◐	○	●	○	○
Nadon and Payeur (2019)	Automatic Selection of Grasping Points for Shape Control of Non-Rigid Objects	2019	○●○○○	◐	○	○	○	○	●	●
Alvarez and Yamazaki (2016)	An Interactive Simulator for Deformable Linear Objects Manipulation Planning	2016	○○●○○	○	○	○	○	●	○	○
Jiao et al. (2018)	Vision Based Cable Assembly in Constrained Environment	2018	○○●○○	●	○	○	○	○	●	○
Gao et al. (2019)	Vision-Based Grasping and Manipulation of Flexible USB Wires	2019	●●○○○	●	○	○	○	○	●	●
Chang and Padir (2020)	Sim2Real2Sim: Bridging the Gap Between Simulation and Real-World in Flexible Object Manipulation	2020	○●●●○	●	●	●	●	●	○	○
Palli et al. (2020); Palli and Pirozzi (2019); Pirozzi and Natale (2018); Homepage ECHORD++ (2020)	Publications in context of the WIRES project	2018-2020	●●●◐○	●	●	●	○	○	●	◐

* addressed steps are indicated   ● fully covered   ◐ partially covered   ○ not covered

only requires the point cloud data of the string from multiple viewpoints. The paper focuses on the shape recognition of unknown deformable linear objects (DLOs) in uncertain conditions.

In Wnuk et al. [9], a conceptual design for an experimental setup for bin picking of deformable objects (DOs) is presented. First, the possibilities of modeling and simulating DOs are presented. The depiction of large deformations with simultaneous reduction of the calculation time was identified as the main requirement for the models. Finite elements methods and mass-spring systems do not match these requirements, whereas multi-body systems do so and therefore are considered the most appropriate. Secondly, a draft conceptual overview for localization, gripping, and manipulation of DOs is presented. Linking existing approaches from the research field of bin-picking with the above modeling approaches are currently being investigated.

In Nadon and Payeur [10], a strategy to control the shape of a DO within a gripper is presented. A depth-sensing camera is positioned almost vertically above a three-finger robot hand which is opened upwards. The user can initially specify the desired contour of the object. Thus, the model-free algorithm automatically generates grasping points and deforms the object into the desired shape.

In Alvarez and Yamazaki [11], an approach is presented that automatically parameterizes physical characteristics of DLOs through visual analysis in order to provide a simulation for manipulation planning. The method takes a set of pictures of a cable before and after a specific manipulation as input. From the pictures of both states, virtual representations are created based on a mass-spring system. The physical behavior of the DLOs is characterized by four synthetic parameters, which values are randomly assigned to each virtual representation of the DLO before the manipulation. Following, a specific manipulation is applied to the parameterized simulation. Based on the geometric distance between the resulting DLO of the manipulation task and the virtual representation of the second state, a difference score is computed. After several iterations with newly randomized parameters, the system selects the parameters with the best fitting score for application in manipulation planning.

In Jiao et al. [12], a model-free method for optically controlled cable routing in structured environments is presented. A cable is routed around pre-known obstacles to achieve a defined shape. Therefore, one cable end is gripped by the robot, while the other cable end is already joined. An industrial collaborative robot with a depth-sensing camera and a two-finger-gripper is used. The cable deformation is detected by markers attached to the cable.

In Gao et al. [13], the soldering of USB wires is described. Visual feedback is applied for the cable identification and sorting of the four cores of a USB cable. Based on the color, the cores are identified and the grasping points are computed, to move them in pre-defined grooves for soldering. In case a wire abandons the groove due to its elasto-plastic behavior, a compensation algorithm is triggered to re-grasp and re-place the wire in the grove.

In Chang and Padır [14], a plug-in task of a household plug is addressed. The problem is approached using a 6 DOF assistive robot arm, two RGB-D cameras, and April tags. A new three-step procedure called Simulation-to-Real-to-Simulation is presented in order to address the gap between simulation and the real world. Firstly, a rough simulation model is created for the plug-in task. Secondly, the performance of the simulation is compared to the behavior in the real world. Lastly, based on the difference between both, the simulation is updated.

In Palli et al. [15], Palli and Pirozzi [16], Pirozzi and Natale [17], a holistic approach and prototype for automated assembly of cables in control cabinets are presented. For this purpose, an industrial robot with a custom end-effector, with multiple integrated sensors, was developed. The system utilizes a tactile sensor between the gripper jaws, force-torque sensing, and multiple optical sensors. 2D-cameras are used to detect and classify cables during the *supply* phase and terminals in the control cabinet. The tactile and optical sensor both estimate the pose of the cable tip, while the gripper manipulates it, to apply a pose correction for the *joining*. Besides, the system has a screwing unit and two linear actuators to screw down cables in the terminals. This work, as well as the prototype, has been developed in the context of the WIRES project [18] of the University of Bologna until the end of 2019. This project shows the largest coverage of the process chain among the compared approaches. The steps *supply*, *grasping*, and *joining* are fully addressed. *Manipulation* is considered with a reduced level of detail because mainly short cable segments are manipulated. An outlook on the *inspection* is given without going into further detail. The EU project REMODEL [19] takes up the topic and is going to continue it until 2023.

2.3 Evaluation of the State of the Art

Two main conclusions can be drawn from the state of the art. Firstly, there is currently no research approach that fully covers the five-step process chain. Moreover, half of the approaches presented address only one process step without aiming at production technology. Secondly, most of the industrial solutions apply a predicament on the cables, which means that the shape of the cable is restricted by the system technology. For instance, a predicament can be created through form closure, e.g. by feeding cables in tubes, or force closure, e.g. by multiple gripping points while the object is kept permanently under tension. Applying a predicament restricts unintentional movements of the cable during assembly. Nevertheless, cable feeding in tubes limits the system in terms of flexibility of variants and customized production. Also, multi-robot-systems with several gripping points stand out due to their high system complexity and effort for system tuning and therefore offer less flexibility, e.g. regarding supply.

3 Concept

To address the limitations identified in 2.3, we propose an approach for robot-based cable assembly that covers the entire process chain and does not apply a predicament in order to increase the flexibility by extending the range of possible cable diameters and connector types. Therefore, a stereo vision sensor and a specifically designed end-effector will complement a 6 DOF industrial robot. The main goal is to increase the product range, in the form of different cable diameters and connectors, of a robot-based cable assembly system in all five steps of the process chain and to reduce system complexity concurrently. This results in a system setup, which aims to be easily adaptable to multiple cable variants, diameters, connectors, and joining technologies. In the upcoming sections, we present our concept with respect to the five-step process chain. Thereby we will discuss the benefits of the proposed system as well as arising challenges and multiple solutions to address such.

3.1 Supply

The way goods are provided for assembly processes has to be adapted to the capabilities and degree of automation of the overall system. Especially semi-automated systems often require assistance, e.g. by a defined provision of the assembly products [15–17]. Compared to existing systems, the proposed approach uses only one grasping point. Therefore, the *supply* can be designed more flexibly, such that the cable can be provided in different ways, e.g. in an undefined shape on a planar surface, hanging, or in a fixture. The flexibilization of the *supply* offers two opportunities. On the one hand, the effort for the *supply* can be reduced by adjusting it to upstream process steps. On the other hand, a decoupling of upstream process steps such as pre-assembly is possible.

A flexible provision through an unknown location and form poses additional challenges to the system technology. When using optical sensor technology, challenges such as occlusion need to be considered, especially for cables with a knot-like structure and overlapping cable course. The occlusion can hide characteristic features of a cable such that pose estimation, object recognition, and gripper accessibility are complicated. The usage of 3D vision technology, as applied in Jiao et al. [12], allows a determination of the cable pose in space. Data acquisition from multiple points of view can inhibit occlusion with additional effort [8]. Optical markers and tools such as April tags can support object identification, but generate extra effort [14]. CAD-matching can be applied for pose measurement using the non-shape-labile components of a cable, such as connectors.

3.2 Grasping

Following the initial recognition in the *supply*, the downstream process *grasping* can be addressed. While two-finger parallel grippers are used in many approaches (e.g. 4, 6, 13, 15, 16, 17, 20), to the authors' knowledge no commercially available gripper exists, which is aimed at robot-based cable assembly. Furthermore, the design process of gripper systems for robot-based cable assembly has received little consideration in the literature so far. Thus, the following section describes the functionalities a gripper system must offer with regard to the further process steps.

Firstly, a gripper system must be able to grasp and later join a wide variety of cables and connectors. Secondly, the gripping pose depends on the shape of the cable or the type of connector as well as the type of joining connection in the downstream joining process. Unforeseeable changes in shape, which can occur e.g. during the gripping process or the manipulation, require once again the detection of the cable end or connector in order to enable a pose correction for the end-effector for the joining. So far, this requirement has mainly been addressed by the WIRES project [18] in which tactile and optical sensor technology is applied. Tactile sensor technology is integrated into the gripper jaws. Thus, the cable shape within the gripper jaws can be detected, and the protruding cable end or connector end can be extrapolated. As an alternative, 2D image data can be used to measure the shape of the protruding cable end or connector end. However, a stationary mounting of the vision sensor limits the flexibility or even the usability of the vision sensor for further functions such as the detection of cables and connectors during supply [15–17]. In order to be able to control the existing diversity of variants with a single gripper system, it is to be investigated how the insights gained in the WIRES project [18] for single core wires can be transferred to connectors and used for a comprehensive solution.

3.3 Manipulation

After the cable has been successfully grasped, it has to be *manipulated* into the joining pose. If the cable is not kept under permanent tension by multiple grippers like, e.g., in [6], but just by one gripper, there is a risk that the freely moving cable end will swing up and collide. Collisions can occur with the robot system, periphery elements, or the cable itself. In our approach, we aim to compensate for additional actuators, such as a second gripper, by using simulations and optical sensor technology.

Therefore, one suitable option is simulation-based offline trajectory planning in advance of the manipulation. Physical models are commonly used to describe the dynamical behavior of deformable objects, as e.g. in Wnuk et al. [9], Alvarez and Yamazaki [11], Arriola-Rios and Wyatt [21], Boonvisut and Cavuşoğlu [22]. Some of these approaches can be applied for robot-based cable assembly. The main disadvantage of this, however, is that model parameters of cables must be provided.

Another possibility is to track the deformable object, e.g. with an optical sensor, as in Gao et al. [13], Matsuno et al. [23], or Leizea et al. [24]. This information could then be used to adjust the robot trajectory during the manipulation. This approach makes the determination of material parameters superfluous but imposes additional requirements on the system technology. On the one hand, there is high data processing effort and the need for real-time capability. On the other hand, it has to be ensured that the sensor has a clear view of the cable and that the permissible distance and angle between the sensor and the cable is maintained.

Therefore, another possibility is to automatically identify the material parameters of a cable in a first step by manipulating it along a specific trajectory and tracking its behavior with an optical sensor. In a second step, material parameters of the cable can be extracted from this data and used in a third step for offline path planning.

3.4 Joining

After the cable has been manipulated into the correct joining pose collision-freely, the cable end or connector has to be *joined* to the product. The goal of the proposed approach is to be able to assemble as many different cable variants as possible with one gripper system, as far as the joining processes are similar. An example of this is the joining of a single wire with an ultrasonically compressed strand into a spring-cage terminal as well as the joining of a coaxial plug into a corresponding socket using the same gripper system. Thus, a major challenge in joining is the filigree character. As the pose accuracy of commercially available, conventional industrial robots is in the range of several millimeters [25], they may be too imprecise for joining specific cable variants. In addition, the cable shape, at least the cable end or the connector end, respectively, has to be determined precisely before joining. Integrating tactile sensor technology directly into the gripper jaws or applying optical sensor technology to measure the protruding cable end [15–17] or connector end, as described in *grasping*, seem to be the most promising approaches. However, it has to be considered that optical sensory can compensate the pose inaccuracy of the robot, whereas this is not possible for tactile sensory. Therefore, the exclusive use of the latter may still be too imprecise for some specific cable variants.

3.5 Inspection

Although joining is the final process step that directly contributes to value creation, quality *inspection* is very important for automating cable assembly to check and record production quality. Manual joining can be prone to errors since many electronic products, such as control cabinets, have a highly complex structure. In practice, employees execute the quality control during the assembly process by pulling each cable to testing if the cable is sufficiently fixed. In scientific papers, both completeness control and quality inspection have been treated very little for automatic cable assembly. To automate this step, two main possibilities are seen. Firstly, 2D or 3D camera technology can be used for a visual completeness check for electronic products. Secondly, an automated quality control of the cable connection during the joining process is feasible. Similar to the quality control carried out by employees, a combination of optical data and force data can be used to assess the correctness of the joining process automatically.

4 Conclusion and Outline

In this paper, a holistic five-step process chain for robot-based cable assembly in industrial production was presented. Based on this process chain, existing approaches for automated cable assembly were examined and for the first time analyzed from a production technology perspective. Furthermore, a concept for robot-based cable assembly was presented, which increases the degree of assembly flexibility by waiving the predicament of the cable during assembly. The focus of the concept was on challenges and possible solutions regarding the five steps. Our future work will aim to implement the possible solutions identified in the concept.

Acknowledgements The research leading to this publication has received funding from the Bavarian Ministry of Economic Affairs, Regional Development, and Energy (StMWi), as part of the project "RoMaFo—Roboterassistenzsystem und maschinelles Sehen zur Montage von formlabilen Bauteilen bei kundenindividuellen Produkten" (DKI0109/01).

References

1. Alvarez, N., Yamazaki, K.: An interactive simulator for deformable linear objects manipulation planning. In: IEEE International Conference on Simulation, Modeling, and Programming for Autonomous Robots (2016)
2. Arriola-Rios, V.E., Wyatt, J.L.: A multimodal model of object deformation under robotic pushing. IEEE Trans. Cogn. Dev. Syst. (2017)
3. Boonvisut, P., Cavuşoğlu, M.C.: Estimation of soft tissue mechanical parameters from robotic manipulation data. IEEE/ASME Trans. Mechatron. (2013)
4. Chang, P., Padır, T.: Sim2Real2Sim: Bridging the Gap Between Simulation and Real-World in Flexible Object Manipulation (2020)
5. Cordis Homepage: Robotic Technologies for the Manipulation of Complex DeformablE Linear objects. https://cordis.europa.eu/project/id/870133/de. Accessed 6 Sept 2020
6. Fraunhofer IPA Homepage. https://www.pitasc.fraunhofer.de/en/application.html. Accessed 6 Sept 2020
7. Gao, Y., Chen, Z., Liu, Y.-H.: Vision-based grasping and manipulation of flexible USB wires. In: Proceeding of the IEEE International Conference on Robotics and Biomimetics (2019)
8. Homepage ECHORD++: WIRES—Wiring Robotic System for Switchgears. https://echord.eu/wires.html. Accessed 6 Sept 2020
9. IFR Statistical Department: World Robotics 2018 (2018)

10. Jiao, C., Jiang, X., Li, X., Liu, Y.: Vision based cable assembly in constrained environment. In: IEEE International Conference on Robotics and Biomimetics (2018)
11. LSP RUB Homepage. https://www.lps.rub.de/forschung/projekte/roboschalt. Accessed 6 Sept 2020
12. Leizea, I., Mendizabal, A., Alvarez, H., Aguinaga, I., Borro, D., Sanchez, E.: Real-time visual tracking of deformable objects in robot-Aasisted surgery. IEEE Comput. Graph. Appl. (2017)
13. Lotter, B., Wiendahl, H.-P.: Montage in der Industriellen Produktion. Springer-Verlag, Berlin (2012)
14. Matsuno, T., Tamaki, D., Arai, F., Fukada, T.: Manipulation of deformable linear objects using knot invariants to classify the object condition based on image sensor information. IEEE/ASME Trans. Mechatron. (2006)
15. Nadon, F., Payeur, P.: Automatic selection of grasping points for shape control of non-rigid objects. In: IEEE International Symposium on Robotic and Sensors Environments (2019)
16. Palli, G., Pirozzi, S., Indovini, M., Gregorio, D. de, Zanella, R., Melchiorri, C.: Automatized switchgear wiring: an outline of the WIRES experiment results. In: Grau, A., Morel, Y., Puig-Pey, A., Cecchi, F. (eds.) Advances in Robotics Research: From Lab to Market (2020)
17. Palli, G., Pirozzi, S.: A tactile-based wire manipulation system for manufacturing applications. Robotics (2019)
18. Pirozzi, S., Natale, C.: Tactile-based manipulation of wires for switchgear assembly. IEEE/ASME Trans. Mechatron. (2018)
19. Rittal AG Homepage. https://www.rittal.com/ch-de/content/de/unternehmen/presse/presse-meldung/pressemeldung_detail_30615.jsp (2014). Accessed 6 Sept 2020
20. Shirakawa, T., Matsuno, T., Yanou, A., Minami, M.: String shape recognition using enhanced matching method from 3D point cloud data. In: IEEE/SICE International Symposium on System Integration (SII) (2015)
21. System Robot Automazione Srl Homepage. https://www.systemrobot.it/en/crlines/wiring-syndy . Accessed 6 Sept 2020
22. Tempel, P., Eger, F., Verl, A.: Reichlich Potential zur Effizienzsteigerung. https://www.schaltschrankbau-magazin.de/artikel/reichlich-potential-zur-effizienzsteigerung/2/. Accessed 7 Sept 2020
23. Ulrich, M.: 3D-Image-Stitching für roboterbasierte Messsysteme. Dissertation, Technical University of Munich (2018)
24. Wnuk, M., Pott, A., Xu, W., Lechler, A., Verl, A.: Concept for a simulation-based approach towards automated handling of deformable objects—A bin picking scenario. In: International Conference on Mechatronics and Machine Vision in Practice (2017)
25. Yumbla, F., Yi, J.-S., Abayebas, M., Shafiyev, M., Moon, H.: Tolerance dataset: mating process of plug-in cable connectors for wire harness assembly tasks. Intel. Serv. Robotics (2020)

Human-Machine Interaction

Improving the Understanding of a Remote Environment by Immersive Man-Machine Interaction

David Böken, Michael Schluse and Jürgen Rossmann

Abstract

In a changing world, the way we interact with machines must change as well. Teleoperation becomes more important. This poses its own set of challenges. To solve these a new Human-Machine interface must be developed. By developing this HMI around the concept of immersion, these challenges can be solved. This new kind of HMI can be applied to different fields. Examples using forestry or remote robot operations are demonstrated.

Key words:

Man-maschine interaction • Human-machine interface • Remote operation • Intuitive visualization

D. Böken (✉) · M. Schluse · J. Rossmann
Institute for Man-Machine Interaction RWTH Aachen University, Ahornstrasse 55, 52074 Aachen, Germany
E-mail: boeken@mmi.rwth-aachen.de
URL: https://www.mmi.rwth-aachen.de/en/

M. Schluse
E-mail: schluse@mmi.rwth-aachen.de

J. Rossmann
E-mail: rossmann@mmi.rwth-aachen.de

T. Schüppstuhl et al. (eds.), *Annals of Scientific Society for Assembly, Handling and Industrial Robotics 2021*,
https://doi.org/10.1007/978-3-030-74032-0_11

1   Introduction

Industry 4.0 combined with the ideas of the Internet of Things (IoT) is changing the way machines and entire factories are operated. Everything is interconnected, is getting more automated, and thus is becoming more complex [1]. The way humans interact with machines is changing. As with remote control, humans don't have to be in the same room as the physical machine, human control can be centralized, control is facilitated and eventually, the number of human operators can be reduced.

Current Human-Machine Interfaces (HMI) are not yet well prepared for this claim. A well suited, modern HMI should provide an intuitively comprehensible overview and intuitively operable interaction metaphors to interact with the physical system.

Modern technologies can be used to create intuitive HMIs for these use cases. We present two examples leveraging such technologies for the benefit of the user.

In the first example, modern simulation and visualization techniques are used to enable the intuitive understanding of a forest and of the effects, different harvesting measures will have. We will show that based on the HMI-principles of "feedback" and "affordance", basic features without futuristic technologies already make a big difference.

In the second example, Virtual Reality (VR) technologies are applied to create an intuitive user interface concept for a remotely controlled robot, see Fig. 1. It connects the Human on the left with a robot on the right.

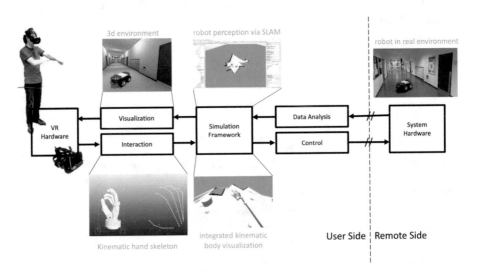

Abb. 1 Structure for connecting a real robot to a virtual reality headset

2   Current Developments in Man-Machine Interaction

2.1   Remote Operation

The way machines are operated today is changing [1]. While previously the operator would stand next to the machine or operate it from the next room, nowadays the physical distance between man and machine as well as the cognitively relevant distance increases. This poses several challenges.

The further away the operator is, the longer it takes for the information to reach him and for his input to be relayed back to the machine. Depending on the application, this latency can be critical.

Another problem lies within the amount of information being relayed via the existing man-machine interface. Current machine interfaces rely mostly on the concept of providing the user with information the machine can interpret itself. But the user also pulls information from different sources.

Take a CNC-Router for example. While the router's interface provides the user with an operational status such as its progress it normally does not provide additional information like vibration or the noise level of the machine. This information is vital as trained operators rely on this additional information. If the machine is too loud or vibrates too much it could be an indicator that the stress on the machine is too high and settings need to be adjusted in order to avoid damaging the machine or the workpiece.

A different problem occurs when considering information that may not be intuitive for a human. This problem is not particular to remote operation but through detaching the operator from his target system, it becomes more pronounced. Previously the human could look at the machine itself to try and make sense of its' data. Without that possibility, he needs to understand the situation solely relying on the data. An example is 2D laser scans. When controlling a mobile robot platform it is easy to show a camera view with which an operator can survey the surrounding. But 2D laser data is often presented as a 2D view from the top. So the operator needs to transfer that knowledge from a 2D view into his 3D awareness. Often he needs to look at several monitors to get all the information and most have different ways of providing that information. This is a challenge for the operator and imposes additional mental strain.

2.2   Reliability

Today's machines are getting more and more complex. In itself, this is not a problem [2]. The problem arises through the reliability of these systems. Grieves [2] defines four categories of system behavior. Predicted Desirable (PD), Predicted Undesirable (PU), Unpredicted Desirable (UD), and Unpredicted Undesirable (UU). The first three classes (PD, PU, and UD) are not problematic, the last class is (UU). With rising complexity, the risk of a failure to be catastrophic also rises [3]. In his book "Normal Accidents", Perrow describes the idea of Normal Accidents, or "system accidents". Small unpredicted events or failures can cascade trough a system unpredictably and cause large events with severe consequences. One prominent example of a small event causing a big failure is the "Space Shuttle Challenger disaster" [3]. Here a simple failure of an O-Ring caused the loss of vehicle and crew.

This should be taken into account when creating a new HMI for remotely operated systems. A new HMI should not be more complex as it needs to be, and if possible complexity should be reduced. At the same time, it has to enable the operator to manage such complex machines without himself becoming part of the complexity.

3   Immersive HMI

To meet these requirements a new way of interaction between a human and the operated machine must be developed. One that does not rely on the human being present at the scene of operation, but one that does not necessarily introduce more complexity into the interaction.

Recent studies have shown, that the sense of vision is the most important in English speaking cultures [4]. While the study suggests that this old hypothesis does not appear to be true for every culture, it still is true for the English one.

The findings of the study suggest that a system should primarily rely on the human sense of vision.

3.1   The Concept

The new HMI needs to address the problems described in Sect. 2. The idea is to "virtually" bring people to the area of interest by using an intuitive visual representation of information collected from the target area. This immerses the operator into the situation of the target system. Using the real sensor data and presenting it to the operator helps him go get a sense of the situation.

By chaining different systems the potential of an "Unpredicted Undesirable" event is raised. By keeping it low, the probability of "Normal Accidents" can be reduced. One possibility to mitigate this problem might be to present the user with all the raw data that is processed for him. The idea is that by making the raw information accessible, the user can verify that the system is working as intended. He can see what is really happening and not only what the system is interpreting. Otherwise, the user would just be another chain in the link of systems and can himself cause a "Normal Accident" based on the wrong information preprocessed by the target system.

This concept does not apply to all types of information. Some information can only be understood by the user if they are preprocessed for him. In those cases, other methods to avoid an "Unpredicted Undesirable" event need to be developed.

At the same time, the system needs to provide a benefit over existing solutions. This is done by giving the user access to as much information as available. Not just as numbers with a label but by actually putting them into a context which he can immediately understand.

Imagine an IoT dashboard within a 3D representation not just with pictograms but actually showing what their values are and showing where the data was collected.

4 Application: Forestry

This concept can be deployed in several different fields. The first example will be a more basic one which shows where this concept is already achievable. This example is taken from forestry. To be more specific privately owned small forest property.

In Germany, there are many small privately-owned forest property. Due to the demographic change, a lot of those properties are inherited by younger generations. Oftentimes they have migrated to cities and lack a personal connection to the forest and knowledge about it.

This results in forest parcels sitting dormant in the forest without being used or cultivated. This leads to problems.

One problem lies withing the economic usage of the surrounding forest. If some owners want to use their forest, they can only do so in an ecological way by combining their property and treating it together, but if someone is not interested in his forest where will be spots which can not be used and this forestry will be more difficult.

4.1 Visualizing the Information

One idea to mobilize the younger generation into getting more involved with their forest is to give them easy access to what it means to own part of a forest and what you can do with it. They often don't know what it means to own a forest. Part of that is that they do not know what they own. They never visited the forest and caused by their disinterest, do not plan to change that.

Abb. 2 App view showing forest at different growth state

The idea is that by using an intuitive visualization you can immerse the user into what he personally owns. This helps to better transfer the knowledge of what they own, and what they can and should do with it. Only through that understanding will the owner be able to make an educated decision based an all facts and not just disinterest.

Through providing the user with a 3D representation the new forest owner is more likely to identify himself with his property and increase interest. See Fig. 2 for an example of a 3D representation of a forest.

Actual engagement can be encouraged by simulating different forest treatments and presenting the results to the user in an intuitive way. This is feedback to the user. He can set different parameters for the simulation and look at the result. This engagement helps fortify the relation to the forest and its possibilities.

Even though all these materials are generated from flyover imaging or satellite pictures they still provide a much higher value especially to the uneducated user who maybe never thought about any properties a forest might have. The user does not have to think about the information he is seeing because it is mostly just like a view he would have in real life. This uses the principle of "affordance". The user might not know what a specific amount of wood is, in numbers. But showing him the trees instead gives him a better sense of the situation.

By not displaying graphs about the forest growths, but instead showing a slide show or a video of the forest growing, the facts are conveyed to the viewer in a simpler and more accessible way.

Figure 3 shows the structure of the system required to visualize the forest. To reach as much people as possible a mobile app is used as a gateway for the users to access and visualize their forest. As mobile devices are not capable of executing all tasks necessary to generate a 3D representation of the forest, this task is outsourced to a server. Interaction with the forest is handled in a way that is compatible to current structures. The app takes instructions from the user and generates tasks for the worker inside the forest to execute.

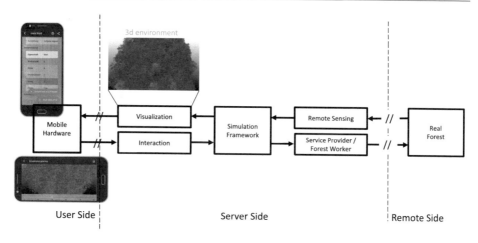

Abb. 3 Structure of a system presenting the real forest to the user

5   Application: Mobile Robotics

Another example where this principle can be applied is remote-controlled robots. While the above example is more of a current use case, this one is more futuristic. Inspiration for this can be taken from different sources. For example from "The Matrix" movies, specifically their flight control room, see [5].

Another source of inspiration for this futuristic interface can be taken from "Bret Victor" and his "Seeing Spaces" [6]. His Vision describes the next generation of "Maker-Spaces" or Workshops. See [7] He describes tools to allow someone to get an insight into a system. He proposes to record and archive lots of data from a system, including a lot of internals and making them accessible. By visualizing them and providing means of "seeing" them inside his "seeing space", the human can get a better idea of why a system is behaving like it is.

When dealing with teleoperated robots one of the difficult tasks is to provide the operator with an understanding of the operated robots surrounding. While cameras and 2D Maps from the top provide an insight into the situation of the robot, the operator needs to stitch these information together in his mind to extend his understanding of the situation.

This can be improved by using a system that helps the operator to immerse into the situation of the operated robot. To achieve this a system needs different components. A 3D representation is the best fit for such a task. But instead of using a computer or a smartphone, Virtual Reality (VR) is the best fit. Because a real-world is represented, the operator just needs to look around and can have most information put into context for him.

But visualization is just one aspect of a remote control system. Actually providing input to it, is another.

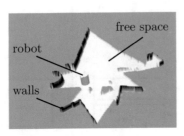

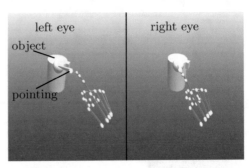

(a) Virtual robot shown in virtual surrounding representation, Live generated using SLAM

(b) Stereo image of operator pointing at target

Abb. 4 Different visualizations

Just like Fig. 3 for the forestry application Fig. 1 shows the structure used to connect a robot to a Virtual Reality (VR) system. As the VR hardware is already powerful, there is no need for an additional server in between the user and remote side. The data from the hardware system is taken and visualized to the user. The input from the user is taken, combined with the information from the robot, and translated into control commands the hardware can understand.

5.1 3D Visualization

As visualization the information provided by the robot is taken and displayed around the user. Lidar data for example is displayed as walls on the ground around the robot, see Fig. 4a. Other information like Camera pictures are integrated into the world and rendered at the position where the image is captured so the operator has the context where the information is coming from. Other information like the current target of the robot or the path the robot will take are overlayed on the floor within the map of walls.

5.2 Input Devices

Due to the challenging way of making interactions with a Virtual Reality (VR) intuitive, different kinds of hardware are needed.

For 3D visualization, an HTC Vive was chosen. Instead of using the provided controllers, which have to be held to provide input, a mixture of a LeapMotion sensor and a Kinect V2 was chosen.

The LeapMotion sensor was mounted to the VR headset, while the Kinect was positioned in front of the user. The LeapMotion was used to track the individual fingers, while the Kinect was chosen to track the Body Position including Hands. The Kinect tracking data made it possible to survey the Hands while they were not in front of the LeapMotion Sensor.

The information from the different devices are combined to provide better coverage for different types of input from the user.

5.3   Interaction

To simplify the interaction and to provide better immersion, the tracking information from the user is shown in the 3D representation. Thus the user can identify when there is a problem with the VR Hardware and can act to avoid causing further problems.

The actual interaction is separated into several parts. One part is navigating in and manipulation of the visualization, the other is operating the target system. A third way of interacting is interaction with a window style of UI elements displayed floating in front of the user, either fixed in the world or fixed to the user, depending on what information is shown or requested from the user.

Interactions with the 3D representation revolve around the way information is displayed to the user. You can move the map around you, scale it, and manipulate what is shown. This can help the user to get a different perspective of a situation. He can also zoom into the scene to inspect details. Trough zooming out he can get a better overview and view the surroundings.

The motion the user uses for these actions is comparable to a drag and drop action. By closing his hands (recognized by the Kinect Sensor) the user starts the action. By opening them again the user ends it. The actions are translated into the world 1 to 1. So "grabbing" something and it around is translated into a map movement. Moving hands closer together is translated into a Zoom gesture. The same goes for moving hands further apart. The user can select different modifiers for this operation. If he wants to move faster or more precisely he can change the factor which links his actions to the movement of the Virtual Reality (VR). The same goes for zooming and rotation actions.

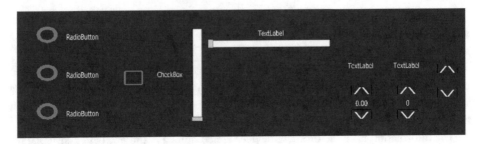

Abb. 5 Virtual window UI element shown to the user in VR, every button and slider is controllable via the virtual hand

Interaction with the real system is operated by the same input capabilities. But instead of manipulating the 3D representation, the real system is manipulated and the results are shown to the user.

One possible action is pointing at the target and telling the robot to move there. The action of pointing is mostly known, but the problem is detecting when the user is pointing where the robot should go, or if he is still deciding. Another problem lies within the accuracy of determining what the user is pointing at. Probably everybody knows this problem, someone is pointing at something and it still turns into a guessing game sometimes. Only when you try to stand where the person pointing person is can you see what they point at. To avoid miscommunication between the system and the operator, a line is drawn where the system thinks the user is pointing. That way the operator only needs to move his index finger, and make sure the line goes where he wants it to go. See Fig. 4b.

Events or special ways of requiring input from the user are handled trough popup windows in the 3D representation. This way of displaying or inquiring information from the user is familiar, or easy to understand. The user simply needs to press the buttons with his finger. Windows can be moved around and positioned where the user wants/needs them. See Fig. 5.

The concept of using windows as a UI element has another benefit on the development side. Developers adapting to this new style of HMI can use a proven way of interacting with the user and don't need to relearn the complete way of interaction. This can also reduce the possibility of mistakes. A new system always has a higher likelihood of the user misunderstanding what is wanted. The developer might think that it is clear what the system might want from the user, but while operating the user misinterprets this information and that might lead to a mistake. Using proven methods as a base for operations can function as a base for new interfaces.

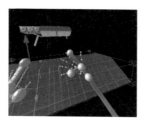

(a) A constellation of two satellites next to each other

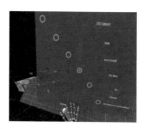

(b) UI presented to the user in front of a satellite, prompting him to select a specific component of the satellite

Abb. 6 View using Virtual Reality (VR) googles, including a visualization of the arms and hands

Most times when a window System would be useful will be when the system has to operate outside of the normal limits. Here a common interaction way is most important which proven UI elements like buttons can provide.

5.4   Different Area of Application

Space is another field where the mobile robotics approach can be utilized. Visualizing the current state of a system using Virtual Reality (VR) can help the user understand the current state of a system. See Fig. 6a. This can help the operator to make educated decisions from the ground station. On the other hand, satellites are complex systems containing lots of parts. Lots of parts are not directly visible to the user, so their state needs to be observable differently. This is where the UI system can help. For example when trying to select a specific component within the satellite, to get further information or execute commands. The user can be presented with a UI listing of all components and can select the desired one. See Fig. 6b.

6   Conclusion

To solve the challenges presented by for example industry 4.0, a new HMI is needed. By using the concept of immersion with intuitive user interaction, this new HMI type can be created. The idea of immersion can be applied to different fields in varying degrees of complexity. Forestry and robotics were given as examples.

The more basic forestry example demonstrates what is already possible with widely available hardware, while the more futuristic robotics example provides a direction in which this concept might develop.

While some problems may be solved with this approach there are still more problems to consider, like suppressing "Unpredicted Undesirable" behavior.

By combining an immersive interface with a digital twin (DT), the DT can act as a mediator between the interface and the actual system [8]. This link can offer more benefits. It allows representing the complete system. Using this representation "Unpredicted Undesirable" behavior can be mitigated [2]. The DT can also provide a standard interface to connect to other DTs and provide a unified interface to immersive interact with multiple systems at the same time.

This leads to the conclusion that integrating digital twins is the next step to further push the concept of an immersive HMI.

Literatur

1. Dorst, W. (Ed.).: Umsetzungsstrategie Industrie 4.0: Ergebnisbericht der Plattform Industrie 4.0. Bitkom Research GmbH. (2015)
2. Grieves, M., Vickers, J.: Digital twin: mitigating unpredictable, undesirable emergent behavior in complex systems (2017). https://doi.org/10.1007/978-3-319-38756-7_4
3. Perrow, C.: Normal accidents: living with high-risk technologies (1999). ISBN: 978-0-691-00412-9
4. Nadja Podbregar: Die Hierarchie der Sinne. (2018). https://www.wissenschaft.de/gesellschaft-psychologie/die-hierarchie-der-sinne-2/
5. Grime, T.: Matrix reloaded: virtual control design (2003). http://www.kazumichi.com/Matrix-Reloaded-Virtual-Control-Design
6. Victor, B.: Seeing spaces. In: Talk at EG Conference (2014)
7. Victor:, B.: Seeing spaces (2014). http://worrydream.com/SeeingSpaces/
8. Cichon, T.: The digital twin: mediator for man-machine interaction (2019). https://doi.org/10.18154/RWTH-2020-01645

Towards a Modular Elbow Exoskeleton: Concepts for Design and System Control

Samet Ersoysal, Niclas Hoffmann, Lennart Ralfs
and Robert Weidner

Abstract

In industrial workplaces, strenuous, repetitive, and long-term tasks at head level or above as well as carrying heavy loads may lead to musculoskeletal disorders of different task dependent body parts. With an increasing trend towards wearable support systems, there is already a large quantity of exoskeletons that may support the user during movements, or stabilize postures, in order to reduce strain on various parts of the body. However, most commercially available exoskeletons mainly focus on the back and shoulder support. Only a few of them address the elbow joint, despite it being prone to injury. Therefore, this paper discusses different possible design and control concepts of modular elbow exoskeletons. The modular architecture potentially enables coupling to existing commercial- and research-associated systems, through appropriate interfaces. Different morphological structures and control mechanisms are assessed in respect to their ability to extend common exoskeletons for back and shoulder support.

S. Ersoysal (✉) · N. Hoffmann · L. Ralfs · R. Weidner
Chair of Production Technology, University of Innsbruck, Institute of Mechatronic, Innsbruck, Austria
e-mail: Samet.Ersoysal@uibk.ac.at

N. Hoffmann
e-mail: Niclas.Hoffmann@uibk.ac.at; Niclas.Hoffmann@hsu-hh.de

L. Ralfs
e-mail: Lennart.Ralfs@uibk.ac.at

R. Weidner
e-mail: Robert.Weidner@uibk.ac.at; Robert.Weidner@hsu-hh.de

N. Hoffmann · R. Weidner
Laboratory of Manufacturing Technology, Helmut-Schmidt-University/University of the Federal Armed Forces Hamburg, Hamburg, Germany

© The Author(s) 2022
T. Schüppstuhl et al. (eds.), *Annals of Scientific Society for Assembly, Handling and Industrial Robotics 2021*,
https://doi.org/10.1007/978-3-030-74032-0_12

141

Based on these considerations, a first functional passive prototype is presented, which supports the flexion of the elbow joint and can be coupled to an existing exoskeleton. In future work, the prototype may be used for further elaboration and practical investigations in laboratory settings to evaluate its technical functionality and biomechanical effects on the user.

Keywords

Exoskeletons · Modular design · Support systems · Elbow · Wearable technologies · Human–machine interaction

1 Introduction

Despite Industry 4.0 and increasing usage of semi or fully automated systems in production lines or logistics, employees still face non-ergonomic work conditions [1, 2]. Tasks in strenuous positions, with repetitive movements, and/or heavy-load-handlings, are associated with work-related musculoskeletal disorders [3]. A feasible solution to support manual tasks may be provided using exoskeletons that physically support different bodily regions of the user [4]. According to a systematic review of exoskeletons with near market maturity [5], there are 25 systems for the upper extremities, 10 systems for the lower extremities, and 26 systems for the back. It is shown, that only 34% of these systems address the elbow joint, because many exoskeletons for the upper extremities only support the neck-and-shoulder-region, with the path of force ending at the upper arm (e.g., exoskeletons like Lucy [6], Airframe [7], and Paexo Shoulder [8]). Any extension to the forearm/wrist implies taking multiple joints with their specific motion patterns and degrees of freedom into consideration, resulting in a more complex kinematic chain, that must be technically reproduced [9]. To bypass this issue, other systems feature a distal path of force, like third-arm-solutions (e.g., [10]), or direct draglines from the shoulder mount to the fingers/wrist (e.g., [11, 12]), which both results in systems being limited in their application, and often tailored to one specific use case. Furthermore, exoskeletons for back support, such as Laevo [13], CrayX [14], or Rakunie [15], often only support work in a bent-over position, or just lifting but not carrying of heavy loads.

Consequently, there is a demand for modularly extensions for existing exoskeletons to support the elbow joint in an additional way, since it is particularly burdened during manual repetitive tasks, as well as the carrying, and holding of heavy goods (e.g., loads and tools). In case of an overload, it is prone to an array of injuries, such as arthritis, biceps bracii tendon rupture, or lateral epicondylitis [16]. Here, the support particularly required in vertical direction, against gravitational forces in respect to support movements elevation and stabilize during static tasks is required. This paper shows and assesses different variants of proximal elbow support systems. For illustration, the preferred solution is presented as a first functional prototype.

2 Determination of Main Requirements

In general, support situations are influenced and determined by several dimensions, some of them are illustrated in Fig. 1.

For the design of a support system, the interaction of user, task, technology, and environment must be considered. Technical design principles for exoskeletons do exist (e.g., [17] and [18]), that address, e.g., safety, ergonomics and comfort, technical affinity, and usability. Additionally, the modular elbow support system follows the idea of Human-Hybrid-Robot [19], in order to reduce the harmful overload on certain body parts, by adapting support systems to different support demands. Thus, an analysis of existing support systems, in terms of their morphology and technical function, is essential to determine a suitable system structure and to ensure the later ability for connection. Furthermore, it is necessary to consider beforehand, the characteristics of the intended tasks and the end-user.

2.1 Analysis of Connectable Exoskeletons

Since various systems on the market, or in research-fields exist [5], only the proximal systems Lucy [6], Airframe [7], and Paexo Shoulder [8], for the upper extremities as well as Laevo [13], CrayX [14], and Rakunie [15] for the back are further analyzed here, in terms of their morphology (see Table 1). These systems feature physical interfaces such as

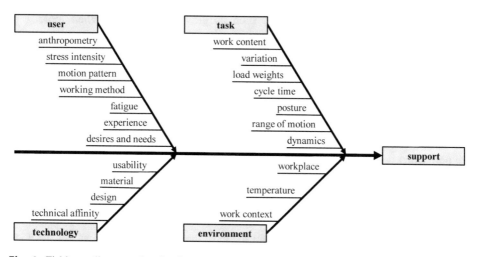

Fig. 1 Fishbone diagram of major factors affecting support situations

Table 1 Overview of the mentioned exoskeletons with main characteristics

Exoskeleton	Lucy	Airframe	Paexo shoulder	Laevo	CrayX	Rakunie
Supported bodily region	Upper extremities			Lower back		
Type of actuation	Active, pneumatic	Passive, springs	Passive, springs	Passive, springs	Active, electric	Passive, elastic bands
Physical interface	Circular	Half-open	Circular	Half-open	Circular	Circular
System weight	5 kg	2.5 kg	1.9 kg	2.8 kg	8 kg	0.25 kg
Structure	Rigid	Rigid	Rigid	Rigid	Rigid	Soft

shoulder straps, and most commonly a belt around the pelvis, for both carrying the systems' weight, and applying and transferring supporting forces to stronger, stabilizing body parts. These forces are often transferred via a proximal back structure, that can be composed of either rigid or soft elements. In some cases, this structure is shifted to the sides (see Paexo Shoulder [8]) or to the abdomen (see Laevo [13]).

2.2 Analysis of Intended Tasks

The intended tasks that should be supported by the elbow system are related to carrying heavy loads or working with heavy tools, in particular, since they are factors contributing to work-related musculoskeletal disorders [2]. It must be considered, that the work environment may be in confined, moist, or rough conditions as well as require the user to wear mandatory protective equipment, such as gloves, or jackets. When looking at working techniques for carrying loads, it becomes obvious, that humans often use the lower part of their inner forearm, either as an additional shelf space to relieve stress on the wrist, or as a clamp, to decrease the grasping force of the fingers. According to the working position for this task, it can be said, that the forearm is normally bent at a 90° angle or an almost fully extended position. When holding heavy tools with both hands, it is also possible, that one elbow may be hyperextended (ca. 135°) and the other in flexion (less than 90°), meaning that the joint angle varies between both arms. Also, a lot of tasks are performed with both hands in use, which requires manual control of the exoskeleton.

2.3 Analysis of Human Anthropometry and Kinematics

The human anthropometry is important for the fit of the system. Different studies for body proportions do exist. For instance, Winter et al. [20] estimate the length of the upper arm and the forearm, in relation to the users' height. Alternatively, Dempster et al. [21] published different data for an array of different body parts, for length, weight, and the center of mass, which may give an appropriate foundation for further development concerning dimensions of the system. In order to maintain the users' natural movement while wearing the system, human biomechanics should also be considered. The elbow joint complex consists of the elbow joint and the radioulnar joints and has two degrees of freedom: flexion/extension and supination/pronation [18]. The flexibility of healthy individuals is 0° to 140° (flexion/extension), 85° (supination) and 75° (pronation). A hyperextension of 5° to 10° in flexion/extension is not seldom [16]. According to comfort, the design, position, and clamp of the interfaces are also important. Principally, interfaces should be designed more broadly, to distribute contact pressures better to the human skin. Avoidable areas of contact are principally the joints themselves (elbow, wrist) and their immediate vicinity (especially medial and lateral epicondyle) [17, Fig. 5.21]. The contact pressure also needs to be limited, because it may significantly constrict the blood supply to the arm [22]. In order to avoid interaction forces, any misalignment of the joints between the exoskeleton and human limb should also be neglected [17].

3 Design and System Control Concepts

Different approaches for supporting the elbow joint are possible. The systems may morphologically vary, mainly in terms of path of force, rigidity of the structural elements, and type of actuation [6], which naturally determines different technical properties and effects on the wearer [4]. In the following, different basic morphological concepts are derived or inspired from already existing approaches, as well as assessed in terms of their suitability for the introduced application field.

3.1 Design Approaches

In general, it is advisable for exoskeleton development to begin with the actuation of only one degree of freedom, in order to maintain users' flexibility, and to reduce the construction complexity [9]. Thus, elbow support systems often only address the more exhausting flexion. Figure 2 clusters different ways of design and control for proximal exoskeletons supporting the elbows joints' flexion. Each group can naturally have individual configurations by experts, but the basic principle stays the same:

- Approach 1: According to the design, approach one uses tensile forces with a circular interface on the forearm and a rope guide on the upper arm, e.g., [23, 24]. Here, the actuator is often placed on the upper back and the force is transmitted by a Bowden cable. Alternatively, the actuator can be directly attached to the upper arm. A biomimetic design with artificial muscles is also possible.

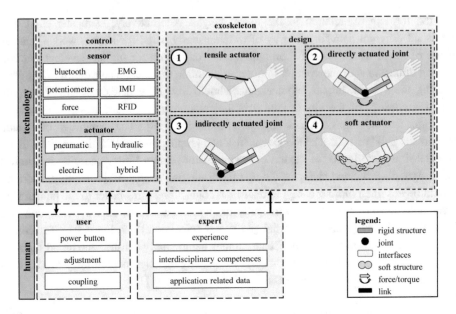

Fig. 2 Considerable design and control strategies for elbow support systems

- Approach 2: A possible way is using a directly actuated joint, consisting of two rigid bars that are respectively fixed to the upper arm and forearm, e.g., [25]. Here, the actuator can generate a direct torque, to support the flexion/extension of the elbow.
- Approach 3: An indirectly actuated joint could be realized by transferring the actuation principle from, e.g., Lucy [6] to the elbow joint, whereas the actuator may be attached to either the upper arm, or the forearm. In this case, the torque, or support, is generated according to the lever arm principle.
- Approach 4: A soft alternative are pneumatically inflatable air cushions, e.g., [26, 27]. A design with soft cylindrical actuators arranged in an array, generating a torque when inflated. A different approach of a soft support system are chamber elements for stabilizing body postures, e.g., [28].

3.2 System Control Strategies

The control mechanisms determine the steering of the systems' support, by either the user, the system itself, or an expert. It depends on whether the system is actuated in a passive or active way. Passive exoskeletons use actuators like springs with by experts pre-determined relations, between the supporting force and the arm angle, or the position of the actuator. Here, the user can sometimes decouple the actuation from the system for subsidiary tasks or pauses, as well as make some adaptions to the curve of force by, e.g., changing the preload of the springs (see, e.g., Paexo Shoulder [8]), or exchanging the actuator (see, e.g., Airframe [7]). Alternatively, active exoskeletons determine the arm position with sensors, like inertial measurement units (IMU) [29] or linear potentiometers [6]. The support demand, dependent on behavior or situation, can be identified by force sensors in gloves [25], by measurements of the muscular activity for grabbing via EMG-sensors [30], or by the determination of the load (e.g., product, ware, tool) via RFID or Bluetooth [31, 32]. Based on this information, experts can pre-determine the necessary support of the system. Also, the user can, adjust the level of support directly with a control element, or by actuating a power button [6].

3.3 Assessment in View to Modularity and Intended Tasks

In reference to the analyzed morphology of the exoskeletons that should be modularly extended and the described application field (see Chap. 2), a qualitative evaluation and comparison of the four described design approaches is made (see Table 2).

It is shown that a support mechanism working with tensile forces (mechanism 1 in Fig. 2) seems to be beneficial for a modular design. In comparison to mechanism 2 and 3, the first mechanism can easily be realized in a soft execution. The soft structure enables wearing the elbow system underneath other systems and tensile forces can easily be proximally extended, to place the actuator to regions, where it does not disturb existing

Table 2 Qualitative evaluation of the presented design concepts according to various criteria

properties	approach 1	approach 2	approach 3	approach 4
use cases				
support static posture	◕	●	●	◑
support dynamic movements	◕	◑	◕	○
support characteristic				
level of support	◑	◕	◕	◔
force adjustment	◑	●	●	◔
ergonomic aspects				
weight at distal end	●	○	◔	◕
wearability/comfort	◕	◑	◔	◑
construction				
modular coupling	●	◔	◑	◑
individual adjustment	●	◕	◕	◔
averaged rating	◕	◑	◑	◔
legend: ○ 1 (low) ◔ 2 ◑ 3 ◕ 4 ● 5 (high)				

structures, and keeps the additional mass on the arm as low as possible [33]. This is technically not possible for the principles of directly (2) and indirectly (3) actuated joints, since rigid structures are required for torque transmission. Soft actuators (4) could have potential as well, but often imply as pneumatically driven active systems both complex control schemes and external energy supplies, e.g., [26]. Here, the modular elbow support system should better be passive in order to keep the modular extension as simple as possible, to reduce the total weight and the amount of needed system elements, and to ease the general connectivity. Active systems simply influence existing systems greater, by changing total system characteristics, particularly from passive systems when adding sensors, energy storages, and control units.

4 Realization of Functional Prototype

The functional prototype (see Fig. 3) is inspired by, e.g., [11, 23, 24] for the structure, elements, and path of force, but changes some vital characteristics for a better modular fit, since these systems are principally designed for being autonomous. In this context, it is essential to ensure that the elements of the developed system do not interfere with other systems. Thus, the path of force on the upper arm is longer proximal than, e.g., [11] to use the interface of the analyzed systems for the upper extremities without any problems. However, this reduces the effectiveness of the tensile forces to induce a supporting torque on the elbow joint [34]. The actuator also moves to the pelvis belt, by extending the tensile lines, due to most systems already having one pelvis belt, interferences with the back structure are avoided, and the pelvis region is appropriate for taking tensile forces. Furthermore, the dual use of possibly existing pelvis belts leads to a reduction of the number of elements and the systems' weight.

The general path of the tensile forces is orientated by the line of non-extension [35] for better effectiveness. A steel rope is used that can be easily adjusted to different body dimensions, by wire rope clips, both near the elbow and on the back. The path proceeds

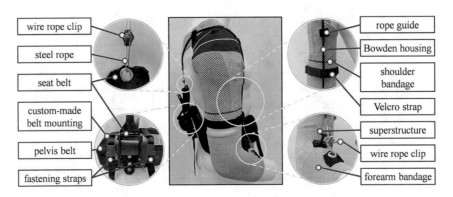

Fig. 3 Overview and central elements of the modular soft elbow support system

from a loop on the forearm bandage over the shoulder bandage, to the pelvis adapter. An installed Bowden housing on the shoulder bandage, as in [29] reduces rubbing and keeps the rope in position. The medical shoulder bandage is extended by a Velcro strip on the upper arm to avoid a protruding of the Bowden cable. The point of force origin on the forearm is also more orientated towards the elbow joint, to keep the lower forearm free and to maintain the typical working techniques. The prototype uses a medical arm bandage, since the use of neoprene interfaces help to reduce contact pressures [36]. Due to the human soft tissue and the soft structure of the connecting element between the bandage around the upper and the forearm, a movability in supination/pronation is preserved. A new-designed superstructure is connected to the forearm bandage, and maintains an induced support torque, by the tensile forces even when the arm is almost fully extended [34].

The actuator is fixed to the pelvis belt via a custom-made mounting. Here, a lock mechanism is preferred, which can lock in different length, in order to support the user in various elbow joint angles, or arm positions. The mechanism is firstly represented by a seat belt of a car, as seen in [34]. Thus, the mechanism locks if the belt is quickly dragged out and disengaged with a little move in the other direction, which means that the user can trigger it on demand. On the back, the belt passes into the steel rope with a hook and eye principle. The custom-made mounting can be easily mounted and tightened to pelvis belts of various widths, by using fastening straps and side-release buckles.

5 Conclusion and Outlook

Although the elbow joint is burdened in many industrial tasks, many exoskeletons for back and shoulder support do not cover it. The presented modular elbow support system may close this gap, by being able to be coupled to such systems and support the flexion of the elbow joint, with induced tensile forces. In the future, a comprehensive laboratory study should be conducted, in order to estimate its technical functionality and biomechanical effects on the wearer. Furthermore, the systems' connectivity to various existing exoskeletons should be checked. Effort could also be made to improve the lock mechanism in its sensitivity, and to merge the elements of the modular system, e.g., both arm bandages.

However, the lock mechanism is limited to static tasks, although many tasks are dynamically performed with different arm positions. Here, an active actuated modular extension of active support systems like, e.g., Lucy [6] could be beneficial and extend possible application cases. Consequently, it is recommended to persist with this concept as well.

References

1. Bogue, R.: Exoskeletons - a review of industrial applications. Ind. Robot. Int. J. **45**(5), 585–590 (2018)
2. Eurofound: Sixth European Working Conditions Survey - Overview report (2017 update). Publications Office of the European Union, Luxembourg (2017)
3. Sluiter, J.K., Rest, K.M., Frings-Dresen, M.H.: Criteria document for evaluating the work-relatedness of upper-extremity musculoskeletal disorders. Scand. J. Work Environ. Health **27**, 1–102 (2001)
4. Fox, S., Aranko, O.: Exoskeletons: Comprehensive, comparative and critical manufacturing performance. J. Manuf. Technol. Manag. (2019)
5. Weidner, R., Linnenberg, C., Hoffmann, N., Prokop, G., Edwards, V.: Exoskelette für den industriellen Kontext: Systematisches Review & Klassifikation. Digitaler Wandel, digitale Arbeit, digitaler Mensch?, 66. Kongress d. Gesellschaft für Arbeitswissenschaften (2020)
6. Otten, B., Weidner, R., Argubi-Wollesen, A.: Evaluation of a novel active exoskeleton for tasks at or above head level. IEEE Robot. Autom. Lett. **3**(3), 2408–2415 (2018)
7. Spada, S., Ghibaudo, L., Gilotta, S., Gastaldi, L., Cavatorta, M.P.: Investigation into the applicability of a passive upper-limb exoskeleton in automotive industry. Procedia Manuf. **11**, 1255–1262 (2017)
8. Maurice, P., Ivaldi, S., Babic, J., Camernik, J., Gorjan, D., Schirrmeister, B., Bornmann, J., Tagliapietra, L., Latella, C., Pucci, D., Fritzsche, L., Ivaldi, S., Babic, J.: Objective and subjective effects of a passive exoskeleton on overhead work. IEEE Trans. Neural Syst. Rehabil. Eng. **28**(1), 152–164 (2020)
9. Asbeck, A.T., Dyer, R.J., Larusson, A.F., Walsh, C.J.: Biologically-inspired soft exosuit. IEEE Int. Conf. Rehabil. Robot, 1–8 (2013)
10. Lockeed Martin: Fortis. Homepage. https://www.lockheedmartin.com/en-us/products/exoskeleton-technologies/industrial.html. Last accessed 30 Aug 2020
11. Li, N., Yang, T., Yu, P., Chang, J., Zhao, L., Zhao, X., Elhajj, I.H., Xi, N., Liu, L.: Bio-inspired upper limb soft exoskeleton to reduce stroke-induced complications. Bioinspiration & Biomimetics **13**(6), (2018)
12. Strongarm Tech: V22 ErgoSkeleton. Homepage. Last accessed 25 June 2020
13. Koopman, A.S., Kingma, I., Faber, G.S., de Looze, M.P., van Dieen, J.H.: Effects of a passive exoskeleton on the mechanical loading of the low back in static holding tasks. J. Biomech. **83**, 97–103 (2019)
14. German Bionic (2020). Cray X. Homepage. Last accessed 09 Sept 2020
15. N-ippin. Rakunie. Homepage. https://www.morita119.com/en/products/supportwear/rakunie/001.html (2020). Last accessed 09 Sept 2020
16. Vavken, P., Rosso, C.: Der schmerzende Ellbogen in der Praxis. Swiss Medical Forum - Schweizerisches Medizin-Forum **17**(44), 953–959 (2017)
17. Pons, J.L.: Wearable Robots: Biomechatronic Exoskeletons (2008)
18. Gopura, R.A.R.C., Kiguchi, K.: Mechanical designs of active upper-limb exoskeleton robots. In: IEEE 11th International Conference on Rehabilitation Robotics, pp. 178–187 (2009)
19. Weidner, R., Kong, N., Wulfsberg, J.P.: Human hybrid robot: a new concept for supporting manual assembly tasks. Prod Eng. **7**(6), 675–684 (2013)
20. Winter, D.: Biomechanics and Motor Control of Human Movement, Fourth Edition (2009)
21. Dempster, W.T., Gaughran, G.R.L.: Properties of body segments based on size and weight. Am. J. Anat. **120**(1), 33–54 (1967)

22. Holloway, G.A., Daly, C.H., Kennedy, D., Chimoskey, J.: Effects of external pressure loading on human skin blood flow measured by 133Xe clearance. J. Appl. Physiol. **40**(4), 597–600 (1976)

23. Dinh, B.K., Xiloyannis, M., Antuvan, C.W., Cappello, L., Masia, L.: Hierarchical cascade controller for assistance modulation in a soft wearable arm exoskeleton. IEEE Robot. Autom. Lett. **2**(3), 1786–1793 (2017)

24. Chiaradia, D., Xiloyannis, M., Antuvan, C.W., Frisoli, A., Masia, L.: Design and embedded control of a soft elbow exosuit. In: IEEE International Conference on Soft Robotics (Robosoft), pp. 565–571 (2018)

25. Otten, B., Stelzer, P., Weidner, R., Argubi-Wollesen, A., Wulfsberg, J.P.: A novel concept for wearable, modular & soft support systems used in industrial environments. In: Proceedings of the 49th Hawaii International Conference on System Sciences, pp. 542–550 (2016)

26. Otherlab. Homepage. https://www.otherlab.com (2015). Last accessed 22 Sept 2020

27. Thalman, C.M., Lam, Q.P., Ngyuen, P.H., Srider, S., Polygerinos, P.: A novel soft elbow exosuit to supplement bicep lifting capacity. In: IEEE/RSJ Iternational Conference on Intelligent Robots and Systems (IROS), pp. 6965–6971 (2018)

28. Weidner, R., Meyer, T., Argubi-Wollesen, A., Wulfsberg, J.P.: Towards a modular and wearable support system for industrial production. Appl. Mech. Mater. **840**, 123–131 (2016)

29. Lessard, S., Pansodtee, P., Robbins, A., Trombadore, J.M., Kurniawan, S., Teodorescu, M.: A soft exosuit for flexible upper-extremity rehabilitation. IEEE Trans. Neural Syst. Rehabil. Eng. **26**(8), 1604–1617 (2018)

30. Koopman, A.S., Toxiri, S., Power, V., Kingma, I., van Dieën, J.H., Ortiz, J., de Looze, M.P.: The effect of control strategies for an active back-support exoskeleton on spine loading and kinematics during lifting. J. Biomech. **91**, 14–22 (2019)

31. Weidner, R., Matthiesen, S., Bruchmüller, T., Mangold, S., Wulfsberg, J.P.: Systematische Entwicklung von Einheiten aus Power-Tools und anziehbaren Unterstützungssystemen – Ansatz einer integrativen Entwicklung. In: Weidner, R. (ed.) (Hg.): Band zur 2. Transdisziplinäre Konferenz „Technische Unterstützungssysteme, die die Menschen wirklich wollen", pp. 527–534 (2016)

32. Bances, E., Schneider, U., Siegert, J.T.B.: Exoskeletons towards industrie 4.0: benefits and challenges of the IoT communication architecture. Procedia Manuf. **42**, 49–56 (2020)

33. Browning, R.C., Modica, J.R., Kram, R., Goswami, A.: The effects of adding mass to the legs on the energetics and biomechanics of walking. Med. Sci. Sports Exerc. **39**(3), 515–525 (2007)

34. Hoffmann, N., Weidner, R., Schubert, T., Wulfsberg, J.P.: Towards a soft shoulder support system. In: Schüppstuhl, T., Tracht, K., Roßmann, J. (eds.) Tagungsband des 4. Kongresses Montage Handhabung Industrieroboter. Springer, Heidelberg (2019)

35. Iberall, A.S.: The use of lines of nonextension to improve mobility in full-pressure suits: AMRL-TR. Aerosp. Med. Res. Lab. (6570th), 1–35 (1964)

36. Xiloyannis, M., Chiaradia, D., Frisoli, A., Masia, L.: Characterisation of pressure distribution at the interface of a soft exosuit: towards a more comfortable wear. In: Carozzo, M., Micera, S., Pons, J. (eds.) Challenges and Trends. WeRob 2018. Biosystems & Biorobotics 22, Springer (2019)

Adaptive Motion Control Middleware for Teleoperation Based on Pose Tracking and Trajectory Planning

Andreas Blank⊙, Engin Karlidag, Lukas Zikeli, Maximilian Metzner and Jörg Franke

Abstract

Concurrent with autonomous robots, teleoperation gains importance in industrial applications. This includes human–robot cooperation during complex or harmful operations and remote intervention. A key role in teleoperation is the ability to translate operator inputs to robot movements. Therefore, providing different motion control types is a decisive aspect due to the variety of tasks to be expected. For a wide range of use-cases, a high degree of interoperability to a variety of robot systems is required. In addition, the control input should support up-to-date Human Machine Interfaces. To address the existing challenges, we present a middleware for teleoperation of industrial robots, which is adaptive regarding motion control types. Thereby the middleware relies on an open-source, robot meta-operating system and a standardized communication. Evaluation is performed within defined tasks utilizing different articulated robots, whereby performance and determinacy are quantified. An implementation sample of the method is available on: https://github.com/FAU-FAPS/adaptive_motion_control.

Keywords

Industrial Robots · Teleoperation · Motion Control

A. Blank (✉) · E. Karlidag · L. Zikeli · M. Metzner · J. Franke
Institute for Factory Automation and Production Systems (FAPS), Friedrich-Alexander-Universität Erlangen-Nürnberg (FAU), Erlangen, Germany
e-mail: andreas.blank@faps.fau.de

J. Franke
e-mail: joerg.franke@faps.fau.de

© The Author(s) 2022
T. Schüppstuhl et al. (eds.), *Annals of Scientific Society for Assembly, Handling and Industrial Robotics 2021*,
https://doi.org/10.1007/978-3-030-74032-0_13

1 Motivation

While conventional robot applications mainly automate repetitive tasks in known environments, the relevance of robotics in applications with existing uncertainties is gaining importance in a variety of industries. This includes among others bin-picking tasks in production or intralogistics, nuclear waste handling in decommissioning, mining or oceanographic operations, assistance in healthcare as well as aeronautics and space missions [1–5] Autonomous robot capabilities are potential enablers to accomplish these tasks [6], but due to uncertainties, still failures and system downtimes occur frequently [7]. In contrast, teleoperation already provides a more reliable robot application in complex scenarios with existing uncertainties [8, 9] and also enables efficient remote intervention for autonomous systems [7]. In addition, teleoperation increases safety at work in hazardous areas, as operators can act at a distance [5, 10]. Due to technological advances in most robotic fields, the demand for teleoperation applications have increased significantly in the last decade [4].

The appropriate translation of operator inputs into robot movements has a key role in teleoperation. Common teleoperation systems for industrial robots either control the robot in cartesian space, through joint angles or by six degrees of freedom pose goals with subsequent trajectory planning. Often these approaches share a lack of reactivity [11–13]. In contrast, approaches optimizing this characteristic through a reactive pose tracking motion control are often proprietary, vendor-specific solutions, resulting in a restricted interoperability [14]. However, for teleoperation in individual remote handling tasks with varying characteristics, interoperability as well as providing different motion control modes is beneficial. This includes reactive pose tracking, jogging control as well as robot movement based on collision-free trajectory planning. While pose tracking allows precise operator control in confined spaces, the latter enables task-specific programming, supervised automation and thus the relief of operator workload.

Our paper focuses motion control for teleoperation of industrial robots especially in scenarios with limited pre-defined circumstances, such as intervention of failed autonomous robot systems. We present a situation-dependent adaptive motion control middleware, featuring reactive pose tracking and collision-free trajectory planning for teleoperation. Thereby interoperability, a standardized communication as well as a simple extension by open-source robot libraries is addressed. The middleware is designed to support different Human Machine Interfaces (HMI) as input, while more intuitive and immersive experiences are achieved in combination with Virtual Reality (VR).

2 Related Work: Motion Control in Teleoperation Systems

Early teleoperation systems linked the controlling master and the controlled slave robot system mechanically [15]. Later, systems became computer-aided and software-intensive [16]. The increasing performance of robot controllers therefore provided a powerful base. Consequently, teleoperation tasks were performed through network-based information and command transmissions [17].

An interface utilized for remote control in industrial applications is the so-called jog-interface, enabling positioning of the manipulator based on incremental applied cartesian vectors or joint angles [18]. Joysticks or teach pendants are common as input devices. Disadvantages are the requirement of specific interfaces respectively an access to the robot controllers processing, limitations regarding an intuitive control operation and risk of missing the targeted pose. The latter results due to difficulties for the operator to recognize reaching the target and due to the time required by the robot to stop.

Remote control approaches of the last decades can often be characterized as proprietary solutions. Reasons therefore are mainly vendor-dependent differing motion control modes, communication interfaces and programming command sets. A distinction can be made between high-level and low-level approaches. While high-level approaches are based on standard robot controller instructions and communication interfaces [12], low-level ones utilize even more vendor-specific procedures and require more advanced programming [14]. These low-level approaches by contrast, enable a reactive and dynamic robot pose control, a requirement for appropriate teleoperation. Alternative methods for teleoperation are based on the well-established Robot Operating System (ROS), which offers middleware-typical functionalities as well as a simple extension of functionalities through open-source algorithms [13, 19, 20]. In particular, the MoveIt path planning extension including the Open Motion Planning Library allows defined target pose reaching. However, limitations derive from a lack of a dynamic and reactive pose tracking due to the time necessary for trajectory planning and the inability to redirect the robot during automated trajectory execution.

The above considered approaches differ on a lower layer regarding communication techniques. Network-based communication via Ethernet and the Internet Protocol is thereby common. At Transport Layer the User Datagram Protocol and the Transmission Control Protocol are utilized. Differentiation between approaches is possible in terms of data encoding standardization. Mentioned should be the Interoperable Teleoperation Protocol (ITP) [21] and the Simple Message Protocol (SMP) of ROS-industrial Edwards and Lewis [22]. While ITP is especially suitable in a standalone setup for relative motions in cartesian space, SMP supports a more complete range of motion types, is ROS-compliant and provides time-efficient communication with robot controllers.

In addition to the robot slave system, the HMI as input system also influences the motion control experience. For a suitable middleware it is important to support a wide range of HMIs. These include input devices as master arms, keyboards, teach pendants and gamepads in combination with single- or multi-monitor setups as well as VR-systems. Especially the latter ones show potential for increased operating intuitiveness and immersivity [23]. Contrary to approaches based on displacement of a master device within a small input area [19], these also enable remote guidance of the Tool Center Point (TCP) based on absolute poses, resulting in an more accurate target pose reaching. Furthermore, VR allows the augmentation by motion relevant information.

For teleoperating industrial robots in not fully defined environments, a middleware with certain characteristics, uniting and extending existing approaches is required. First, interoperability needs to be ensured through hardware abstraction and standardized communication. Due to the variety of expected tasks, different motion control types during teleoperation should be provided for an improved situation-dependent operation. Available should be a reactive, dynamic pose tracking to fulfill more complex movements, collision-free trajectory planning for relieving workload from the operator and as base for automated operations as well as a cartesian jogging for precise linear motions for performing assembly operations. Finally, the middleware should be complementable with modern HMIs for intuitive operation.

3 Adaptive Motion Control Middleware for Teleoperation

In the following sections our method of an adaptive motion control middleware for teleoperation of industrial robots is described. First, an overview on architectural level is given, followed by a description of the implemented motion control modes. Finally, the systematics for a situation-dependent adaptive switching of modes is explained.

3.1 Overview: HMI, Middleware and ROS-Adaptive Control Driver

The teleoperation system consists of three main components: HMI, Teleoperation Middleware and ROS-Adaptive Control Driver (ROS-ACD). An overview of the middleware architecture and its mechanisms is given by the following Fig. 1.

An HMI consists of an operator input and a feedback channel. Thereby the input device is used for target pose control and the feedback channel allows multimodal integration of the operator. For both channels, our middleware supports a variety of devices. In our architecture the HMI is linked to the slave robot system via the Teleoperation Middleware which is integrated to a computer system running ROS. The latter is chosen since it provides a comprehensive set of open-source software libraries and tools for robotic applications. Thereby, the ROS-Industrial extensions provide hardware compatibility through drivers for a wide range of industrial robot systems.

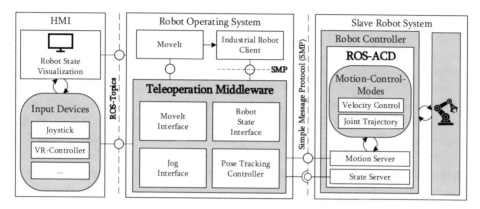

Fig. 1 Teleoperation middleware integration as technical architecture modeling

An essential element of our approach is the ROS-ACD, which provides different motion control modes for a specific robot system. The driver itself is an extension of existing SMP-based ROS drivers. It includes a motion server for performing motion instructions and a state server for feedback. A key feature is enabling reactive velocity control in contrast to common provided motions based on planned trajectories.

Our proposed middleware pursues to retain the existing ROS architecture for ensuring interoperability. It serves as an additional abstraction layer between ROS and the ROS-ACD, enabling functional extension, especially for providing different motion control modes: jogging, collision-free trajectory planning and pose tracking. Different interfaces are used for communication, e. g. the MoveIt-Interface serves to interchange target pose goals and planned trajectories. For dynamic pose tracking a specific controller is implemented. The Robot State Interface records the actual robot position for monitoring and control purposes, whereby all messages are also tunneled to ROS. The middleware is designed for usage in combination with the ROS Industrial Robot Client. Instead of a direct linkage to the robot driver, the client interconnects with the middleware. Thereby, data traffic is bi-directionally passed through the middleware.

3.2 Motion Control

The following section describes the motion control modes. Since jogging and trajectory planning are integrations of common methods, pose tracking will be described in detail.

Collision-Free Trajectory Planning. Collision-free trajectory planning is suited for safely approaching target positions, especially in environments with potential collision. This is of importance for teleoperation since it enables to reduce operator workload. It also serves the teleoperation system to provide autonomous robot capabilities, such as supervised grasping of objects. As described, our approach is based on MoveIt for planning trajectories. The target pose thereby is specified via the HMI (e. g. VR-controller). As feedback from MoveIt the planned trajectory with intermediate robot settings is spatially visualized for the operator through the HMI. If released, the middleware performs the motion control for carrying out the planned trajectory.

Pose Tracking Control. Pose tracking is based on cartesian velocity control implemented within the ROS-ACD. It allows reactive remote guidance of the TCP based on a dynamic target pose. First an initial target pose is defined near to the actual robot TCP through the HMI. The subsequent operator movements are tracked, so highly individual splines can be performed. The actual implementation depends on the functions available through the vendor-specific Application Programming Interface (API). For reactive motion control, a deterministic real-time capable low-level function is required, which is preferably executed in each interpolation cycle T_{IP} of the robot controller.

If the API provides a native function for cartesian velocity control, this is preferably used to take advantage of existing safety features, such as limiting acceleration and velocity or singularity handling. Otherwise, position based incremental motion functions are used for the implementation. This is achieved by conversion of the velocity commands into incremental displacement for an interpolation cycle. A velocity command v_{cmd} is a 6D vector consisting of a linear and an angular component. Multiplying the vector by T_{IP} results in a relative, incremental motion vector Δx_{inc}.

$$v_{cmd} = \begin{bmatrix} v_x & v_y & v_z & \omega_x & \omega_y & \omega_z \end{bmatrix} \tag{1}$$

$$\Delta x_{inc} = v_{cmd} \cdot T_{IP} \tag{2}$$

For a closed loop, the pose controller requires the actual pose of the TCP as feedback. It is necessary to adapt the driver to transmit feedback messages synchronized to the interpolation clock which are used to trigger the pose control loop. This ensures the deterministic processing as well as the necessary frequency matching of the velocity commands. Since the driver only sends joint positions q as feedback, the actual pose p_{fbk} is calculated through the forward kinematics. The target pose p_{cmd} is provided by the HMI. A PID controller compensates the tracking error e. The output can be limited to avoid unwanted high accelerations and velocities. Finally, the resulting velocity commands v_{cmd} are transmitted to the robot controller. The block diagram of the control loop is shown in Fig. 2.

Jogging Control. In addition to the two described methods above, the middleware provides a velocity-command based jogging of the robot. This serves to perform simple movements. For realization, one or more input-device joints of the HMI are accordingly mapped to robot joints. Motion control is either based on the individual joints of the robot via joint angles or in cartesian reference to the TCP.

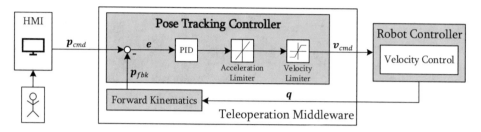

Fig. 2 Block diagram of the pose tracking controller

3.3 Adaptive Control-Mode Switching

Depending on the task, it is beneficial to switch between motion control types while teleoperation. For this purpose, a state machine is implemented within the ROS-ACD. The switching is triggered by the reception of corresponding message types defined for the specific motion command. Provided modes can be used immediately without having to switch manually. This enables an adaptive, situation-dependent switching. For example, in combination with a VR-HMI, the distance between the actual TCP pose and the target pose serves as a threshold for the system to switch automatically. Furthermore, the middleware ensures a fail-safe switching through idle state monitoring.

4 Experimental Setup and Evaluation

In this chapter, our demonstrator system setup and experiments are described. A qualitative evaluation is performed based on a real-world teleoperation handling task. The results retrieved are backed up by a quantitative evaluation.

4.1 System Setup and Qualitative Evaluation

The system is realized according to Fig. 1. As HMI an existing VR-interface is used (cf. [8, 22]. The Teleoperation Middleware is integrated into ROS Melodic running on Ubuntu 18.04 LTS. The robots utilized are a Yaskawa HC10 with YRC1000 controller and a Stäubli TX60L with CS8C controller. While the Stäubli controller has a native function for velocity control, the ROS-ACD for the Yaskawa YRC1000 is extended according to our method. Within this implementation only the proportional part of the PID controller is used. Both robot controllers run with an interpolation cycle T_{IP} of 4 ms. For safety reasons during evaluation, velocity commands are limited to a maximum velocity v_{max} of 250 mm/s.

For qualitative evaluation, different operators perform handling tasks (see Fig. 3). Thereby, industrial components are picked out of small load carriers (SLC) and

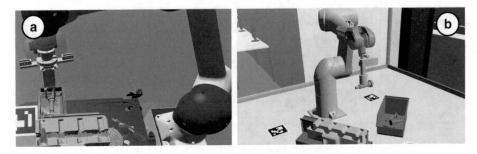

Fig. 3 Teleoperated handling of components: Yaskawa HC10 (**a**) and Stäubli TX60L (**b**)

subsequently are positioned within an assembly fixture. As depicted, the HC10 is equipped with a two-finger parallel gripper and the TX60L with a magnetic gripper.

Due to the intuitive operation, the sensitive and accurate positioning, even in confined spaces (e. g. within the SLC or fixture), pose tracking is the most preferred motion control mode. After a few minutes of handling, the operators begin to use the collision-free trajectory planning mode as well to reduce time and required physical body movements. This mode is thereby used to bridge greater distances and for automated reaching of defined pose goals. The jogging mode in cartesian space is seen as suitable for linear joining operations. One reason for this is the possibility to simply lock other movement directions. The velocity control of the CS8C is perceived as smoother compared to our implementation for the YRC1000. In terms of reactivity the latter performs slightly better. Both results from the more intensified filtering by the proprietary function.

4.2 Quantitative Evaluation of Pose Tracking Motion Control

In quantitative terms, pose tracking motion control is evaluated regarding reactiveness, processing determinacy, dynamics and accuracy. Analyzed are the velocity control and the higher-level pose tracking controller of the middleware. Evaluations are performed on the Yaskawa YRC1000/HC10 to verify the feasibility of our own velocity control implementation. The analyses are performed by recording target and actual states for position and velocity of a single axis (x-axis), whereby the behavior is valid for all axes.

The velocity control step response and trigger response time are shown in Fig. 4. The commanded velocity $v_{x,cmd}$ of 100 mm/s is reached after about 200 ms, while the settling time (with 2% error) is already reached at 175 ms. The observed feedback velocity $v_{x,fbk}$ results from the processing of the velocity commands on the robot controller as incremental motion. It is obvious that the robot controller limits the acceleration, which is reflected in the gradient of the graph. In addition, a small delay at the beginning is noted, which is due to communication, execution time and the internal dynamics of the robot system. However, the reactivity of the velocity control can be evaluated as sufficient for the addressed use cases.

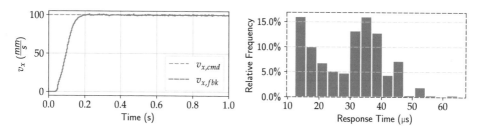

Fig. 4 Velocity control step response (left) and trigger response time (right)

The evaluation of the right plot allows conclusions about reliable and deterministic command execution. The response time represents the duration between feedback arrival from the robot controller as trigger signal and the consequent transmission of the velocity command. On the robot controller side, feedback streaming as well as the execution of the last received velocity command is synchronized with the interpolation clock with a cycle time T_{IP} of 4 ms. Thus, the velocity command must be sent as soon as possible to be available on-time at the robot controller until the next cycle. Since the execution of velocity commands is delayed by one interpolation cycle, T_{IP} at the same time represents a system dead-time. The results show a maximum response time of 64.7 µs with an average value of 29.7 µs. The on-time provision of commands is also reflected by the velocity profile. As a result, it can be concluded that the velocity control ensures a robust and reactive motion control.

The pose tracking controller is evaluated with respect to dynamics. The step response and the dynamic behavior for different controller settings are investigated (see Fig. 5). It is obvious that higher controller gains K_P results in a faster controller behavior. However, values chosen too high ($K_P \geq 4$) tend to overshoot, which is not desirable in teleoperation. The linear gradient of the step response graphs at the beginning is characteristic, as it results from the limitation of v_{max}. While $x_{fbk,1}$ cannot reach the steady state after 5 s, the others stabilize within the range of 2–3.5 s. The noticeable minimal stationary error is due to the resolution of the increments, which is limited to 1 µm. Since such a small error would result in an increment smaller than the resolution, it cannot be applied. However, depending on K_P the observed errors are within a range of 18–96 µm. Consequently, these can be neglected for our application.

The right graph shows a trajectory performed via pose tracking. Here x_{cmd} corresponds to the target position and $x_{fbk,n}$ corresponds to the individual feedback position. Here v_{max} represents a limiting factor. Thus, the tracking error can be minimized by a balanced parameterization of the two values. Performed experiments show, that a teleoperation with a too small resulting tracking error is perceived as hypersensitive. For example, the resulting control behavior of $x_{fbk,1}$ was preferred by operators.

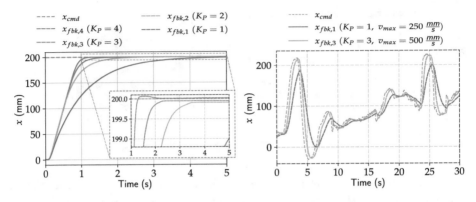

Fig. 5 Step response (left) and dynamic tracking (right) with different control settings

5   Summary, Conclusion and Future Work

In this work, a method of an adaptive motion control middleware for teleoperation of industrial robots has been described. The resulting middleware improves interoperability and offers a variety of motion control modes to be chosen adaptively depending on the situation. This has proved to be beneficial for tasks in environments with existing uncertainties. The integration within ROS enables among other advantages the support of a wide range of robot systems. The evaluation results prove reactive pose tracking capabilities suitable for dynamic teleoperation in complex tasks. The performed handling experiments show the potential for intuitive operation, especially in combination with a VR-based HMI. Thereby, bandwidth requirements, system latencies and the suitability for long-distance teleoperation are detailed in Kohn et al. [23]. Future work will further improve the control behavior by using the entire PID-controller and low-pass filtering of HMI-inputs. In addition, we will implement collision avoidance in pose tracking mode. These and further research activities are addressed within the close to application research project VIRERO on volume-optimized conditioning of radioactive waste [5].

Acknowledgements Parts of this research were conducted within the project FORo-botics (AZ-1225-16), funded by the Bavarian Research Foundation (BFS). The authors would like to thank the foundation and the expert reviewers for their valuable advice.

References

1. Althoefer, K., Konstantinova, J., Zhang, K.: Position and velocity control for telemanipulation with interoperability protocol. In: TAROS 2019. Lecture Notes in AI. Springer, Berlin (2019)
2. Basso, A., Hlaváč, V., Hulka, J., Jilich, M., Krsek, P., Malassiotis, S., Molfino, R., Smutný, V., Wagner, L., Zoppi, M.: Towards intelligent autonomous sorting of unclassified nuclear wastes. Procedia Manuf. **11**, 389–396 (2017). https://doi.org/10.1016/j.promfg.2017.07.122

3. Blank, A., Hiller, M., Zhang, S., Leser, A., Metzner, M., Lieret, M., Thielecke, J., Franke, J.: 6DoF pose-estimation pipeline for texture-less industrial components. In: Bin Picking Applications IEEE European Conference on Mobile Robots (ECMR), Prague, Czech, pp. 1–7 (2019)

4. Blank, A., Havenith, A., Kohn, S., Querfurth, F., Zwingel, M., Metzner, M., Franke, J.: Robotic technologies for volume-optimized conditioning of radioactive waste—VIRERO. In: AiNT (Hrsg) 9th International Conference on Nuclear Decommissioning, Aachen (2020)

5. Blank, A., Berg, J., Zikeli, L., Lu, S., Sommer, O., Reinhart, G., Franke, J.: Intervention strategy for autonomous mobile robots. wt Werkstattstechnik online (Bd. 110 (Nr. 9)):601–606 (2020)

6. Buschhaus, A., Blank, A., Ziegler, C., Franke, J.: Highly efficient control system enabling robot accuracy improvement. Procedia CIRP **23**, 200–205 (2014). https://doi.org/10.1016/j.procir.2014.03.200

7. Buschhaus, A., Blank, A., Franke, J.: Vector based closed-loop control methodology for industrial robots. In: 2015 International Conference on Advanced Robotics (ICAR), pp. 452–458 (2015)

8. Dalvand, M.M., Nahavandi, S.: Improvements in teleoperation of industrial robots without low-level access. In: IEEE International Conference on Systems, Man, and Cybernetics (SMC), San Diego, USA, pp. 2170–2175 (2014)

9. Edwards, S., Lewis, C.: ROS-industrial: applying ROS to industrial applications. In: IEEE International Conference on Robotics and Automation, ECHORD Workshop (2012)

10. Furuya, T., Yamashita, T., Kurisaki, T., Ikeda, H., Monji, T., Katoh, R.: Supervised robot system and computer aided teleoperation. In: IEEE International Workshop on Intelligent Robots, pp. 685–689. IEEE (1988)

11. Global Information Inc. Teleoperation and Telerobotics: Technologies and Solutions for Enterprise and Industrial Automation 2020–2025. Report. MindCommerce (2020)

12. Goertz, R.C.: Remote-control manipulator. US Patent No. 2,632,574 (1953)

13. Kawatsuma, S., Fukushima, M., Okada, T.: Emergency response by robots to Fuku-Shima-Daiichi accident: summary and lessons learned. Ind. Robot. Int. J. **39**(5), 428–435 (2012). https://doi.org/10.1108/01439911211249715

14. King, H.H., Hannaford, B., Kwok, K.-W., Yang, G.-Z., Griffiths, P., Okamura, A., Farkhatdinov, I., Ryu, J.-H., Sankaranarayanan, G., Arikatla, V., Tadano, K., Kawashima, K., Peer, A., Schauß, T., Buss, M., Miller, L., Glozman, D., Rosen, J., Low, T.: Plugfest 2009: global interopera-bility in telerobotics and telemedicine. In: IEEE International Conference on Robotics and Automation, pp. 1733–1738. https://doi.org/10.1109/robot.2010.5509422 (2010)

15. Kohn, S., Blank, A., Puljiz, D., Zenkel, L., Bieber, O., Hein, B., Franke, J.: Towards a real-time environment reconstruction for VR-based teleoperation through model segmentation. In: IEEE International Conference on Intelligent Robots and Systems, Madrid, Spain (2018)

16. Lima, A.T., Rocha, F.A.S., Torre, M.P., Azpurua, H., Freitas G.M.: Teleoperation of an ABB IRB 120 robotic manipulator and barrettHand BH8–282 using a geomagic touch X haptic device and ROS. In: Joint Conference on Robotics: SBR-LARS, Joao Pessoa, Brazil, pp. 188–193 (2018)

17. Obermeier, B., Treugut, L.: Bericht Lernende Systeme in lebensfeindlichen Umgebungen Plattform Lernende Systeme, München (2019)

18. Pires, J., Sá da Costa, M.G.: Object-oriented and distributed approach for programming robotic manufacturing cells. Robot. CIM **16**(1), 29–42 (2000)

19. Prexl, M., Zunhammer, N., Walter, U.: Motion prediction for teleoperating autonomous

vehicles using a PID control model. In: 2019 IEEE Australian and New Zealand Control Conference, Auckland, New Zealand, pp. 133–138 (2019)

20. Trevelyan, J.P., Kang, S.-C., Hamel, W.R.: Robotics in Hazardous Applications Springer Handbook of Robotics, Bd 19, pp. 1101–1126. Springer, Berlin (2008)

21. Wang, M., Liu, J.: A novel teleoperation paradigm for human-robot interaction. In: Turau, V., Weyer, C. (eds.) (Hrsg) 2004 IEEE Conference on Robotics, Automation and Mechatronics, pp. 13–18. IEEE, Piscataway, NJ (2004)

22. Yamada, S., Nomura, T., Kanda, T.: Healthcare support by a humanoid robot. In: ACM/IEEE International Conference on Human-Robot Interaction, Daegu, Korea, pp. 1–2 (2019)

23. Yuan, F., Zhang, L., Zhang, H., Li, D., Zhang, T.: Distributed teleoperation system for controlling heterogeneous robots based on ROS. In: 2019 IEEE International Conference on Advanced Robotics and its Social Impacts (ARSO), pp. 7–12. IEEE, Piscataway, NJ (2019)

Approach of a Decision Support Matrix for the Implementation of Exoskeletons in Industrial Workplaces

Lennart Ralfs, Niclas Hoffmann and Robert Weidner

Abstract

Despite the advancing trends in automation, workers in industrial workplaces often face repetitive tasks with heavy workloads. Whenever methods or adaptions in both technology and organization are insufficient to improve working conditions, personal-related interventions as exoskeletons come into question. They may prove successful in alleviating musculoskeletal disorders and relieving physical strain. The rising number of market-ready exoskeletons often challenges users or companies to select an appropriate system for their applications. In order to address this issue, this paper presents a generic approach for supporting both the selection and evaluation of exoskeletons. With respect to the task, user, and technical system, the decision support matrix (DSM) merges work profiles, motion patterns, and postures into one schematic representation. It aims to suggest exoskeletons with inherent properties matching these external requirements. In summary, the DSM may help users and companies to assess the fundamental suitability and select appropriate support devices for specific applications.

L. Ralfs (✉) · N. Hoffmann · R. Weidner
Chair of Production Technology, Institute of Mechatronic, University of Innsbruck, Innsbruck, Austria
e-mail: Lennart.Ralfs@uibk.ac.at

N. Hoffmann
e-mail: Niclas.Hoffmann@uibk.ac.at; Niclas.Hoffmann@hsu-hh.de

R. Weidner
e-mail: Robert.Weidner@uibk.ac.at; Robert.Weidner@hsu-hh.de

N. Hoffmann · R. Weidner
Laboratory of Manufacturing Technology, University of the Federal Armed Forces Hamburg, Helmut-Schmidt-University, Hamburg, Germany

© The Author(s) 2022
T. Schüppstuhl et al. (eds.), *Annals of Scientific Society for Assembly, Handling and Industrial Robotics 2021*,
https://doi.org/10.1007/978-3-030-74032-0_14

Keywords

Exoskeleton · Industrial workplace · Decision Support · Human–machine interaction · Expert system

1 Introduction

In recent times, advancing trends in automation and mechanization afford and affect the restructuring of workplaces, as well as the implementation of new technologies in industrial applications (e.g., systems for human–machine cooperation [1], exoskeletons [2], or augmented reality systems [3]). Nevertheless, workers often face heavy workloads, repetitive tasks, or need to perform in ergonomically improvable positions. Almost one in three people manipulates heavyweight goods, 43 percent daily work in either tiring, exhausting, or painful postures [4]. As a result, workers are physically and psychologically stressed and exposed to a risk of developing musculoskeletal disorders (MSD) [4, 5]. One viable option to prevent future and reduce existing effects is the use of exoskeletons [6]. They have shown to support workers at industrial workplaces in repetitive tasks that cannot be fully automated or are physically demanding [7].

Regarding industrial applications, exoskeletons are externally wearable mechanical devices [8] that either empower, facilitate, stabilize, or add movements [9]. Depending on their respective functional and morphological configuration, exoskeletons do not only potentially relieve physical stress on users, but also enhance their performance, strength, and endurance [10, 11]. Frequent capabilities and purposes of exoskeletons are lifting and manipulation assistance, body stabilization during the execution of manual tasks, and positioning correction [7, 10]. In this sense, exoskeletons amplify users' capabilities and build a bridge between fully manual work and tasks demanding the implementation of industrial robots [7].

2 Approaches for Classification of Support Situations

As a wide range of requirements and market-ready exoskeletons exist, selecting the most suitable system for the respective application is complex and challenging. In current science and practice, different approaches exist to characterize support situations for exoskeletons. According to crucial criteria (e.g., power supply mechanism, supported part of body, purpose of use), exoskeletons can be classified into different types or groups, respectively [8, 11]. In this respect, distinctive features such as affordable dynamics, motion ranges, or postures are of varying importance [12]. Hitting the same notch, checklists from, e.g., assurance associations [13] and calculation tools from manufacturers [14], indicate relevant factors to characterize the support situation. A similar but more illustrative approach is a periodic classification table for technical support systems. It puts

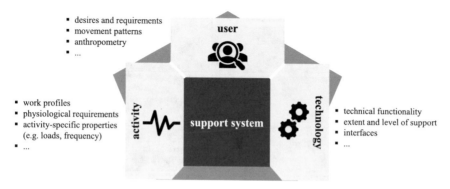

- desires and requirements
- movement patterns
- anthropometry
- ...

user

- work profiles
- physiological requirements
- activity-specific properties
  (e.g. loads, frequency)
- ...

activity

support system

technology

- technical functionality
- extent and level of support
- interfaces
- ...

Fig. 1 Characterization of the solution space of support systems, based on [12]

together the two dimensions relational patterns of human and exoskeleton, as well as supported movements of body parts [9]. In this way, the table enables to assign appropriate types of exoskeletons to given support situations. Another visual approach uses a path diagram to classify support situations by the three fundamental determinants spatial–temporal relation, form of coupling, and location of control [15].

All in all, the described approaches characterize and classify different support situations, but there is no comprehensive tool going even further and supporting the selection and evaluation of exoskeletons for specific activities. Therefore, this paper addresses these demands by presenting a generic approach, bringing together the dimensions of technology, users' needs, and tasks in a united concept (see Fig. 1).

3   Influencing Dimensions of the Decision Process

Several factors in different dimensions influence the decision-making process of selecting appropriate exoskeletons for specific tasks. The main characteristics of applications in industrial workplaces and technical core properties of exoskeletons are introduced.

3.1   Characteristics of Tasks in Industrial Workplaces

Based on preventing MSD, analyses of work profiles are an expedient way to identify and determine (potential) risk factors in industrial applications. Two categories group these factors: the first consists of those addressing the workplace environment and organization (e.g., frequencies, repetition of tasks, or pace of work), while the second type focusses on physical aspects of the activities (e.g., physical exertion, postures, loads, repetition of movements) [16]. According to the causes of MSD, the methods aim to identify exposure factors and evaluate either repetitive movements, strained postures, or the handling of loads [16]. Even though the rapid upper limb assessment (RULA), the Ovako Working

		environment and organization of work	physical aspects of work
RULA (repetitive movement)		range of movement, repetition	working postures and positions of different anatomical regions: flexion, extension and twist, static muscle work, force
OWAS (strained postures)		work sampling (variable or constant interval): frequency and time spent in each posture	weight of loads, classification of postures: straight, bent, twisted, kneeing, on or above/below shoulder level
NIOSH (load handling)		travel distance, cycle time, duration of task	physical exposure factors: object weight, horizontal/vertical location, lifts or lowers per minute, type of grasp, asymmetric angle

Fig. 2 Characteristic aspects of work based on selected examples

Posture Analyzing System (OWAS), or National Institute for Occupational Safety and Health equation (NIOSH) cover different application scenarios or bodily regions, all of them provide exemplary factors as relevant characteristics for further considerations concerning the analysis of tasks. Figure 2 exemplarily shows aspects to characterize work profiles.

RULA examines factors implying a risk for upper limb disorders [17]. OWAS helps identify, classify and evaluate working postures and enables a structured analysis of frequencies and times spent in strained poses [18]. NIOSH assesses the risks of manual material handling, aiming to determine whether an object or weight is unsafe to handle in lifting and lowering tasks [19]. Even though these methods primarily aim to evaluate existing working conditions and are not designed to select a technical support system, they highlight essential factors in characterizing work profiles.

Another approach to identify characteristics of work profiles is the use of distinguishing features. These generic factors also describe the variance in industrial activities. Factors like dynamics, precision, additional non-task related activities, external process forces, or the variance of tasks complement the aspects of work mentioned above [12, 20].

3.2 Technical Properties of Exoskeletons

Besides the characteristics of tasks in industrial workplaces, the specific knowledge about exoskeletons and their technical properties is essential for selecting an appropriate system. In order to provide crucial support in the respective field of application, the importance of different functionalities of exoskeletons varies with their requirements. Differentiating characteristics are, e.g., the mechanism of action, supported bodily region, intended use, power requirements, and construction material [6, 7, 8, 21]. Table 1 shows the core characteristics and associated explanations.

Moreover, the characteristics of the system may restrict the possible use [9]. E.g., the design of an exoskeleton possibly influences the potential application, as its size or weight

Table 1 Core characteristics of exoskeletons (in alphabetical order)

characteristics	explanation
actuation and control	method of actuation (e.g., active, passive), recognition of intention, sensor technology
extent of support	supported bodily regions (e.g., upper extremities, back)
form of support	enable/empower/stabilize/facilitate/add movements
level of support	performance support (torque, weight), degree of support (0–100%)
morphological structure	force transmission, stiffness (e.g., rigid, soft, hybrid), material (e.g., lightweight)

may hinder the user in executing his task. Considering the distance between the exoskeleton and the human body is also paramount, as it could be impossible to access narrow passages. According to the resemblance to human anthropometry, exoskeletons have different ranges of motion and, thus, enable diverse movements [8].

4 Development of a Tool Providing Decision Support

A decision-supporting tool including multicriterial factors may aid the selection of an appropriate exoskeleton in a structured manner. Before the approach is presented in detail, relevant characterization dimensions are determined, and the display form is specified. According to the human hybrid robot concept, the tool should consider the influencing factors user, technical system, and task [22].

4.1 Determination of Decision-supporting Dimensions

Even though there is a large variety of tasks appearing in workplaces of different industries [17], three elementary dimensions generally characterize the activities:

(1) Properties of tasks: Relevant properties of activities for selecting an exoskeleton, e.g., dynamics, bodily region, posture, weights, variance (concerning weight and tools), travel distances, and ranges of motion.
(2) Work profiles: Deriving specific requirements with respect to the properties, based on the work profiles to describe the activities/workplaces.
(3) System characteristics: Each exoskeleton inheres its own functional and morphological characteristics and, thus, differently supports the user.

The main challenge is to link the ambient requirements of work systems with the technical properties of exoskeletons. In order to be purposively applicable in various industries, the approach needs to meet the following criteria: (a) the decision support tool should be generic to cover a wide range of application scenarios, (b) it should address different loading cases, and (c) it should leave individual margins of discretion, as the importance of parameters differ for the respective applications.

4.2   Selection of a Display Method

For the three main influencing dimensions, a form of presentation needs to visualize the potential suitability of exoskeletons for specific applications in one compact figure. In this regard, a visual display method addresses the various dimensions and raises awareness of the complex selection process best. As a result, a multidimensional matrix has been chosen. It depicts the three dimensions and contrasts the selection alternatives in a structured way for specific applications. Effectively, the matrix enables a generic task modeling and selection of exoskeletons for respective scenarios.

4.3   Creation and Application of the Decision Support Matrix

The top of Fig. 3 illustrates the concept of the decision support matrix (DSM). The three main dimensions are: (1) core overall properties with possible characteristics, (2) modeled tasks from industrial workplaces, as well as (3) properties of exoskeletons. The lower part of Fig. 3 shows the four stages to create and apply the DSM in a step-by-step guide.

Step 1 – Selection of relevant properties: The horizontal axis represents the general characteristics as requirements for the system. By the use of rating scales, the properties can be specified. The multicriteria selection process requires the consideration of relevant factors for describing the work profiles. For the specific application, attributes may be individually chosen to characterize activities.

Step 2 – Modeling of activities: This step adds the ambient activities to the matrix. Vertical pillars in each column visualize work profiles for concrete application scenarios. Comparable to a box plot, the ranges of the properties are qualitatively modeled and set the requirements for a successful implementation of appropriate exoskeletons.

Step 3 – Implementation of exoskeletal features: Available exoskeletons with their technical properties are introduced to the matrix. Different lines indicate inherent characteristic values for each system (idealized course with exemplary variance). Before an exoskeleton is selected for implementation, the requirements should be considered thoroughly. The line progressions in the matrix represent real data of different exoskeletons.

Step 4 – Selection of an exoskeleton: Based on defined requirement profiles, the match between the line progressions for different exoskeletons with the modeled pillars of the activities allows a well-founded selection of a system. This indication enables decision-makers to assess whether the implementation of an exoskeleton in the respective application scenario may be sufficient or not.

The matrix may add value by enabling to systematically link the ambient requirements of work systems and the technical properties of the exoskeletons. Thus, it helps to both

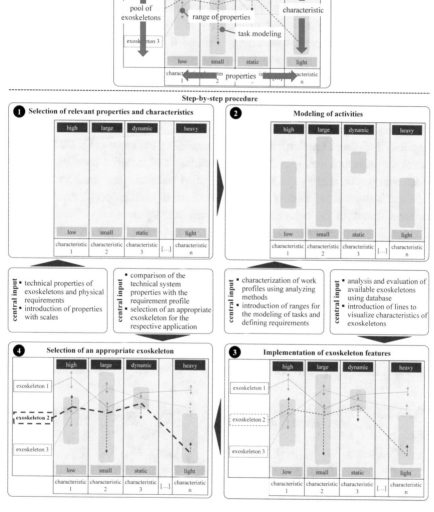

Fig. 3 Basic systematic as well as a step-by-step guide to the creation and application of the decision support matrix

select and evaluate exoskeletons for given tasks. If the matrix becomes confusing due to too many lines, it is highly recommended to distinguish different loading cases and, thus, to create different matrices.

5 Application Example

For a real application scenario, the step-by-step procedure for creating and applying the DSM is explained. This scenario considers manual screwing, assembling, and material handling activities in different workplaces and postures.

Step 1 – Selection of relevant properties: In order to describe the loading cases, technical properties incl. scales need to be defined first. In this example, the properties dynamics, bodily region, (arm) posture, weight, variance, distance, and range of motion are chosen and displayed on the abscissa (see Fig. 4). Except for the bodily region and (arm) posture (discrete scale), the factors can be distributed continuously as possible expressions. Referenced papers, checklists, or calculation tools may support the identification of relevant properties.

Step 2 – Modeling of activities: The three exemplary use cases (1) screwing overhead, (2) screwing with tool changes, and (3) assembling and material handling activities in extended arm positions are considered. Grey-scaled pillars in each column of the matrix model these scenarios. Depending on the expressions of each property, the corridors take up different spaces in the columns. The screwing activity overhead of use case 1, e.g., mainly strains the upper extremities. Conversely, work profile 2 requires different postures and variances due to necessary tool changes. In contrast to the previous cases, scenario 3 includes combined assembly and manual handling activities and, thus, strains different body parts. A high variation concerning dynamics, weights, distances, movement ranges, and postures is characteristic. Vertical pillars in the respective columns of the DSM illustrate these requirements (see Fig. 4). Methods as OWAS, RULA, or NIOSH might help to determine the exact expressions.

Step 3 – Implementation of exoskeletal features: In this example, two different exoskeletons are implemented into the DSM: an active and rigid exoskeleton, as well as a

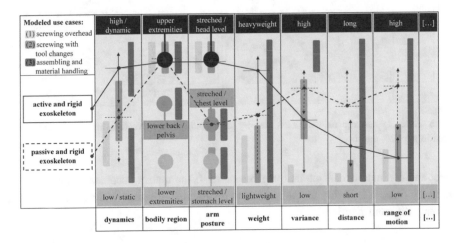

Fig. 4 Decision support matrix for three exemplary loading cases

passive system. They are visualized by full (active exoskeleton) and dashed lines (passive system), respectively (see Fig. 4). These lines represent the properties of commercial exoskeletons. Due to objectivity and the methodological focus, they are not explicitly named in this example. The vertical arrows illustrate the corridors to cover occurring variance. Datasheets or measured data may help to determine and compare the expressions for each exoskeleton.

Step 4 – Selection of an exoskeleton: The last step schedules the comparison of the exoskeletal properties to the requirements of the work profiles. For an accurate interpretation, minimum requirements should be considered. These implications lead to the following suggestions for this exemplary scenario:

(1) The minimum requirements of use case 1 are met and even surpassed by the active exoskeleton. As a result, this system is well-suited to support users in this use case. The passive exoskeleton is not suitable, as it does not support overhead activities.

(2) Neither the active nor the passive exoskeleton can adequately support all the different postures. But all other minimum requirements are met by both systems. The appropriateness of the corresponding exoskeletons for this use case should be considered in more detail.

(3) Neither the active nor the passive exoskeleton meets the requirements of use case 3 (e.g., concerning varying dynamics, back support, high variance). As a result, both systems cannot fully support all tasks. Nevertheless, each system can assist single activities of the work profile.

In short, the possible eligibility of exoskeletons for a task is given, if the line characterizing the specific system matches the range of the highlighted pillars of the respective work profiles in the DSM (see Fig. 4).

As the example shows, the matrix may indicate the suitability of different exoskeletons and support the selection process. However, it is always necessary to consider the specific application case detailly. In general, higher expressions for the technical features are more advantageous than lower ones, as the minimum requirements are easier met. Moreover, the chosen aspects underlying the recommendations for the selection should not be the only decisive factor. Other factors, such as acceptance and usability, should also be considered. In order to reduce physical strain and support the workers, it is also crucial to choose an appropriate level of support.

6 Conclusion and Outlook

Exoskeletons should be carefully selected for the respective applications, as an inappropriate selection and use may also affect unintended consequences, like restricting the freedom of movement or potentially causing or enhancing MSD [11]. Therefore, this paper has presented a generic and schematic approach to assess the fundamental suitability of exoskeletons. Additionally, it may also lead to a higher sensibility and help to identify characteristics of activities.

However, the matrix cannot represent the complex decision problem in its entirety. There is potential left to improve and specify the approach of a decision support matrix, e.g., by adding another dimension considering the specific technical properties of exoskeletons. It may also help distinguish between ambient work profiles and inherent core properties of exoskeletons. Interrelations between the external and internal levels could therefore be addressed more clearly.

In summary, this generic approach may support decision-makers to select an appropriate exoskeleton for their specific industrial application. Further critical evaluation of the decision support matrix may be reached by incorporating it into practical selection processes of exoskeletons. Despite this generic approach presented in the spotlight of exoskeletons, the application of the DSM is not limited to this technology. It is also possible to classify any other kind of (technical) support system in the matrix. In this respect, the matrix aims at a similar effect as the approach of a periodic table for technical support systems does [9]. Even though the DSM cannot represent the complexity of the selection problem, it aids in determining a starting point for a well-founded selection of suitable support devices.

Acknowledgements This article was written as part of the "Exo@Work" research project, sponsored by the German Employer's Liability Insurance Association for Trade and Logistics (BGHW). The authors are solely responsible for the content of the article.

References

1. Bloss, R.: Collaborative robots are rapidly providing major improvements in productivity, safety, programing ease, portability and cost while addressing many new applications. Industrial Robot **43**, 463–468 (2016)
2. Weidner, R., Linnenberg, C., Hoffmann, N., Prokop, G., Edwards, V.: Exoskelette für den industriellen Kontext: Systematisches Review und Klassifikation. In: 66. Kongress der Gesellschaft für Arbeitswissenschaften: Digitaler Wandel, digitale Arbeit, digitaler Mensch (2020)

3. Saggiomo, M., Longé, G., Gloy, Y.-S.: Augmented-Reality-basierte Applikation zur Assistenz von Webmaschinenbedienern. Melliand-Textilberichte **96**, 22–24 (2016)
4. Eurofound: Sixth European working conditions survey—overview report (2017 update). Publications Office of the European Union, Luxembourg (2017)
5. Bogue, R.: Exoskeletons—a review of industrial applications. Ind. Robot Int J **45**, 585–590 (2018)
6. Hensel, R., Keil, M., Mücke, B., Weiler, S.: Chancen und Risiken für den Einsatz von Exoskeletten in der betrieblichen Praxis: https://www.asu-arbeitsmedizin.com/chancen-und-risiken-fuer-den-einsatz-von-exoskeletten/chancen-und-risiken-fuer-den-einsatz-von. Accessed 25 Aug 2020
7. Kara, D.: Industrial exoskeletons: new systems, improved technologies, increasing adoption. https://www.therobotreport.com/industrial-exoskeletons/. Accessed 22 Sep 2020
8. De Looze, M.P., Bosch, T., Krause, F., Stadler, K.S., O'Sullivan, L.W.: Exoskeletons for industrial application and their potential effects on physical work load. Ergonomics **59**, 1–11 (2015)
9. Weidner, R., Karafillidis, A.: Distinguishing support technologies. A general scheme and Its application to Exoskeletons. In: Karafillidis, A., Weidner, R. (Hrsg.): Developing support technologies—integrating multiple perspectives to create assistance that people really want, pp. 85–100 (2018)
10. Bogue, R.: Robotic exoskeletons: a review of recent progress. Ind. Robot Int. J. **42**, 5–10 (2015)
11. Fox, S., Aranko, O., Heilala, J., Vahala, P.: Exoskeletons: comprehensive, comparative and critical analyses of their potential to improve manufacturing performance. J Manuf Technol Manage **31**, 1261–1280 (2019)
12. Weidner, R., Hoffmann, N.: Technische Unterstützungssysteme – Menschen gewollt. In Hartard, S., Schaffer, A. (Hrsg.): Mensch und Technik – Perspektiven einer zukunftsfähigen Gesellschaft, pp. 225–246 (2020)
13. DGUV: Checkliste für den betrieblichen Einsatz von Exoskeletten: https://publikationen.dguv.de/regelwerk/publikationen-nach-fachbereich/handel-und-logistik/physische-belastungen/3909/fbhl-020-checkliste-fuer-den-betrieblichen-einsatz-von-exoskeletten?c=54. Accessed 03 Feb 2021
14. Japet: Ergonomic evaluation tool. https://en.japet.eu/calculator/. Accessed 03 Feb 2021
15. Weidner, R., Argubi-Wollesen, A., Karafillidis, A., Otten, B.: Human-machine integration as support relation: individual and task-related hybrid systems in industrial production. I-Com **16**, 143–152 (2017)
16. Gomez-Galan, M., Perez-Alonso, J., Callejon-Ferre, A.-J., Lopez-Martinez, J.: Musculoskeletal disorders: OWAS review. Ind. Health **55**, 314–337 (2017)
17. McAtamney, L., Corlett, E.N.: RULA: a survey method for the investigation of work-related upper limb disorders. Appl. Ergonomics **24**, 91–99 (1993)
18. Karhu, O., Kansi, P., Kuorinka, I.: Correcting working postures in industry: a practical method for analysis. Appl. Ergonomics **9**, 199–201 (1977)
19. Garg, A., Hegmann, K.T., Moore, J.S., Kapellusch, J.M., Bhoyar, P., Thiese, M.S., Merryweather, A., Deckow, G., Bloswick, D., Malloy, E.J.: The NIOSH lifting equation and low-back Pain, Part 1: association with low-back pain in the back: works prospective cohort study. Hum. Factors **56**, 6–28 (2014)
20. Argubi-Wollesen, A., Wollesen, B., Leitner, M., Mattes, K.: Human body mechanics of pushing and pulling: analyzing the factors of task-related strain on the musculoskeletal system. Saf. Health Work **8**, 11–18 (2017)

21. Otten, B., Weidner, R., Argubi-Wollesen, A.: Evaluation of a novel active exoskeleton for tasks at or above head level. IEEE Robot. Autom. Lett. **3**, 2408–2415 (2018)
22. Weidner, R., Wulfsberg, J.P.: Concept and exemplary realization of human hybrid robot for supporting manual assembly tasks. Procedia CIRP **23**, 53–58 (2014)

An Approach to Integrate a Blockchain-Based Payment Model and Independent Secure Documentation for a Robot as a Service

Rainer Müller, Ali Kanso and Fabian Adler

Abstract

Robots for hire instead of purchase are developing an increasing interest among customers. This article will present a concept developed at ZeMA for the integration of a blockchain-based payment module and manipulation-proof documentation of process-specific data. This should improve the business model of hiring robots not only in terms of technical components but also in terms of economic considerations. In this way, the billing process can be automated to a certain extent, and a legally secure basis with the manipulation-proof storage of process-specific data can be created. The advantages and disadvantages of blockchain, in relation to Robot as a Service, will be highlighted and it will be shown how the disadvantages can be negated. The current limits of blockchain will also be shown. The blockchain technology IOTA is used.

Keywords

Robotic · Robot as a service · Rent-a-robot · Blockchain · Human–machine-interaction

R. Müller · A. Kanso · F. Adler (✉)
Centre for Mechatronics and Automation gGmbH (ZeMA), Group of Assembly Systems and Automation Technology, Gewerbepark Eschberger Weg 46, 66121 Saarbrücken, Germany
e-mail: f.adler@zema.de

T. Schüppstuhl et al. (eds.), *Annals of Scientific Society for Assembly, Handling and Industrial Robotics 2021*,
https://doi.org/10.1007/978-3-030-74032-0_15

1 Introduction[1]

By 2022, up to 4 million industrial robots are expected in industrial companies worldwide with an annual growth rate of 12%. In addition, there is an increasing demand for rentable industrial robot systems or robots as a service, which means that the number of suppliers is also rising. ABI research [2] forecasts that by 2026 up to 1.3 million "*Robot as a Service*" (*RaaS*) systems will be in use. Due of the temporarily shut down of entire production facilities in the context of the Covid-19 pandemic, robot systems and automation solutions are gaining more attention for several industries. This could lead to an increase in the number of and demand for robotic systems. [2, 3, 12, 17].

In case of *Robot as a Service*, the customer will no longer buy the robot system, but will rent it for a certain period of time in order to counteract bottlenecks, staff shortages or increased demand for a product. The customer only wants to pay for the robot service time, or per part produced, that is actually used, in order to ensure a return on investment for the customer. Especially in small and medium sized enterprises (SMEs) a robot system in assembly does not produce continuously like in a large company [7].

The company benefits from renting instead of buying with lower initial investment costs, no long-term commitment to the system, no costs incurred when the robot system is shut down and no maintenance work or costs. To a certain extent, the robot rental business can be compared to a car rental business. One major difference is the commissioning. Whereas a car only requires a key and driver to drive off, the robot system in addition requires considerable programming effort. Therefore, robot rental companies usually offer various applications (palletizing, material handling, welding or machine tending) that are compatible with the system [1, 9].

A further issue for the hiring company is the clarification of the question of responsibility in the event of failure, downtimes and ensuring that the contractually agreed working hours or produced parts are observed. On the one hand, the lessor of the robot system does not want the hirer to work with the robot more than contractually negotiated. On the other hand, the lessor must be assured that in the event of a standstill or failure, the fault for the failure or the standstill is attributable to the lessor. Only then, a tenancy agreement be concluded in a legally secure and holistic manner.

The question of responsibility could, for example, be clarified by collecting and storing process-relevant data, logs and events. This process-relevant information is stored on a data storage device (PLC, hard disk, database, cloud) and can be used to determine the invoice amount. This point requires mutual trust between tenant and lessor, as the entered and generated data is not stored in a manipulation-proof manner. This means there is a chance that the data on the data storage device will be changed after the conclusion of the contract or, for example, in the event of a fault.

---

[1]The results are part of the project "Robotix-Academy" funded by the Interreg V A Großregion (no 002–4-09–001). Furthermore the research is part of the "COTEMACO"-project and funded by the European Union's Horizon 2020 research and innovation program at ZeMA.

In the context of this article, a concept developed at ZeMA is presented, which shows how this trust could be generated with a data storage device without reducing the accessibility of the data. An approach of a manipulation-proof database with the coupling of a payment model is presented. Blockchain offers and combines these two functionalities in one platform, but will not considered as the superior solution at this state in the areas of manipulation-proof database and payment models.

For implementation, the potential advantages of blockchain as a tool are applied to the use case *Robot as a Service*, also considering the disadvantages of blockchain. Advantages of blockchain are data and process integrity for risk reduction, potential for transaction processing and verification and automated business transactions using smart contracts. Further advantages are transparency and pseudonymity, which can also be seen as a disadvantage of blockchain. E.g.: This means that every participant has an encryption key when uploading transactions transparent but encrypted to the blockchain. If the key is stolen by a stranger, he can accordingly trace what and when the participant uploaded something to the blockchain (pseudonymity) [4].

Due to a possible transaction processing, the data, events and logs of a robot can be directly linked to a digital currency, which will not be discussed in detail in this article. The focus is on the manipulation-proof storage of process-specific data and a payment concept for the control and release of a robot system. The service business model "*Rent-a-Robot*" is regarded as given and not further researched.

2 State of the Art and Initial Situation

This article links the blockchain topics with the business model of renting a robot. In the following, blockchain and *Rent-a-Robot/Robot as a Service* will be examined more closely in order to define the basics and points of contact of the article with the state of the art.

2.1 Rent-a-Robot and Robot as a Service

When researching the literature, it is particularly noticeable that the topic of *Robot as a Service* has a strong industrial relevance, but only few scientific articles have been published on it so far. Considerably more articles have been published in scientific journals or blogs on this topic. These are examined in more detail below.

Rent-a-Robot or *Robot as a Service* is part of the "*as-a-Service*" philosophy and consists of the components of a service system as well as a service business model to reflect the rental business. Service systems can be industrial robots, driverless transport systems or even entire robot cells. During the loan transaction, fees can vary from a fixed continuous value to a variable value based on the service duration or number of products. This ensures that no further costs are incurred if the rented system stops operating and the service provider is responsible for maintaining and ensuring availability. This is warranted within the framework of a *Robot as a Service*—Service Level Agreement (SLA) [3, 14, 19].

The idea or vision of *Robot as a Service* is to integrate cloud-based "robot rental" systems at the user's premises for the required assembly processes with on-site robot hardware and cloud-based programming. This allows systems to be adapted to changing requirements without having the necessary infrastructure in advance. Users are therefore able to quickly expand or redesign their production and reduce pre-production costs. In addition, the cloud connection allows, for example, the robot operating time to be monitored and data to be stored [1, 14].

In the project *"Rent-a-Robot"*, the consortium around Schuh et al. [3] has dealt with the consistent conceptual design for the temporary automation of assembly activities. On the one hand, a modular robot assembly system suitable for industrial use is described and. On the other hand, methods for evaluating, planning and designing the required operator models are developed. This should ensure a quick cost calculation and selection of an operator model when a system is requested. Flexible costs during the runtime of the system as well as data collection and storage on a manipulation-proof medium in order to have a basis for decision making in case of downtimes were not considered [3].

READY Robotics, RobotWorx, Hirebotics or Essert are examples for companies that offer robots for rent. They offer robot cells with or without pre-defined tasks for rent. The collection and storage of data on service lifetime and workload is done sporadically. Payment methods are pay-per-hour, pay-per-week or pay-per-use. The problem and starting point of this article is a concept for the manipulation-proof storage of process-relevant data for a well-founded assessment of the costs incurred [1, 6, 9].

2.2 Blockchain and the Alternative IOTA

Blockchain technology is a type of Distributed Ledger Technology (DLT), in which transactions are bundled together in blocks (see Fig. 1. DLT forms the basis for innovative applications such as crypto-currencies, identification possibilities, machine communication or smart contracts, with which orders can be automatically triggered or contracts concluded. Blockchain represents a distributed, manipulation-proof, time-stamped database. The transactions and data are protected against manipulation using cryptographic procedures and consensus algorithms. Each transaction is encrypted by a so-called hash.

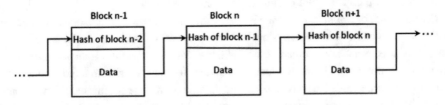

Fig. 1 Data structure of a blockchain [16]

In addition, each network participant is regularly provided with an updated transaction history, which means that 51% of the network would first have to be "convinced" in case of attempted manipulation. This is because each block in the blockchain refers to the previously completed block (see Fig. 1) [8, 10, 18].

The blockchain technology is based on five basic principles: Distributed data bank, peer-to-peer data transmission, transparency through pseudonymity, irreversibility of recordings and computerised logic. Not each of these basic principles bring only advantages. Once something is uploaded in the blockchain, it is difficult to remove the data or messages. Furthermore, not every company wants to have company related data stored in a public accessible database. These problems are taken into account when implementing a blockchain based payment model and independent secure documentation for a *Robot as a Service*[1, 11].

Based on the characteristics of the blockchain technology, Chowdhury et al. [5] analysed under which aspects a blockchain or a conventional database is more reasonable. Blockchain is considered to be the better solution for confidence building, robustness and validity of the data. For confidentiality and performance reasons, the traditional database is better. [5].

The low performance is due to the straightness of most blockchain technologies. To achieve better scalability and thus performance, IOTA is used as alternative with their Directed Acyclic Graphs (see Fig. 2). IOTA is not an acronym for IoT, but is a proper name and based on the smallest Greek letter in the alphabet. Instead of bundling transactions into blocks and chaining them together, IOTA links transactions as nodes in a

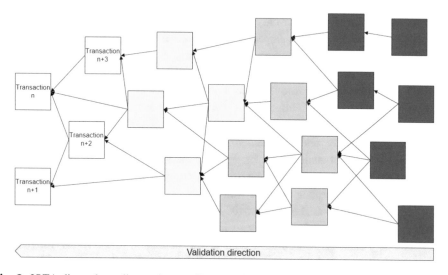

Fig. 2 IOTA directed acyclic graph according to [13]

graph. IOTA is subject to the basic idea, advantages and characteristics of blockchain and is also often perceived as a blockchain. Accordingly, IOTA is seen as a kind of modification of the blockchain.

Furthermore, IOTA allows zero-value data transactions, which means that the transactions can be used for pure data exchange or data storage. Masked Authenticated Messaging (MAM), which is a data communication protocol and publishes encrypted data streams, called channels, in transactions in the tangle, is used for this purpose. The Tangle is shown in Fig. 2 [13, 15].

3 Concept for Expanding the Business Model *Robot as a Service* with Blockchain

In order to create a fundamental trustworthy basis for the hiring of robot systems, existing concepts and business models of *Rent-a-Robot/Robot as a Service* will be combined with blockchain. IOTA is used as "blockchain technology" to counteract the disadvantages of scalability and performance of other blockchain technologies. By using IOTA, the disadvantages of the previous business model of *Robot as a Service*, such as validity and the manipulation-capable storage of data, are to be avoided. The concept is intended to map *Robot as a Service* more economically.

In the case of rental transactions, static and dynamic costs can arise during the rental period. The static costs can be calculated in advance, which Schuh et al. have already researched in the project "*Rent-a-Robot*" [3] for robot rental. Until now, the dynamic costs and the effects of failures, malfunctions or non-compliance of contract numbers on the final cost calculation were not or only partially considered.

This is where the concept developed at ZeMA takes effect. In order to offer a dynamic rental model, process-specific and -relevant characteristics, data as well as logs and events, e.g. in case of malfunction or failure, must be validly stored. Process-relevant data are operating time, production turnover, sensor data as well as safety data in case of an emergency stop. Furthermore, information which is used as a basis for decision-making, e.g. for cost calculation, must be stored in a more manipulation-proof manner. This way, well-founded and legally binding statements can be made if the contractual objectives are not reached. In case of malfunction or damage, the question of responsibility can be clarified via the data. Figure 3 shows the necessary components and links between them to extend the *Robot as a Service* use-case with IOTA.

Starting from the robot, the concept for the extension of existing technologies with blockchain using the example of *Robot as a Service* uses a single-board computer (e.g. Raspberry Pi), a database, a web dashboard and a connection to the IOTA Tangle. The single board computer is the centre point and forms the interface between the blockchain technology (IOTA Tangle), the robot and the database. The single board computer should be a universal interface to IOTA and should also be useable for other systems and databases. This creates a certain modularity and theoretically allows to use any

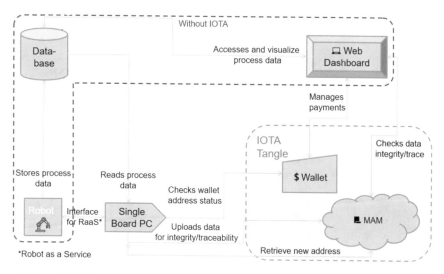

Fig. 3 Extended *Robot as a Service* use-case with IOTA

infrastructure (machine, database). Through the connection, each robot gets its own virtual wallet, which can be used to implement a pay-per-use model and to store process-relevant data in a manipulation-proof way. These two scenarios are presented in more detail below. Finally, the web dashboard serves as a control centre for the systems in use by managing and monitoring process data and payments.

3.1 Pay-per-Use Via Crypto Currency

Pay-per-use in itself exists and already functions in the rental business. The novelty is that the blockchain technology allows to automate the rental process. The robot is equipped with an IOTA wallet through the single-board computer, which is used for payment. By means of activities with a value, the costs can be tracked in an automated, manipulation-proof and transparent way. If the hirer wants to use the robot, he has to pay accordingly with the certain valence. This could be implemented time-based as well as product number-based. The renter pays only as long as he produces with the rented machine. Figure 4 shows how this model could be implemented on basis with IOTA.

Initially, payment via the web dashboard is managed by the lessor. In the web dashboard, transactions to the robot can be viewed. Moreover, addresses for handling payments can also be managed in the web dashboard. The transmission of a new address is triggered with the web dashboard and communicated to the robot via a MAM channel. This ensures an automated transmission. The hirer can use the address associated with the robot to pay accordingly. After payment, the hirer is able to move and produce with the robot. If no payment has been made, the rented robot system is on-hold and waits either

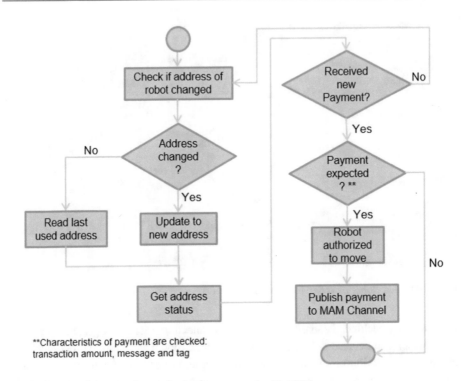

Fig. 4 Payment for production authorization executed with IOTA

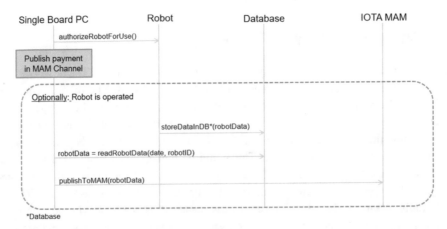

Fig. 5 Robot authorized to move through IOTA

for a new address or payment. This results, the hirer only pays if the robot is used. In the meantime, it is also checked whether a payment is expected. If this is the case, the robot is authorised to move. It sends and documents a confirmation to the IOTA MAM channel (see Fig. 5).

For the payment history, further data on the executed process is optionally recorded after the robots movement has been authorised. In the first step, the process-relevant data (robot data, sensor data) are stored in the database. In the second step, the single-board computer polls the database at regularly definable intervals and writes the data in a specific MAM channel in the IOTA Tangle. In this way, the payment history can be stored directly linked to the process-relevant data in a manipulation-proof manner.

3.2 Independent Data Certification for Manipulation-proof Documentation of Process Data, Events and Logs

A disadvantage of this storage method for process- and company-relevant data is the transparency and pseudonymity of the block chain technology. Not every company is willing to store sensitive data in a publicly accessible database. Therefore, two options are presented below, one using the blockchain technology as pure data certification and the other using blockchain as a database. Figure 6a shows the two options.

When information is uploaded, a unique non-changeable hash of this information is formed. If the information would be uploaded again at a later time, the hash is still the same as when it was first uploaded. If the information is manipulated, the two hashes would be different. Accordingly, the hash can be regarded as the unique signature of the information. With option one, the single board computer retrieves the corresponding data from the database, generates a hash and stores it in the blockchain. Whereas with option two, the corresponding data from the database is stored directly in the blockchain. Both options ensure that the data is stored in a certified way. The advantage of the first option is that the data is not uploaded. The advantage of the second option is that the data could theoretically be released directly, as they are already available in the IOTA tangle.

For maintenance, checking or billing purposes, the logged data can be accessed via the Web Dashboard (see Fig. 6b). The Web Dashboard reads the process data from the database and from the IOTA MAM channel. By comparing the two data sets from the

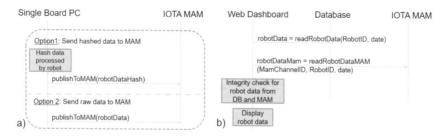

Fig. 6 **a** Robot logging activity; **b** Access to logged data and data integrity check; with IOTA

different storage media, an integrity check can be performed to guarantee the authenticity and that the data is not manipulated. The data and the integrity check are displayed in the Web Dashboard. In addition, the stored hash from the IOTA MAM channel with the newly created hash of the data from the database could also be used for the integrity check.

4   Conclusion and Outlook

In the article a concept was presented that with blockchain a technology under development can add value for already existing use cases, in this case *Robot as a Service*. A modular approach was chosen to apply the presented concept to other industries and systems. The aim of the implementation was to increase the confidence in the generated process specific data of *Robot as a Service* without reducing the accessibility of the data. An approach to combine the resulting dynamic costs and pay-to-use with the blockchain crypto currency was also presented. This was implemented with IOTA, as it offers advantages in terms of scalability and performance compared to other blockchain technologies. Other blockchain technologies also offer features such as the manipulation-proof storage of data and could be applied.

Although blockchain has promising properties, more attention must be paid to the development of the technology in the coming years in order to realise an industrial application of the technology. At the moment, there are good approaches, but some questions and undesired peculiarities for the industry, such as transparency and pseudonymity of blockchain as well as strongly fluctuating exchange rate of the crypto currency, have not been clarified. Blockchain still has to find its benefit to existing technologies. Furthermore, Blockchain and IOTA should always be seen as a tool to solve a problem.

Especially the point that many blockchains have a fluctuating price of its cryptocurrency must be considered. Therefore, from an economic point of view, it is extremely difficult to implement an automated payment or pay-per-use model via blockchain technology. At the moment, it makes more sense to use the cryptocurrency as a fictitious currency like a token system to link process data, events and logs (e.g. operating time, produced parts) with a valence. The links between the token and the data can be used for automated invoice generation.

Nevertheless, the concept offers a high potential to extend existing technologies and to create added value. The next step is the further concretization of the concept and its application to a use case and a demonstrator. Thereby, further approaches and weak points shall be defined in order to make it usable for industrial application. Furthermore, the modular aspect is to be extended to cover a larger number of use cases, especially for assembly activities.

References

1. Anandan, T.M.: Robots for rent—why RaaS works. (2018)https://www.robotics.org/content-detail.cfm/Industrial-Robotics-Industry-Insights/Robots-for-Rent-Why-RaaS-Works/content_id/7665 Accessed 22 Sep 2020
2. Bay, O.: Robotics-as-a-Service is the key to unlocking the next phase of market development. (2018) https://www.abiresearch.com/press/robotics-service-key-unlocking-next-phase-market-development/ Accessed 22 Sep 2020
3. Brecher, C., Müller, R., Schuh, G.: Rent-A-Robot—Abschlussbericht des MBMF-Verbundsprojektes. Aachen, 2.Auflage, Apprimus Verlag (2008). ISBN: 978-3-940565-22-8
4. Buhl, H.U., Urbach, N., Schweizer, A.: Blockchain—Technologie als Schlüssel für die Zukunft? Zeitschrift Für Das Gesamte Kreditwesen **2017**, 596–599 (2017)
5. Chowdhury, M.J.M., Colman, A.W., Kabir, A., Han, J., Sarda, P.: Blockchain Versus Database: A Critical Analysis. In: 17th IEEE International Conference On Trust, Security And Privacy In Computing And Communications/12th IEEE International Conference On Big Data Science and Engineering, (2018). https://doi.org/10.1109/TrustCom/BigDataSE.2018.00186
6. Essert Robotics: https://www.essert.com/en/robot-as-a-service Accessed 22 Sept 2020
7. ETCIO agency: The rising trend of Robots as a Service (RaaS). (2019). https://cio.economictimes.indiatimes.com/news/strategy-and-management/the-rising-trend-of-robots-as-a-service-raas/70553000 Accessed 22 Sep 2020
8. Fill, H.G., Meier, A.: Blockchain kompakt: Grundlagen, Anwendungsoptionen und kritische Bewertung. IT komapkt, Springer, Wiesbaden, (2019). https://doi.org/10.1007/978-3-658-27461-0
9. Hirebotics: https://www.hirebotics.com/botx Accessed 22 Sep 2020
10. Hosp, J.: Kryptowährungen: Bitcoin, Ethereum, Blockchain, ICOs & Co. einfach erklärt. München, 2. Auflage, FinanzBuch Verlag, (2018). ISBN: 9783960922483
11. Iansiti, M., Lakhani, K.R.: The truth about blockchain. (2017). www.hbr.org/2017/01/the-truth-about-blockchain Accessed 22 Sept 2020
12. IFR – International Federation of Robotics: Robot Investment reaces record 16.5 billion USD—IFR presents World Robotics. (2019). https://ifr.org/ifr-press-releases/news/robot-investment-reaches-record-16.5-billion-usd Accessed 22 Sep 2020
13. IOTA Wiki: https://www.iota-wiki.com/de/ Accessed 22 Sep 2020
14. Robotic Industries Association, Robotics Online Blog: The rise of Robots-as-a-Service. (2020). https://www.robotics.org/blog-article.cfm/The-Rise-of-Robots-as-a-Service/259 Accessed 22 Sep 2020
15. Schiener, D., Snsteb, D., Popov, S.: What is IOTA? IOTA Foundation. (2019). https://www.iota.org/getstarted/. Accessed 22 Sep 2020
16. Shahsavari, Y., Zhang, K., Talhi, C.: Performance modeling and analysis of the Bitcoin inventory protocol. IEEE International Conference on Decentralized Applications and Infrastructures (DAPPCON 2019), San Francisco, California, USA. (2019). https://doi.org/10.1109/DAPPCON.2019.00019

17. Thomas, Z.: Coronavirus: Will Covid-19 speed up the use of robots to replace human workers? (2020). https://www.bbc.com/news/technology-52340651 Accessed 22 Sep 2020

18. World Bank Group: Distributed Ledger Technology (DLT) and blockchain. Washington International Bank for Reconstruction and Development/The World Bank, FinTech Note No. 1, (2017)

19. Yates, G.: 2019 State of the world for Robot-as-a-Service Companies. (2020). https://insights.rlist.io/p/report-robot-as-service-companies.html#nutshell Accessed 22 Sep 2020

Human-Robot Collaboration

Human Body Simulation Within a Hybrid Operating Method for a Safe and Efficient Human-Robot Collaboration

Kai Lemmerz and Bernd Kuhlenötter

Abstract

The planning and integration of production systems with a direct human-robot collaboration (HRC) is still associated with various technical challenges. This applies especially to the realization of the operation methods speed and separation monitoring (SSM) as well as power and force limiting (PFL). Due to the limited consideration of the human motion behaviour, the required dynamic separation distance in SSM is frequently oversized in practice. The main consequences are wasted space as well as cycle time and performance losses within the corresponding HRC application. In PFL a physical contact between the operator and robot is permissible, taking into account specified biomechanical thresholds. However, there is still a lack of suitable use-cases since the maximum permissible speeds are on a very low level. Moreover some thresholds regarding the transient contact case are still non-applicable for critical body areas (e.g. temple, middle of forehead). The study of this paper is related to a kinematic state determination of the human operator within a new hybrid collaborative operation. In this method the SSM type is extended regarding the description of the operator and coupled with the two-body contact model of the PFL. Using a planning and simulation tool for HRC, the kinematic states of different body regions are derived from an integrated and parameterized digital human model. Afterwards, these body regions are mapped to the characteristic body areas of the ISO/TS 15066, whereby the resulting information will be applied in an adaptive robot speed control. The performance of the presented concept will be evaluated using an exemplary simulated HRC scenario.

K. Lemmerz (✉) · B. Kuhlenötter
Ruhr-University Bochum, 44801 Bochum, Germany
E-mail: lemmerz@lps.rub.de
URL: http://www.lps.ruhr-uni-bochum.de/

© The Author(s) 2022
T. Schüppstuhl et al. (eds.), *Annals of Scientific Society for Assembly,
Handling and Industrial Robotics 2021*,
https://doi.org/10.1007/978-3-030-74032-0_16

191

Keywords

Human-robot collaboration • Speed and separation monitoring • Power and force limiting •
Modeling and simulation

1   Introduction

In order to integrate and certify a direct human-robot collaboration (HRC), four types of
operations are permissible according to [1, 2]. The paper at hand deals with the research
on the collaborative operations *speed and separation monitoring* (SSM) as well as *power
and force limiting* (PFL). With the ISO/TS 15066 [3], system integrators are supported by
easy to use principles for the estimation of the dynamic separation distance in SSM and the
configuration of permissible robot speeds in PFL. However, problems arise in the practical
implementation of these two operations. In SSM, the safety zones are often oversized due
to different simplifications such as a punctual and constant directional operator speed as
well as conservative values for the robot stopping distances. While PFL on the other hand
allows the closest level of interaction, it also defines the most stringent safety requirements.
Thereby, the permissible robot speeds depend on the affected characteristic body areas. If no
exact determination of these body areas can be performed during the process, the practical
implementation of PFL usually avoids larger translatory robot motions at the height of
the human head and limitations of the maximum Cartesian speed to 250 mm/s by a time
consuming path planning. In both types of operations the mentioned circumstances can lead
to cycle time losses and wasted space in the respective production system.

1.1   Related Work

The separation distance S_p to be maintained applying the SSM depends especially on the
motion behaviour of the operator and robot [4]. Applying an extended path planning strategy,
advantages can be achieved compared to the classic human-robot coexistence, since the robot
motion can be dynamically adjusted to its environment [5–9]. Despite this, it is noticeable that
the motion behavior of the human operator is either often neglected or facilitated. However
a simplified consideration of the operator has been examined in some approaches [7, 10,
11]. Nevertheless, there are no suitable concepts for the differentiation and localization of
individual body regions and the use of the acquired status information for an adaptive robot
path planning. This results in the loss of usable workspace due to conventional assumptions
such as a constant operator speed of 1.6 m/s. In conclusion it can be established that none
of these approaches consider the complete human musculoskeletal system within an HRC
environment. According to the PFL specification, the forces and pressures acting on the
operator in the event of a collision must not exceed the biomechanical thresholds of the
affected body areas according to [3]. Numerous studies have been carried out in this regard to

correlate factors such as maximum forces or stresses with respect to the onset of injuries and pain perception [12–15]. In a more recent study, new pressure thresholds were determined and assessed with regard to a revision of the ISO/TS 15066 [16]. A Risk reduction can be achieved either by safety control systems or inherently safety features on the robot. In recent years, scientific work has focused on safety-related implementations as well as the investigation of collision consequences and influencing values (e.g. effective masses) in case of contact [14, 17–20]. Nonetheless, the PFL method is still in a research stadium and up to now rarely suitable for industrial practice. A main reason for this are the *non-applicable* areas for transient contact scenarios according to [3]. This complicates the design of safe and efficient HRC scenarios as well as the certification process. Based on the related work, a need for action can be derived with respect to an extended human modeling within the SSM and PFL type.

1.2   Novelty and Objective

The paper at hand deals with the development and investigation of an extended human modeling and the kinematic state consideration of specific body areas. This approach will be embedded into a new hybrid collaborative operation (HCO), which combines the SSM and PFL method [27]. In order to estimate the motion data we will apply a previously developed HRC simulation tool. The main objectives are the reduction of safety zones and the increase of productivity within HRC systems under the early consideration of relevant safety requirements. Section 2 introduces the applied simulation tool and provides an overview of the HCO method. A new approach for the kinematic state determination of the specific body areas based on a digital human body model is discussed in Sect. 3. The evaluation of this approach is carried out by means of a designed and simulated HRC scenario in Sect. 4. Section 5 summarizes the results of this paper and discusses future research topics.

2   Preliminary Work

2.1   Simulation Tool for Human-Robot Collaboration

Subsequently, an HRC planning and simulation tool, which has been developed within the BMBF research project KoMPI (fund number 02P15A060), will be applied for the consideration of the human motion behaviour and the kinematic state estimation of single body regions. Using this tool, both human operators and robots can be analyzed in a collaborative assembly environment and evaluated in terms of ergonomics and automation aspects. The underlying system architecture consists of two main components: A robot and peripheral simulation, and a human and process simulation [21]. The extended concept is based on the integration of evaluation functionalities for the field of automation technology (e.g.

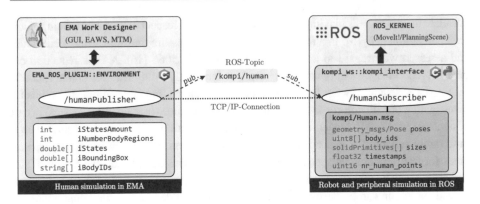

Fig. 1 Transmission of virtual human motion data in the HRC simulation tool

collision-free robot path planning) into the proprietary software solution ema Work Designer (emaWD) of imk automotive GmbH [22, 23]. emaWD enables a temporal evaluation and ergonomic analysis of the designed production processes. A central part of emaWD is a realistic simulation and visualization of human body motions, whereby a digital human model (DHM) can be parameterized and controlled via a simple and intuitive task-based process modeling [24]. The part of the robot and peripheral simulation is based on the Robot Operating System (ROS) [25]. ROS is widely used in robot research and enables a holistic simulation of robot systems, sensors and the environment.

Figure 1 shows the concept of the data transmission between the emaWD-ROS interface (`EMA_ROS_PLUGIN`) and ROS itselfs by means of the animated human motion in emaWD. Hereby, the human model is reduced to $n_{br} = 18$ body regions. A `/humanPublisher`-Node collects the appropriate information about the individual and time-dependent kinematic poses in a user defined ROS message `Human.msg` and published it on the `/kompi/human` topic. In the `kompi_interface` package this message will be received and interpreted by a corresponding callback function. By using the incoming data and simplified geometrical primitives, an approximated human motion is then reconstructed on the `planning_scene` of MoveIt! [26]. Hence, this dynamic environment can be considered in the estimation of a collision-free robot path.

2.2   Method of a Hybrid Collaborative Operation

The simulation based description of the human motion behaviour, discussed in this paper, will be embedded in a new hybrid collaborative operation method (HCO) for HRC applications. Assuming a pre-defined robot trajectory, the main objective of the HCO is to achieve a safety-oriented and efficient configuration of the Cartesian robot speed [27]. With respect to the specified safety strategies of SSM, HCO involves a pure speed adaption and no alternative

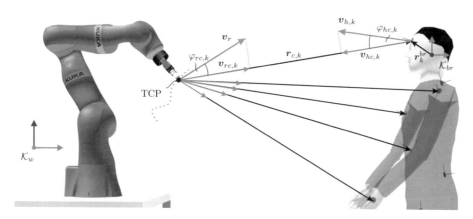

Fig. 2 Kinematic Relation between robot's TCP and a specific body area k

path planning. Thereby, the robot system must comply with the normative requirements of the PFL and be equipped with an external sensor system for workspace monitoring according to the SSM. The procedure of the HCO method is divided into three stages, which are briefly described as follows:

1. **Kinematic state determination**: The kinematic states of the interaction partners are determined at the tool center point (TCP) of the robot (assumption in this paper) and $n_k \in \mathbb{N}$ characteristic body areas k of the operator (cf. Fig. 2). They include the local position vectors $\mathbf{r}_r \in \mathbb{R}^{3 \times 1}$ and $\mathbf{r}_{h,k} \in \mathbb{R}^{3 \times 1}$ as well as linear velocities $\mathbf{v}_r \in \mathbb{R}^{3 \times 1}$ and $\mathbf{v}_{h,k} \in \mathbb{R}^{3 \times 1}$. Based on the kinematic positions, the respective collision vectors $\mathbf{r}_c$ and actual distances S_k between the operator and robot can be formulated as:

$$S_k = \|\mathbf{r}_c\| = \|\mathbf{r}_{h,k} - \mathbf{r}_r\| . \tag{1}$$

For the sake of clarity, the index k does not occur in the further steps unless they have to be used explicitly. A key aspect in the SSM and PFL method is the description of the absolute directional speeds

$$v_{rc} = \mathbf{v}_r \frac{\mathbf{r}_c}{|\mathbf{r}_c|}, \quad v_{hc} = -\mathbf{v}_h \frac{\mathbf{r}_c}{|\mathbf{r}_c|} \tag{2}$$

and the collision angles

$$\varphi_{rc} = \cos^{-1}\left(\frac{\mathbf{v}_r \, \mathbf{r}_c}{|\mathbf{v}_r| \, |\mathbf{r}_c|}\right), \quad \varphi_{hc} = \cos^{-1}\left(\frac{\mathbf{v}_h \, \mathbf{r}_c}{|\mathbf{v}_h| \, |\mathbf{r}_c|}\right) \tag{3}$$

between both interaction partners [6, 8, 9].

2. **Safety measurement according to SSM**: The second stage of the HCO includes the determination of the current protective separation distance

$$S_p(t_0) = \underbrace{v_{hc}\left(T_r + T_s(v_{rc})\right)}_{S_h} + \underbrace{v_{rc}\,T_r}_{S_r} + S_s(v_{rc}) + S_m \qquad (4)$$

according to [3] as well as the proof of the required safety condition $S \geq S_p$, direction of motions via $\{\varphi_{hc}, \varphi_{rc}\}$ and relative speed $v_{rel} = v_{rc} + v_{hc}$ between the human operator and robot. With regard to Eq. (4), T_r denotes the reaction time of the robot system. S_m includes the inaccuracy of the applied monitoring system in the sense of a constant minimum distance. The variables $T_s(v_{rc})$ and $S_s(v_{rc})$ describe a speed-dependent braking behaviour of the robot in Cartesian space regarding category stop 1 (cf. [1, Annex A]). In case all required conditions $S \geq S_p$, $\varphi_{rc} \geq \pi/2$ and $v_{rel} \leq 0$ are violated, a reduced robot speed $\tilde{v}_{rc}(S, S_p(t_0))$ is desired to maintain the separation distance according to the SSM type (cf. [8]).

3. **Integration of PFL Requirements**: In the third stage of the HCO method, the SSM will be combined with the safety-related requirements of the PFL. If a reduction of the robot speed should be performed in stage 2 ($0 < \tilde{v}_{rc} < v_{rc}$), a remaining potential collision speed $v_c = v_{rc} - \tilde{v}_{rc}$ is specified, which will be compared with the permissible collision speed $v_{p,k}$ according to the transient contact model of the PFL [3]. Hence, a modification of the original robot speed would only be required within the HCO method if $v_c > v_{p,k}$ applies. Thereby, $v_{p,k}$ is decisively influenced by the maximum permissible transfer energy and the biomechanical properties (effective mass, stiffness) of the concerned specific body area k, which usually changes with a high probability during the collaboration scenario.

For further details of the HCO method, please refer to [27]. In context of the this method, the present paper deals with the determination and usage of the kinematic human states based on virtual motion data. These informations are necessary in order to firstly get a knowledge about the critical and hazard body areas, secondly to determine the local safety distances $S_{p,k}$ and thirdly to obtain the body area-dependent permissible collision speed over the entire process.

3   Kinematic States of Human Body Areas

As part of the virtual acquisition of human motion data and the kinematic state determination, the DHM ema and the human-specific subtasks created by the planner are applied. In this regard, Sect. 2.1 briefly summarizes the export of the human behaviour simulated in emaWD via the developed ROS interface EMA_ROS_PLUGIN and the reconstruction of the kinematic states within the robot and peripheral simulation. The two parameters poses and velocities of the transmitted message Human.msg provide the spatial positions $r_{br} \in \mathbb{R}^{3 \times 1}$ and rotations $R_{br} \in \mathbb{R}^{3 \times 3}$ as well as the linear and angular velocities $v_{br} \in \mathbb{R}^{3 \times 1}$ and $\omega_{br} \in \mathbb{R}^{3 \times 1}$ of the individual body regions. In the next step, the geometric relationship

Table 1 Assignment of the specific body areas k according to (**author?**) [3] to the body regions br of the reduced EMA model (l = left; r = right)

br	Body region (EMA)	k	Specific body area (ISO/TS 15066)	Displacement r_k^{br}		
				x [mm]	y [mm]	z [mm]
1	HEAD	1	Middle of Forehead	90	0	80
		2	Temple (l/r)	50	±70	40
		3	Masticatory muscle (l/r)	50	±50	40
		4	Neck muscle (l/r)	−40	±30	−80
2	LOWER_BODY	7	Fifth lumbar vertebra	−60	0	−20
		10	Abdominal muscle	80	±40	100
4	UPPER_BODY	5	Seventh neck vertebra	−20	0	60
		6	Shoulder Joint	100	±130	20
		8	Sternum	160	0	−70
		9	Pectoral muscle (l/r)	140	±70	−50
5/6	ARM (r/l)	12	Deltoid muscle	−40	−40	0
7/8	FOREARM (l/r)	13	Humerus	−30	20	0
		15	Forearm muscle	−20	30	50
		16	Arm nerve	0	−30	30

between these body regions br and the specific body areas k according to [3] has to be described in a formulation.

The assignments of the body areas k to the respective body regions br are listed in Table 1. For the sake of clarity, the listing is limited to the body's head, center and the upper limbs. In addition to the designation of the ema body regions, the local displacements $r_k^{br} \in \mathbb{R}^{3\times1}$ of the body areas with respect to the parent coordinate system $\mathcal{K}_{br}$ are given. The Cartesian coordinates of r_k^{br} were determined geometrically using the underlying ISO/TS 15066 specification and the human model type *50th percentile, male* of the applied simulation tool. With the transformation of r_k^{br} into the global frame $\mathcal{K}_w$, the spatial position of the specific body area k is given by

$$r_{h,k} = r_{br} + R_{br}\, r_k^{br}. \tag{5}$$

In addition, considering the linear velocity v_{br} and the rotational velocity ω_{br} in the origin of $\mathcal{K}_{br}$, the following relation applies with Euler's derivation rule:

$$v_{h,k} = v_{br} + \omega_{br} \times r_k^{br}. \tag{6}$$

4   Evaluation

Afterwards the human motion consideration within the HCO method is carried out by means of an exemplary HRC scenario. In the scenario, the TCP of a Kuka LBR iiwa 14 R820 moves along a pre-defined circular path $x(t)$ (cf. Fig. 3). An extended investigation of the entire kinematic chain has been done in a primary work [8]. The characteristics of the SSM measurement and the influence of the extended human modeling discussed in this paper is summarized in Fig. 4.

First of all, Fig. 4a illustrates the current distances S_k and dynamic safety distances $S_{p,k}$ according to Eq. (4) of four selected body areas. The operator initially approaches to the robot's workspace and then walks around the circular and contrary trajectory of the TCP. So a reduction of the actual distances can be be noticed in the beginning of this scenario. After 3.04 s, the minimum distance is reached at 0.57 m by means of the *left shoulder joint* ($k = 6$). Moreover, the operator penetrates the safety zone in the segment $1.55\,\text{s} < t < 3.71\,\text{s}$ with at least one body area. The partial oscillation of $S_{p,k}$ can be explained by the periodically varying speed $v_{h,k}$ during the operator's motion. This effect is intensified especially in the upper limbs (see e.q. $k = 16$). Figure 4b shows the variation of the critical body area k_{cr}

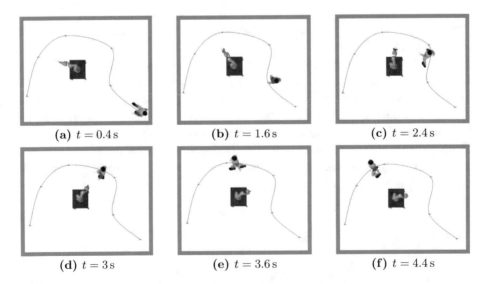

(a) $t = 0.4\,\text{s}$ (b) $t = 1.6\,\text{s}$ (c) $t = 2.4\,\text{s}$

(d) $t = 3\,\text{s}$ (e) $t = 3.6\,\text{s}$ (f) $t = 4.4\,\text{s}$

Fig. 3 Simulated motion of the human operator and the Kuka LBR iiwa 14

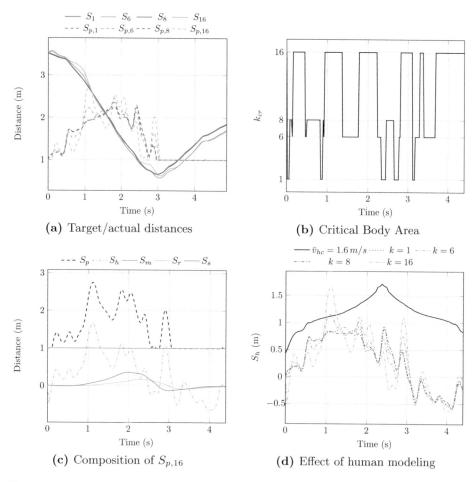

Fig. 4 Safety distance characteristic and influence due to the human modeling

for the proposed scenario. This parameter results from the minimum time-to-collision value TTC between the several body areas and the TCP. For further details please refer to [8]. Even within this short process and the selection of only four body areas, a strong shift of k_{cr} can be recognised. This emphasizes the high relevance for the reproduction of the entire human musculoskeletal structure regarding the SSM method. Furthermore, the extended human modeling will also effect on the permissible collision speed v_p according to PFL, since whose characteristics strongly depend on the properties and thresholds of the respective body areas. With Fig. 4a and b, the *left arm nerv* ($k = 16$) appears to be the most critical area with a share of 57.55 % over the hole process. Accordingly, this area will primarily be observed for the next point. In Fig. 4c the composition of $S_{p,16}$ due to the individual terms of Eq. (4) can be seen. Beside the required minimum distance S_m, the speed-dependent courses of S_r, S_h and S_s are apparent. It can be stated that these terms may become negative, since the direction of motion of the interaction partners is considered. This is significant regarding the objective of a close cooperation and the minimization of the separation distance in a HRC environment. Nevertheless, S_p must not be less than S_m according to [3]. From the displayed courses it can be noticed, that S_h has got the highest influence on S_p with an average of 0.48 m (75.61 %) and a maximum value of 1.66 m at $t = 1.1$ s (95.35 %). In contrast, the terms S_r and S_s resulting from the robot's responsiveness and stopping behaviour have a comparatively small influence, with an average of $\bar{S}_r = 0.05$ m (10.35 %) and $\bar{S}_s = 0.098$ m (14.04 %).

Following this, the part of $S_{h,k}$ both in the conventional interpretation with $\bar{v}_{hc} = 1.6$ m as well as in the speed-dependent variant of the body areas are shown in Fig. 4d. According to Eq. (4) the reaction time T_r and the stopping time $T_s(v_r)$ affect the value of $S_{h,k}$ besides the actual operator speed v_{hc}. Due to the proportionality of the stopping values to the actual robot speed, a non-linear course can be seen even in the conventional interpretation. Thereby, a maximum value is reached at $t = 1.71$ s. When comparing the courses of the different body areas, the obvious potential of a possible distance reduction can be recognized, as long as the local velocities $v_{hc,k}$ of the operator are taken into account. In this scenario an average reduction of 0.88 m which ranges from a minimum of -0.54 m (increase) to a maximum of 1.8 m could be achieved. It should be noted that an exact modeling and consideration of the human motion can also lead to an increase of the dynamic separation distance. This can be seen for example in the time interval 0.97 s $< t < 1.25$ s.

5 Conclusion and Future Work

The paper at hand deals with the investigation of a new concept for the modeling and consideration of a human body motion within HRC environments. A central objective is to assist planning engineers in the design stage of HRC system development. Moreover it is intended to improve hybrid production scenarios between human operators and robots under the consideration of the normative safety conditions. In an exemplary HRC scenario different key factors in the calculation of the separation distance (according to SSM) has been investigated. Especially the extended consideration of the human motion data was evaluated. The results illustrate appropriate potentials for the reduction of oversized safety zones compared to the conventional operating methods. The kinematic state determination of the human body areas, as the first stage of the HCO method, has been integrated into the ROS based planning and simulation tool (cf. Sect. 2.1). Due to the ROS-Industrial initiative, real control interfaces of various industrial robots are provided by different vendors. In this context, future work will include a further evaluation of the HCO method using different robot systems and interaction scenarios. This will be carried out in the learning and research factory of the chair of production systems.

References

1. ISO 10218-1:2011, Robots and Robotic Devices – Safety requirements for industrial robots – Part 1: Robots (2011)
2. ISO 10218-2:2011, Robots and Robotic Devices – Safety requirements for industrial robots – Part 2: Robot systems and integration (2011)
3. ISO/TS 15066:2016, Robots and Robotic Devices – Collaborative Robots (2016)
4. Marvel, J.A., Norcross, R.: Implementing speed and separation monitoring in collaborative robot workcells. Robot. Comput.-Integr. Manuf. **44**, 144–155 (2017). https://doi.org/10.1016/j.rcim.2016.08.001
5. Vicentini, F., Giussani M., Tosatti, L.M.: Trajectory-dependent Safe Distances in Human-Robot Interaction. In: IEEE Emerging Technology and Factory Automation (ETFA), pp. 1–4 (2014). https://doi.org/10.1109/ETFA.2014.7005316
6. Zanchettin, A.M., Ceriani, N.M., Rocco, P., Ding, H., Matthias, B.: Safety in human-robot collaborative manufacturing environments: metrics and control. IEEE Trans. Autom. Sci. Eng. 882–893 (2015). https://doi.org/10.1109/TASE.2015.2412256
7. Dröder, K., Bobka, P., Germann, T., Gabriel, F., Dietrich, F.: A machine learning-enhanced digital twin approach for human-robot-collaboration. Proc. CIRP **76**, 187–192 (2018). https://doi.org/10.1016/j.procir.2018.02.010
8. Glogowski, P., Lemmerz, K., Hypki, A., Kuhlenkötter, B.: Extended calculation of the dynamic separation distance for robot speed adaption in the human-robot interaction. In: IEEE International Conference on Advanced Robotics (ICAR), pp. 205–212 (2019). https://doi.org/10.1109/ICAR46387.2019.8981635
9. Byner, C., Matthias, B., Ding, H.: Dynamic Speed and Separation Monitoring for Collaborative Robot Applications - Concepts and Performance. Robot. Comput.-Integr. Manuf. **58**, 239–252 (2019). https://doi.org/10.1016/j.rcim.2018.11.002
10. Lasota, P.A., Rossano, G.F., Shah, J.A.: Toward safe close-proximity human-robot interaction with standard industrial robots. In: International Conference on Automation Science and Engineering (CASE), pp. 339–344 (2014). https://doi.org/10.1109/CoASE.2014.6899348
11. Vogel, C., Walter, C., Elkmann, N.: Safeguarding and supporting future human-robot cooperative manufacturing processes by a projection- and camera-based technology. Proc. Manuf. **11**, 39–46 (2017). https://doi.org/10.1016/j.promfg.2017.07.127
12. Yamada, Y., Hirasawa, Y., Huang, S., Umetani, Y., Suita, K.: Human-robot contact in the safeguarding space. IEEE/ASME Trans. Mechatrons. **4**, 230–236 (1997). https://doi.org/10.1109/3516.653047
13. Haddadin, S., Albu-Schöffer, A., Hirzinger, G.: Requirements for safe robots. measurements, analysis and new insights. Int. J. Robot. Res. **28**, 1507–1527 (2009). https://doi.org/10.1177/0278364909343970
14. Haddadin, S., Haddadin, S., Khoury, A., Rokahr, T., Parusel, S., Burgkart, R., Bicchi, A., Albu-Schöffer, A.: On making robots understand safety. Embedding injury knowledge into control. TInt. J. Robot. Res. **31**, 1578–1602 (2012). https://doi.org/10.1177/0278364912462256
15. Behrens, R., Elkmann, N.: Study on meaningful and verified thresholds for minimizing the consequences of human-robot collisions. In: IEEE International Conference on Robotics and Automation (ICRA), pp. 3378–3383 (2014). https://doi.org/10.1109/ICRA.2014.6907345
16. Melia, M., Geissler, B., König, J., Ottersbach, H.J., Umbreit, M., Letzel, S., Muttray, A.: Pressure pain thresholds: subject factors and the meaning of peak pressures. Eur. J. Pain **23**(1), 167–182 (2019). https://doi.org/10.1002/ejp.1298

17. Oberer-Treitz, S.: Crashworthiness analysis of robots for the safety assessment in human-robot-cooperation University of Stuttgart, Stuttgart (2017)
18. Vemula, B., Matthias, B., Ahmad, A.: A design metric for safety assessment of industrial robot design suitable for power- and force-limited collaborative operation. Int. J. Intell. Robot. Appl. **2**(2), 226–234 (2018). https://doi.org/10.1007/s41315-018-0055-9
19. Aivaliotis, P., Aivaliotis, S., Gkournelos, C., Kokkalis, K., Michalos, G., Makris, S.: Power and force limiting on industrial robots for human-robot collaboration. Robot. Comput.-Integr. Manuf. **59**, 346–360 (2019). https://doi.org/10.1016/j.rcim.2019.05.001
20. Schiemann, M., Hodapp, J., Zurn, M., Berger, U.: Roboskin: Increased robot working speed within human-robot-collaboration safety regulations. In: 5th International Conference on Control 2019, pp. 85–91 (2019). https://doi.org/10.1109/ICCAR.2019.8813448
21. Glogowski, P., Lemmerz, K., Schulte, L., Barthelmey, A., Hypki, A., Kuhlenkötter, B., Deuse, J.: Task-based simulation tool for human-robot collaboration within assembly systems. In: Tagungsband des 2. Kongresses Montage Handhabung Indusstrieroboter, pp. 155–163 (2017). https://doi.org/10.1007/978-3-662-56714-2_1
22. Lemmerz, K., Glogowski, P., Hypki, A., Kuhlenkötter, B.: Functional integration of a robotics software framework into a human simulation system. In: 50th International Symposium on Robotics (ISR), pp. 1–8 (2018). ISBN:978-3-8007-4699-6
23. Lemmerz, K., Glogowski, P., Miro, M., Kuhlenkötter, B.: Verrichtungsbasierte Planung, Simulation und sozialpartnerschaftliche Potentiale der Mensch-Roboter-Kollaboration. In: Mensch-Technik-Interaktion in der digitalen Arbeitswelt (WGAB 2020), pp. 21–38 (2020)
24. Bauer, S.: Process language based system for controlling digital human models as a software component for planning and visualization of human activities in the Digital Factory. Technische Universistöt Chemnitz, Chemnitz (2015)
25. Quigley, M., Conley, K., Gerkey, B.P., Faust, J., Foote, T., Leibs, J., Wheeler, R., Ng, A.Y.: ROS: an open-source robot operating system. ICRA Workshop on Open Source Software (2009)
26. Chitta, S.: MoveIt!: an introduction. In: Robot Operating System (ROS): The Complete Reference pp. 3–27 (2016). https://doi.org/10.1007/978-3-319-26054-9_1
27. Lemmerz, K., Glogowski, P., Kleineberg, P., Hypki, A., Kuhlenkötter, B.: A hybrid collaborative operation for human-robot interaction supported by machine learning. In: International Conference on Human System Interaction (HSI), pp. 69–75 (2019). https://doi.org/10.1109/HSI47298.2019.8942606

Towards Semi Automated Pre-assembly for Aircraft Interior Production

Florian Kalscheuer, Henrik Eschen and Thorsten Schüppstuhl

Abstract

The growing aviation market puts first tier suppliers of aircraft interior under great pressure. Cabin monuments, not only consist of various assemblies with a wide range of parts, they are also highly customized by the airliners. Historically grown, poorly optimized manual processes offer the required flexibility, but limit the production rate of the individual products. The aviation industry responds with an increased use of automation technology. Recent standardization and automation approaches for efficient manufacturing, lead to an increase in productivity of these low volume products. However, complementary approaches to increase the degree of automation during assembly of aircraft interior components are missing. To reach a higher degree of automation this paper presents a derivation of cabin specific assembly processes with a varying degree of automation. First the range of components and processes in pre-assembly is analyzed with respect to automation. Based on the analysis, components and processes are classified in standardized groups. Fully automated and flexible automation processes are introduced to develop a semi-automated system. Furthermore, the required flow of information is described. Discussion of the results shows that the presented solution allows a flexible pre-assembly of low-volume interior parts and sets a baseline for further digitalization approaches.

Keywords

Pre-assembly · Aircraft interior · Hybrid assembly · Digitalization

F. Kalscheuer (✉) · H. Eschen · T. Schüppstuhl
Institute for Aircraft Production Technology, Hamburg University of Technology, Denickestraße 17, 21073 Hamburg, Germany
e-mail: florian.kalscheuer@tuhh.de
URL: https://tuhh.de/ifpt

1   Introduction

1.1   Motivation and Problem Statement

The aviation market is constantly growing. Until recently, the two biggest OEMs Boeing and Airbus both predicted a demand of new airliners which would increase the existing global fleet to 50,000 aircraft by 2038 [1, 4]. Despite the COVID-19 crisis and the sudden increase in cancellations of aircraft orders in March 2020 global revenue passenger kilometers are predicted to recover at 10% below the pre COVID level in 2025 [2, 5, 18]. During the 25-year lifespan of an aircraft [1], the cabin interior is re-furbished in intervals of 5–10 years [3, 25]. The line fit products, combined with these so called "retrofit monuments" result in an aftermarket "that is typically two to three times bigger than the OEM market" [27].

Although the abovementioned numbers will decrease, first tier suppliers have to remain competitive in an increasingly consolidated market [22]. On the one hand, previous publications primarily address automation approaches for manufacturing processes e.g. robot supported installation of threaded inserts, pressure-controlled potting of honeycomb panels and concepts for the automated manufacturing of flat sandwich panels [8, 9, 14, 19]. These works can be classified as digital aircraft interior production and set a digitalization baseline for downstream processes. On the other hand, suppliers face the assembly of customized, highly individual lightweight panels with a high proportion of manual tasks. These provide to the required flexibility, but cannot be carried out economically in a high wage country. To increase the productivity they either outsource work to low-wage countries or strive for a higher degree of automation. However, the interior assembly does not provide an adequate database and a categorization and parametrization of tasks and data for digital, automated assembly in lot size 1 as well as cabin specific assembly concepts are missing.

Since the automated manufacturing of flat lightweight panels promises a high standardization potential, this paper pursues the approach of the automation of the pre-assembly of flat lightweight panels. The goal is to give a detailed overview of the existing parts and their respective processes, as well as an evaluation and parametrization towards automated assembly in order to set a baseline for digital assembly solutions. At last, a partial demonstrator for a hybrid-assembly station will be described which allows further research in the field of cabin interior assembly.

1.2   Automation Approaches for Interior Production

Besides the automated manufacturing approaches mentioned in Sect. 1.1 only a few other publications regarding the automation aircraft interior assembly exist. Halfmann et al. [12] defined a concept for the final assembly line in aircraft factories which allows the manual pre-assembly of e.g. hatracks, passenger service units and lining panels outside the fuselage to reduce lead-time and working hours. Other concepts focus on a modular, assembly-friendly design of interior components or the application of augmented reality tools for faster cabin conversions [7, 11, 13]. Fette et al. [10] introduced a concept of robot-supported assembly of an overhead storage compartment. The abovementioned publications address the assembly either of complete monuments at the OEMs production line during line fit or the assembly of monuments from single panels and do not contribute to an increased productivity during the assembly of the subassemblies themselves.

Currently, composite sandwich structures made of a honeycomb core and pre-impregnated glass-fiber facesheets show large tolerances and are not optimized for automated assembly processes. The optimized panel design in combination with an automated potting process which was introduced in Eschen et al. [8] allows to produce individual panels with functional properties like solid wooden panels used in the furniture industry. Machine and plant manufacturers such as HOMAG Group offer specific solutions for the furniture industry to produce pieces of furniture in small batch sizes or even batch size one. Together with digital solutions such as manufacturing executions systems (MES) and the connected drilling, milling, sawing and edge banding machines, they provide highly automated production systems [16]. Specific configurations for customized lightweight panels are not addressed. Furthermore, chained, fully automated assembly systems do not meet the custom framework of cabin interior production. Standardized and assembly-oriented design of parts and subassemblies as well as a digital database are missing.

Huang et al. [17] present a skill-based programming system for automated assembly of furniture sets with robots using a library of skills and artificial intelligence. Knepper et al. [20] implement a furniture assembly system which consists of two KUKA youBots to assemble an IKEA Lack table with a planning algorithm which requires only the geometric form of the components as an input. The described concepts are not transferable economically, since the programming effort still exceeds economically justifiable process improvement and the necessary process standardization is missing.

Current research shows, that these limitations can be overcome by a combination of automated systems and worker supporting systems, creating an economically functioning, semi-automated assembly system. Müller et al. [23] introduced a modular production equipment for visual assistance which can implement different projection devices using MQTT and Node-RED for the application in different industries. Müller et al. [24] presents an approach for cognitive assistance to handle deviations from repetitive process flows in assembly. Furthermore, commercial solutions such as "Der Schlaue Klaus" (Optimum) offer visual assistance for assembly and quality assurance using image processing. However, such systems have not been integrated in aircraft interior assembly and the specific requirements for these systems have not been analyzed yet.

1.3   Outline

In order to give an overview of involved components and processes and to identify solutions to reach a higher degree of automation, first the current range of components and the existing assembly processes are analyzed. These results can further be evaluated with respect to automatability and classified into different categories. This allows to develop cabin specific assembly processes with a varying degree of automation. Since existing, manual production systems for highly individual components do not provide a data basis for the integration of automated systems, the flow of information will be analyzed and discussed. This includes the identification of process specific parameters, as data input for the control of automated or semi-automated systems. Since semi-automated (hybrid) systems require a consideration of human tasks, a partial demonstrator of a hybrid assembly station is introduced and described before the results are discussed.

2   Pre-assembly of Aircraft Interior

2.1   Range of Components in Pre-assembly

The production process used for the production of aircraft interior monuments can be divided in the manufacturing and machining of raw panels, the pre-assembly of sub-assemblies and the final assembly of complete monuments. All processes that use a single sandwich panel as a base part are allocated in the pre-assembly e.g. mounting of threaded inserts and dowels by gluing or fastening attachment parts with screws. Joining different panels to create a complete monument as well as the application of e.g. electrical wires, air and water pipes are considered as a part of the final assembly. All processes are currently executed manually.

The pre-assembly of interior panels contains a wide range of aircraft interior specific components. Assemblies usually consist of threaded inserts, edge protection, lock fittings, guiding rails, retainers and intermediate retainers, bumper plates, hinges and lamp housings. The base component which is considered in the following analysis, is a flat honeycomb panel manufactured as described in 1.1. This basic structure is shown in Fig. 1 using the example of a service-unit panel. Three different types of inserts, attachment parts with individual geometries and parts are assembled onto surfaces in all joining directions.

Besides the variation of different attachment parts, the threaded inserts form the numerically largest group of components in pre-assembly. They are used to create joints for screws and can be subdivided in three different categories: cylindrical inserts (potted inserts), cylindrical inserts with a one-sided flange and high-load inserts.

Potted inserts are usually made of metal or fiber-reinforced plastic (FRP) and a stainless-steel thread. They are mounted in drilled holes on the top and bottom sur-face or front surfaces for the attachment of low-, medium- and high-loaded components. In some

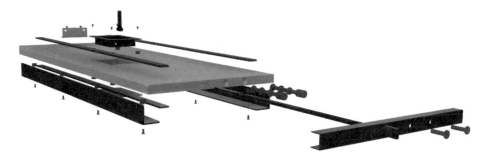

Fig. 1 Basic structure of a service-unit panel

cases, potted inserts can be substituted by so called friction-inserts. They have a thermoplastic coating, that is melting while drilling them into the honey-comb panel to create
a form-fit connection [15]. The cylindrical inserts with a one-sided flange and the
high-load inserts are made of aluminum and a stainless-steel thread. The flange is used to
transfer loads into the face sheets. The core diameter of the high loaded inserts is similar to
the cylindrical inserts. However, its length is about four times higher and it has three shaft
shoulders along its length. Furthermore, other special types of inserts exist. Between insert
and part assembly, a decorative film is applied.

2.2 Range of Pre-assembly Tasks

The pre-assembly tasks are divided in five different operations joining, handling,
inspection, adjustment and additional tasks according to Lotter and Wiendahl [21]. Predominantly, primary assembly tasks such as joining and handling can be found and are
listed in Fig. 2.

	Joining	• Adhesive application • Adhesive bonding	• Screwing • Composing • Friction welding
Assembly tasks	Handling	• Pick-and-place • Orientating	• Placement
	Inspection	• Torque Control	• Presence monitoring
	Adjustment	-	
	Additional tasks	• Highlighting	

Fig. 2 Assembly tasks during aircraft interior pre-assembly

The assembly sequence can be described as follows. First, all inserts are mounted into the panel. Either friction welding or two-component adhesive is used to create a form-fit connection. The remaining parts are joined mostly with screws. A few parts are joined by adhesive bonding. Usually, each part is attached separately to the panel. Only a small amount of parts requires simultaneous handling of more than one component.

Since airlines and OEMs use the cabin as a unique selling point over competitors, most of the time cabin interior is custom made and produced in small lot-sizes. Also, the parts themselves undergo changes in geometry and material. In addition, the number of parts of an assembly as well as joining positions vary.

3   Semi-Automated Assembly of Aircraft Interior

3.1   Automation-Oriented Classification of Parts and Their Corresponding Assembly Processes

For an increased degree of automation during the pre-assembly of aircraft interior, solutions have to be found for a high amount of different parts and individual assembly processes. Therefore, the parts and processes are analyzed with respect to automation-oriented product design based on the guidelines mentioned in Boothroyd et al. [6], Lotter and Wiendahl [21], Ponn and Lindemann [26], Stelzer et al. [28]. The design criteria allow to classify groups of parts (GP) with similar, automation-friendly assembly processes within the pre-assembly (Fig. 3). Because of high safety restrictions in product approval in the aviation industry and since cabin products must be able to be reproduced unchanged over a long period of time, assembly-oriented design changes cannot be considered. The honeycomb panel itself can be considered as an assembly friendly base part. The large support surface ensures stable positioning and no assembly path is geometrically blocked. The front faces are easily accessible.

The first group of parts (GP1) consists of various types of inserts. All can be joined unidirectional in a linear motion and their position is defined by drilled blind holes. GP2 consists of attachment parts that can be described as 2-dimensional or 2, 5-dimensional. They maintain the flat support surface of the base part after they are assembled so the assembly remains automation-friendly. They can be joined unidirectional by adhesive or screws. However, their position on the base part and their orientation is not defined by any geometric feature.

GP3 consists of parts that can be described as 3-dimensional. The parts are joined unidirectional and their position is usually defined by the corresponding inserts. Joining position as well as the screws can easily be accessed. Whereas all these parts can be joined vertically, unidirectional, collision free, are easy to handle and joined separately, some special parts are remaining (GP4). They cannot be considered automation-friendly since they require simultaneous handling of multiple parts, have multidirectional joining paths and have difficult to grasp contact surfaces.

Parts	Process	Parameter		
GP1	Insert handling + friction welding	• Initial position • Final position • Diameter	• Rotational speed • Process time	
	Insert handling + adhesive application	• Initial position • Final position • Orientation	• Insert diameter • Adhesive volume	
GP2	Part handling + (highlighting) + adhesive/force application	• Initial position • Final position • Orientation • Joining face contour	• Part weight • Part dimension • Adhesive volume	
GP3	Part handling + (highlighting) + screwing	• Initial position • Final position • Orientation • Joining face contour • Part weight	• Part dimension • Type of screw • Drill hole position • Torque • Process time	
GP4	Handling of multiple parts + joining task	• Initial position • Final position • Orientation • Assembly sequence • Joining face contour	• Weight/Dimension • Type of screw • Drill hole position • Torque • Process time	

Fig. 3 Groups of parts, respective processes and process parameters

3.2 Parametrized Assembly Processes

Although most of the parts and their assembly operations can be considered as easy to automate, the effort in production planning increases as soon as automation technology is implemented, particularly when integrated in manual assembly systems. To contribute to a digital description of the processes along the digital production process chain the identified groups of parts are analyzed with regard to their required input parameters (Fig. 3).

Due to comparably simple base parts and predominantly one-dimensional joining directions, processes can be found which vary only in a small number of specific input parameters. The insert placement process only varies in its final position whereas for highlighting and handling, the dimensions and joining faces are required to define gripping position and highlighted contours. The remaining parameters are defined once for each component and are not process-specific.

3.3 Research Demonstrator for Hybrid-Assembly

The classification of components and the description of their respective assembly operations yield potential for automation. Numerically predominant and easy to automate,

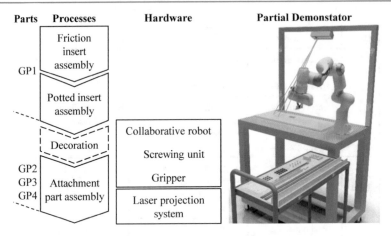

Fig. 4 Research demonstrator for hybrid interior assembly

insert assembly can be fully automated. Numerically low represented attachment parts with different geometries require manual work, which can be supported by assistance systems.

In an ongoing research project, a semi-automated assembly process has been developed. The process chain and a part of the assembly system is shown in Fig. 4. The demonstrator represents the attachment part assembly. Joining and handling is supported by a collaborative robot and allows parallelization of tasks. The orientation of parts or the assembly sequence can be supported by a laser projection system.

4   Discussion

The introduced hardware in combination with the abovementioned processes offer a solution to increase the degree of automation of the current processes. The classification and identification of different groups of parts, with respect to their automation capability contributes to process standardization. High-volume processes such as insert placement and gluing can be fully automated with complete flexibility regarding type, number and position of inserts. Commonly used hardware e.g. gantry kinematics, screwing and adhesive dosing units or grippers can be used for this purpose. However, because of the varying geometry of attachment parts, manual assembly cannot be neglected completely. For assembly tasks which require simultaneous handling of multiple parts in different joining directions, manual assembly is preferred. Nevertheless, the laser projection system or collaborative robot offer flexible worker assistance for material handling, joining or highlighting. This introduced combination of fully automated and semi-automated processes contributes to an increase in productivity during the pre-assembly of aircraft interior components and defines a baseline for future research in the field of cabin interior production.

The analysis of the required process information for the automated assembly and assistance systems is a first step to a digital process description. These information complements the previously developed digital manufacturing process chain. Part and assembly specific parameters can be derived from existing construction data or are defined once during production planning. Process specific parameters can be easily adapted and executed by the proposed hardware. This contributes to a paperless documentation, distribution of information on the shopfloor and allows quick process reconfiguration for new assemblies. However, production management systems and databases have to be implemented to ensure reliability.

The hybrid approach on pre-assembly shows potential for high process flexibility although manual labor is reduced, due to expandable hardware, such as additional assistance systems like industrial cameras, sensors or a moving head spot.

5 Summary and Future Work

In this work, a semi-automated pre-assembly system has been introduced to increase the degree of automation and productivity to enable suppliers to remain competitive with low wage countries. The range of components and assembly processes have been analyzed and classified with respect to their automation capability. Standardized groups of processes have been identified and hardware has been proposed. For further research, a partial demonstrator of a hybrid-assembly station was presented.

In future work, the fully automated solutions for insert placement have to be implemented and the automation of adhesive application has to be investigated. Furthermore, a holistic concept for continuous digital cabin production must be developed which allows to implement further automation solutions in a manual production environment. This requires research regarding efficient assembly planning methodologies for semi-automated assembly systems focusing on flow of information, digitalization and development of data-oriented design criteria to reduce commissioning effort of flexible automation technology. In future work, sensors can be integrated in the hybrid-assembly station to allow e.g. presence or assembly progress monitoring, action recognition or quality ensuring inspection tasks.

Acknowledgements The work presented in this paper is carried out in a project funded by the Federal Ministry of Economic Affairs and Energy of German (BMWi).

References

1. Airbus S.A.S.: Global Market Forecast 2019–2038: Cities, Airports & Aircraft (2020a). https://www.airbus.com/content/dam/corporate-topics/strategy/global-market-forecast/GMF-2019-2038-Airbus-Commercial-Aircraft-book.pdf. Accessed 25 Aug 2020
2. Airbus: Orders and Deliveries (2020b). https://www.airbus.com/aircraft/market/orders-deliveries.html. Accessed 25 Aug 2020
3. Black, S.: Advanced Materials for Aircraft Interiors (2020). https://www.compositesworld.com/articles/advanced-materials-for-aircraft-interiors. Accessed 31 Aug 2020
4. Boeing: Boeing: Commercial Market Outlook (2020a). https://www.boeing.com/commercial/market/commercial-market-outlook/. Accessed 25 Aug 2020
5. Boeing: Boeing: The Boeing Company: General Information (2020b). https://www.boeing.com/company/general-info/index.page#/overview. Accessed 25 Aug 2020
6. Boothroyd, G., Dewhurst, P., Knight, W.A.: Product design for manufacture and assembly. CRC Press, Boca Raton, London (2011)
7. Deneke, C., Oltmann, J., Schüppstuhl, T., Krause, D.: Technology Innovations for Faster Aircraft Cabin Conversion. In: AST 2019 (2019)
8. EJOT Holding GmbH & Co. KG. EJOT-TSSD (2020). https://www.ejot.de/medias/sys_master/Industry_Flyer/Industry_Flyer/h76/h1e/9047841079326/EJOT-TSSD-Flyer-de-02.18.pdf. Accessed 7 Sep 2020
9. Eschen, H., Harnisch, M., Schüppstuhl, T.: Flexible and automated production of sandwich panels for aircraft interior. Proc. Manuf. **18**, 35–42 (2018). https://doi.org/10.1016/j.promfg.2018.11.005
10. Eschen, H., Kalscheuer, F., Schüppstuhl, T.: Optimized Process Chain for Flexible and Automated Aircraft Interior Production. Proc. Manuf. (2020)
11. Fette, M., Büttemeyer, H., Krause, D., Fick, G.: Development of multi-material overhead stowage systems for commercial aircrafts by using new design and production methods. In: SAE Technical Paper Series (2019)
12.. Halfmann, N., Krause, D.: Towards Innovative Assembly Concepts: Integral Product—and Assembly Structure (2010)
13. Halfmann, N., Krause, D., Umlauft, S.: Assembly Concepts for Aircraft Cabin Installation. In: ASME 2010 10th Biennial Conference, pp. 733–739 (2010)
14. Halfmann, N., Elstner, S., Krause, D.: Product and Process Evaluation in the Context of Modularization for Assembly (2011)
15. Harnisch, M., Schüppstuhl, T.: High Quality Automated Honeycomb Potting with Active Pressure Control (2019)
16. HOMAG Group: Die HOMAG Group auf dem Weg zu Industrie 4.0 (2015). https://www.homag.com/fileadmin/systems/brochures/vernetzte-produkton-industrie40-de.pdf. Accessed 10 Jan 2021
17. Huang, P.-C., Hsieh, Y.-H., Mok, A.K.: A skill-based programming system for robotic furniture assembly. In: 2018 IEEE 16th International Conference on Industrial Informatics (INDIN), Porto, July 2018, pp. 355–361
18. IATA: COVID-19: Outlook For Air Travel in the Next 5 Years (2020). https://www.iata.org/en/iata-repository/publications/economic-reports/covid-19-outlook-for-air-travel-in-the-next-5-years/. Accessed 25 Aug 2020
19. Kähler, F., Eschen, H., Schüppstuhl, T.: Automated Installation of Inserts in Honeycomb Sandwich Materials. Proc. Manuf. (2020)

20. Knepper, R.A., Layton, T., Romanishin, J., Rus, D.: IkeaBot: an autonomous multi-robot coordinated furniture assembly system. In: 2013 IEEE International Conference on Robotics and Automation, Karlsruhe, Germany, May 2013, pp. 855–862
21. Lotter, B., Wiendahl, H.-P.: Montage in der Industriellen Produktion. Springer, Berlin (2012)
22. Morrison, M.: AIX: Diehl Aviation Looks at Next Step as Competitors Consolidate (2020). https://www.flightglobal.com/systems-and-interiors/aix-diehl-aviation-looks-at-next-step-as-competitors-consolidate/132027.article. Accessed 1 Sep 2020
23. Müller, R., Hörauf, L., Vette-Steinkamp, M., Kanso, A., Koch, J.: The assist-by-X system: calibration and application of a modular production equipment for visual assistance. Proc. CIRP **86**, 179–184 (2019). https://doi.org/10.1016/j.procir.2020.01.021
24. Müller, R., Hörauf, L., Speicher, C., Bashir, A.: Situational cognitive assistance system in rework area. Proc. Manuf. **38**, 884–891 (2019). https://doi.org/10.1016/j.promfg.2020.01.170
25. Niţă, M.F., Scholz, D.: Business opportunities in aircraft cabin conversion and refurbishing. J. Aerosp. Oper . **1**(1–2), 129–153 (2011). https://doi.org/10.3233/AOP-2011-0008
26. Ponn, J., Lindemann, U.: Konzeptentwicklung und Gestaltung technischer Produkte. Springer, Berlin (2011)
27. Red, C.: Composites in Aircraft Interiors, 2012–2022 (2020). https://www.compositesworld.com/articles/composites-in-aircraft-interiors-2012-2022. Accessed 31 Aug 2020
28. Stelzer, R., Grote, K.-H., Brökel, K., Rieg, F., Feldhusen, J. (eds.): Entwerfen, entwickeln, erleben: Methoden und Werkzeuge in der Produktenentwicklung. TUDpress, Dresden (2012)

An Approach for Direct Offline Programming of High Precision Assembly Tasks on 3D Scans Using Tactile Control and Automatic Program Adaption

Maximilian Metzner, Dominik Reisinger, Jan-Niklas Ortmann, Lukas Grünhöfer, Andreas Handwerker, Andreas Blank and Jörg Franke

Abstract

This contribution defines a methodology for the direct offline programming of robotic high-precision assembly tasks without the need for real-world teach-in, even for less-accurate lightweight robots. Using 3D scanning technologies, the relevant geometrical relations of the offline programming environment are adjusted to the real application. To bridge remaining accuracy gaps, tactile insertion algorithms are provided. As repetitive inaccuracy compensation through tactile search is considered wasteful, a method to automatically adapt the robot program to continuously increase precision over time, taking into account multiple influence sets is derived. The presented methodology is validated on a real-world use case from electronics production.

Keywords

Lightweight robots · Peg in hole · Photogrammetry · Machine learning

M. Metzner (✉) · J.-N. Ortmann · L. Grünhöfer · A. Blank · J. Franke
FAU Erlangen-Nürnberg, 91056 Erlangen, Germany
e-mail: maximilian.metzner@faps.fau.de

D. Reisinger
Technical University of Applied Sciences Nuremberg, 90489 Nuremberg, Germany

A. Handwerker
Digital Industries Division, Siemens AG, 91058 Erlangen, Germany

T. Schüppstuhl et al. (eds.), *Annals of Scientific Society for Assembly,*
Handling and Industrial Robotics 2021,
https://doi.org/10.1007/978-3-030-74032-0_18

1   Motivation

Many manual assembly tasks use the sensory abilities of the human worker to achieve high precision. Automation of such tasks usually involves special high-accuracy manipulators and tedious calibration of the application. Such systems are usually expensive and cannot be included into the material flow of existing manual production lines, in the sense of a hybrid system. To allow a safe robot operation in a manual system without the need for complete encapsulation of the automated process, lightweight robots (LWR) are often chosen. The lightweight construction of these manipulators however, further decreases their inherent accuracy [19]. The manual teach-in of such tasks is also time-consuming, which is why most programs are prepared in simulation tools by offline programming (OLP) and are then adapted to the actual application via re-teaching of key interaction points. This approach however greatly decreases system flexibility, as new product or variant introduction thus requires a manual touch-up on the real system resulting in production downtimes. Furthermore, the teach-in of such processes is often itself not accurate enough by itself, resulting in iterative teach-point optimization.

To address this, we introduce a methodology for the direct OLP of high-precision assembly tasks that still allows for the use of low-accuracy LWRs. The system is also capable of automatic program adaption for increasing process accuracy over time.

First, we introduce the relevant state of the art concerning high-precision robotic assembly, often referenced as peg-in-hole, as well as 3D scanning for production system planning. Based on this, we derive a need for action and requirements for such a method. In the third chapter, the new approach is described, going into detail on main features. We then validate the approach on a real scenario from power electronic production. The results are discussed and the contribution is concluded on an outlook of further planned research activities.

2   State of the Art

As our approach combines strategies of robotic high-precision assembly and 3D scanning, relevant research approaches in both fields are presented. From this, we derive a research gap to allow direct OLP of high precision assembly tasks and detail our contribution.

2.1   Robotic Peg-In-Hole Assembly

Robotic peg-in-hole assembly is categorized in five strategies by Li and Qiao: Sensing-information based, compliant mechanism-based, environment constraint-based, sensing constraint-based and human-inspired [16]. Xu et al. use a differentiation between contact-model-based and contact-model-free strategies [25]. The latter involve learning-by-demonstration techniques and reinforcement-learning approaches, which are custom to every application and thus hardly reusable. Contact-model-based approaches involve environment-constraint-based strategies and sensing-based compliant control. As shown in Metzner et al. [19], sensing-information-based methods offer great potential for economic automation of high-precision assembly even without highly accurate manipulators. Another advantage of such systems is the ability to reuse compensation routines and thus to increase engineering efficiency. In this contribution, we focus on force/torque-sensing (FTS) based systems, as in contrary to vision-based systems, high precision can be achieved without the need for online setup of the vision system and without the need for markers.

FTS-based precision assembly has been demonstrated by Jasim et al. A spiral-shaped search pattern with continuous surface contact is introduced [11]. This concept is abstracted in multiple contributions, e.g. to ambidextrous robots [26]. More general strategies have been implemented through the LIAA project, leading to the PiTaSC library [20]. Lately, commercial software systems, like Drag&Bot or Artiminds RPS have been introduced, that also offer FTS- and vision-based compensation modules [6]. However, all of these approaches are based on development, combination and parametrization on the real application, leading to longer system setup times and an inherent invariability to variant changes, new product introduction or processual changes, as a touch-up of the real system is required, leading to lengthy down-times. Metzner et al. present an approach for offline program synthesis to allow an offline preparation and parametrization of the program [19]. However, the application still needs to be calibrated with the real robot, as the simulation scene for OLP frequently deviates from the real application more drastically than can be compensated with the search strategies. Furthermore, all of the above approaches are static, meaning that a data-driven internal optimization of the robot frames is not performed, leading to wasteful repeated search times under the presence of poorly taught points or changes in the system.

2.2 3D-Scanning of Industrial Settings for Production Planning

3D-Scanning is used mostly for planning of new production systems in brown-field scenarios [18]. Scans are often taken using LiDAR-Scanners, also used for surveying, archaeology and building process documentation. This procedure allows for the capturing of very large areas on a factory scale, are however only partially suited for fine adjustment of small individual components due to resolution restrictions. Such scans are most suited for global planning of layouts, media in- and outlets and clearances.

Precise local scanning is done using either simultaneous localization and mapping (SLAM) approaches or structure-from-motion (SfM) routines. In SLAM approaches, an initial frame from an RGBD-sensor is iteratively complemented with succeeding frames that are fitted into the generated map [3]. These maps are mostly created using handheld RGBD-sensors and offer a time-efficient way to capture 3D scenes. However, such sensors suffer from significant sensor noise, which, complemented with fitting error propagation from the SLAM routine leads to subpar scanning accuracy.

SfM routines base on feature matching on numerous RGB images with identical optical parameters, which are then used to calculate the relative capturing poses for each frame based on a sparse reconstruction on the feature level. In a next step, this information is used to perform a dense reconstruction of the scene using multi-view stereo approaches [22, 23]. Such approaches have shown to offer high accuracy, especially on texture-rich settings, whilst only requiring a simple RGB camera for capturing. However, industrial application and validation for fine adjustment of virtual planning approaches has not been shown.

2.3   Research Gap and Need for Action

This contribution aims to enable a direct offline programming of high-precision assembly by providing a methodological model for deviations influencing accuracy in assembly. A compensation of the different deviations is conducted using CAD calibration from 3D scans and FTS-based tactile insertion strategies. Furthermore, an approach for data-driven optimization of the robot interaction frames is presented. We extend the state of the art by providing a systematic classification of accuracy influences in high-precision assembly tasks and corresponding strategies for their compensation. We furthermore quantitatively evaluate the usage of SfM approaches for the detailed 3D capturing of production systems for precise OLP. We extend existing strategies for FTS-based deviation compensation by data-driven optimization and validate the functionality on a realistic test scenario. We also demonstrate the integrated applicability of the methodology by an offline implementation of a robust sub-millimeter assembly process for an electronics device without any manual touchups.

3   Methodology

The methodology focuses on compensating three general types of accuracy influences: global independent deviations, systemic dependent deviations and random deviations. The global independent deviations are static biases for the entire system and independent of other operational parameters or time. They can be distinguished in macro-deviations and micro-deviations. Macro-deviations stem mostly from differences in geometric dependencies between the virtual ideal planning scene and the real system. Micro global deviations are caused by absolute accuracy errors of the manipulator as well as manufacturing and assembly tolerances of system components.

Dependent deviations occur due to manufacturing and assembly tolerances of the parts, e.g. varying by supplier, the work piece carriers or the material supply channels. They are thus dependent on at least one other parameter. Random deviations include the repeatability precision of the manipulator and other random external influences.

To compensate for global independent deviations, a one-time adaption of the robot interaction points is sufficient. For macro-deviations, this is achieved by integrating the real system geometry into the virtual planning scene through 3D scanning, as detailed in Sect. 3.1. Independent micro-deviations, dependent deviations and random influences are compensated by using tactile assembly strategies, see Sect. 3.2. As the tactile compensation of deviations is as such a wasteful, time-consuming process, a data-driven approach to further differentiate the independent micro-deviations and dependent deviations, as well as their dependent parameter, is shown in Sect. 4.2. The goal is to eliminate all systemic deviations, leaving only the random deviations, which however occur on a much smaller scale than the other influences.

3.1 Precision 3D-Scanning and Surface Reconstruction

To allow a low-cost 3D scanning of the real system, we use a structure-from-motion (SfM)-based approach for photogrammetric reconstruction from high-resolution RGB images. The real system is photographed with static camera parameters from different viewpoints ensuring a high overlap between pictures. Through inter-picture feature matching, the geometric dependency of the picture viewpoints, and thus the 3D structure of the system is derived. The resulting point cloud is then reconstructed into a surface model to allow fine adjusting with the virtual system model in the simulation/OLP tool, in this case Tecnomatix Process Simulate. Most relevant for robot OLP is the robot base, the work piece and the material supply 6D pose, as those are used to define the interaction points in the robot world coordinate system. Fine-tuning can be done manually or aided by e.g. ICP routines [2]. To implement ICP-based approaches for known parts, the CAD model is subsampled into a point cloud. Unknown parts, for which no CAD model is available, are categorized based on their geometry. Elements compounded from a few primitive geometries, such as the floor, tables and pillars, are estimated using a RANSAC approach. Here, the individual elements are fitted with the most likely primitive model based on fitting error minimization [7, 21]. More complex elements of the 3D scan, such as freeform surfaces and cables, are meshed using implicit function approaches, allowing for an efficient processing of large point sets [1, 4, 5, 12–15]. All other reconstructed geometry from the scan, which is not used for interaction point determination is used for collision-free path planning.

3.2 Tactile Assembly Strategies and Data-Driven Program Adaption

The general approach for tactile assembly used in this contribution is detailed in Metzner et al. [19]. The assembly process is divided into three phases: approach (A), search (S) and insertion (I). Dependent on part and assembly-specific parameters, the different variants for each phase are selected and parameterized. For compensation of micro-deviations, the search phase is most important, as the total deviation is determined here.

The results of each individual search phase in every dimension is recorded together with relevant process parameters, such as carrier ID, base part manufacturer ID or material supply ID. The global deviation for each point is determined after a small sample of test runs as the mean of the found positions in each dimension.

The dependent variable compensation requires more data, depending on the magnitude of the impact of each variable. Through a statistical hypothesis test, significant deviations between the actual goal value and the distribution of recent points of one class of one parameter are identified. As variance homogeneity cannot be assumed, the dependent variable may also influence the distribution, and sample size will also differ, a two-sided test statistic for independent samples according to Welch is used [24]. In case a significant deviation is found, the difference of the old and new arithmetic mean is saved as an additional compensation factor linked to the investigated parameter value. A new target position for each parameter set is calculated on the robot controller by adding up all such parameter-linked deviations.

4   Validation

First, each functional part of the methodology is validated individually on a common setup. A fully integrated execution of the entire methodology is also demonstrated. We validate our system on a scenario from electronics production, the assembly of through-hole-devices (THD) into printed circuit boards (PCB). This process is very challenging to automate, as assembly tolerances are usually only tenths of millimeters, to allow the robust soldering, and the parts are inherently fragile. We use a UR10 LWR equipped with an ATI FT AXIA 80 force-torque sensor (FTS) and a Weiss CRG 200–085 gripper for our setup. The system also consists of a workpiece carrier system for the PCBs and a material supply in the form of a tray.

4.1   Evaluation of Scanning Accuracy

The photogrammetric 3D scanning is validated by direct comparison of a point cloud and surface model generated by our SfM pipeline using a Sony Alpha 6000 camera (6000 $\times$ 4000 pixel), see Fig. 1, with those of a FARO Quantum V2/Blue LLP scanner. The Faro Scanner is based on a calibrated coordinate measuring arm with a laser line sensor and has a nominal measuring accuracy below 25 μm. It is thus chosen as a baseline

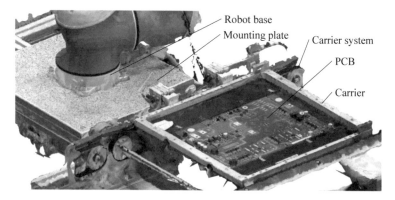

Fig. 1 3D reconstruction of the robot base and PCB carrier system from images using SfM

ground truth for comparison with our scan. Directly using the Faro scanner itself is not feasible in most applications due to long setup and calibration times, reachability issues and very high initial hardware cost in comparison to a simple handheld camera as used by our approach.

The distance measurements, trimmed at 3 mm maximum distance and limited to the robot base and work piece carrier with PCB to account for differences in scanned portions of the system, is shown in Fig. 2. The distance measurement is done in CloudCompare [9], using quadratic function distance calculation (k = 6) to account for cloud resolution differences, see also Girardea-Montaut [8]. Most points fall well under an absolute distance of 1 mm (green) (median distance 0.63 mm). Especially feature-rich areas, such as the edges of PCB, carrier and robot base are very accurately captured. These distinct features are also the most suitable for fine-adjustment of CAD geometries. The same observation is made comparing the surface models.

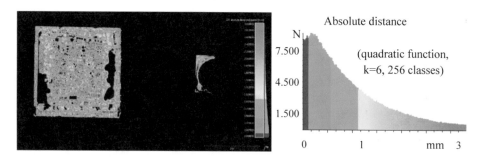

Fig. 2 Cloud to cloud distances of the SfM scan to the FARO scan (reference)

4.2   Performance of Compensation Strategies

As the functionality of our search and insertion strategies has already been shown in Metzner et al. [19], we focus on the validation of the scan-to-execution approach and our data-driven compensation strategies. The scan-based OLP methodology is validated by using the 3D scan to OLP a program implementing our methods. Both material supply and assembly position are derived from the scan and optimized via tactile strategies. The program is transferred to the real setup and executed without further adaption or calibration. As shown in Fig. 3b, the first execution run results in a visible offset (1) which is compensated by the tactile search (2). In the second run (3), this independent deviation is compensated to a degree that an insertion is possible without a search.

To validate our data-driven approach, we conduct a baseline analysis, repeatedly measuring the same assembly position of a mechanically fixed PCB by a four-point tactile approach in positive and negative x-y to quantify the robot's inherent measurement uncertainty. Then, we enhance a robot program for insertion of a THD into a circular hole by specified offsets in the perpendicular plane to the insertion vector. The reference goal location is determined as precisely as possible using the average hole center from the baseline test. The reached compensation values for each input parameter set, determined through a spiral-search approach, are saved. We then compare the resulting calculated offset values from our data-driven approach with the specified ones to evaluate the functionality of our compensation approach.

We recreate both independent as well as dependent deviations. For the dependent deviation, we also log the dependent parameter value set for our analysis. Asymmetries are deliberately added to the induced dependent deviations to simulate a biased parameter population in the initial compensation phase, which is likely in reality.

We implement three parameters, labeled manufacturer (M), carrier ID (C) and material feed (F). M and F each have two parameter values (A/B, 1/2), C has three (1/2/3). The magnitude of influence from each parameter, see Table 1, is derived from actual tolerance

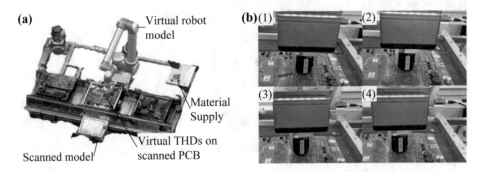

Fig. 3 OLP of a precision assembly process; **a** OLP scene with integrated 3D scan; **b** Program execution; (1) First approach with noticeable deviation; (2) Final position after spiral search; (3) Optimized second approach; (4) Direct assembly without search

Table 1 Magnitude of parameter influence for test series

Parameter	Value	Δx (mm)	Δy (mm)	s_x (mm)	s_y (mm)
M	A	0.1	−0.1	0.025	0.025
	B	−0.1	0.1	0.025	0.025
C	1	0.5	0.5	0.125	0.125
	2	−0.5	0.2	0.125	0.05
	3	0.3	−0.2	0.075	0.05
F	1	0.7	−0.2	0.175	0.05
	2	−0.2	0.7	0.05	0.175

values of a similar application in electronics production. For each parameter in each test run, a random number from a normal distribution using the absolute deviation and its standard deviation s is drawn using the polar method [17]. For each M and F, we conduct 450 test runs, for each C 300, resulting in 900 test cycles plus 75 baseline test for both standard and optimized insertion.

In 1950 insertion tests, no failure is recorded. The distribution of specified and found deviations is shown in Fig. 4. We optimize the search points using the mean deviation calculated for each M, C and F parameter value as well as the asymmetry in global distribution (Δx_{mean} = 0.14 mm; Δy_{mean} = 0.49 mm). The mean search time without optimization is 1.23 s, with 71% of insertions requiring a search. With our approach, on the same data points, no search is ever required.

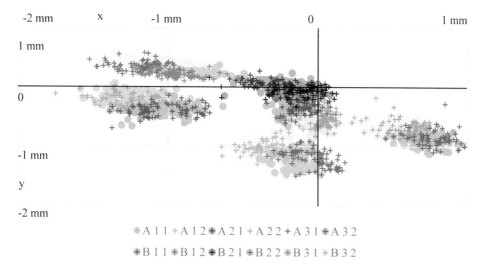

Fig. 4 Insertion test series results colored by parameter combination. Negative specified position (•) and found position (+)

5   Discussion

The findings on the point cloud and surface model accuracy show that our 3D scanning method is capable of reducing deviations of the virtual OLP scene to the real application into the submillimeter area. Even though further influences for deviations, such as the robot absolute accuracy, exist, a successful execution of a high-precision THD assembly process without touchups is demonstrated. We furthermore validate our data-driven approach for compensation of dependent deviations and find a total elimination of search times. We thus conclude that even deviation influences as low as 0.1 mm can be determined and permanently compensated after a reasonable data gathering period, even when an LWR is used.

6   Conclusion and Outlook

In this contribution, we propose a methodology for direct OLP of high-precision assembly tasks through a combination of 3D scanning and tactile assembly routines including data-driven program adaption approaches based on defined deviation influences for continuous, automatic process improvement. We quantitatively evaluate the individual methods as well as the overall methodology. It is found that our approach is capable of robust OLP of processes with high accuracy requirements. Further research topics include the applicability on further use-cases and the validation on real-life industrial production data.

References

1. Agarwala, A.: Efficient gradient-domain compositing using quadtrees. ACM Trans. Graph. **26**, 94 (2007)
2. ArtiMinds Robotics GmbH Artiminds Robot Programming Suite (2020). https://www.artiminds.com/de/. Accessed 24 Jul 2020
3. Arun, K.S., Huang, T.S., Blostein, S.D.: Least-squares fitting of two 3-d point sets. IEEE Trans.. Pattern Anal. Mach. Intell. **9**, 698–700 (1987)
4. Bailey, T., Durrant-Whyte, H.: Simultaneous localization and mapping (SLAM): part II. IEEE Robot. Automat. Mag. **13**, 108–117 (2006)
5. Calakli, F., Taubin, G.: SSD: smooth signed distance surface reconstruction. Comput. Graph. Forum **30**, 1993–2002 (2011)
6. Crane, K., Weischedel, C., Wardetzky, M.: The heat method for distance computation. Commun. ACM **60**, 90–99 (2017)
7. drag and bot GmbH Drag & Bot: Industrieroboter wie ein Smartphone bedienen (2020). https://www.dragandbot.com/de/. Accessed 24 Jul 2020
8. Fischler, M.A., Bolles, R.C.: Random sample consensus: a paradigm for model fitting with applications to image analysis and automated cartography. Commun. ACM **24**, 381–395 (1981)
9. Girardea-Montaut, D.: Distance Computation: Cloud-Cloud Distances (2015). https://www.cloudcompare.org/doc/wiki/index.php?title=Distances_Computation
10. Girardea-Montaut, D: Cloud Compare (2019). https://www.cloudcompare.org/
11. Jasim, I.F., Plapper, P.W., Voos, H.: Position identification in force-guided robotic peg-in-hole assembly tasks. Proc. CIRP **23**, 217–222 (2014)

12. Kazhdan, M., Hoppe, H.: Screened Poisson surface reconstruction. ACM Trans. Graph. (tog) **32**, 1–13 (2013)
13. Kazhdan, M., Hoppe, H.: An adaptive multi-grid solver for applications in computer graphics. Comput. Graph. Forum **38**, 138–150 (2019)
14. Kazhdan, M., Bolitho, M., Hoppe, H.: Poisson surface reconstruction. In: Proceedings of the fourth Eurographics symposium on Geometry processing, vol. 7 (2006)
15. Kazhdan, M., Chuang, M., Rusinkiewicz, S., et al.: Poisson surface reconstruction with envelope constraints. Comput. Graph Forum **39**, 173–182 (2020)
16. Li, R., Qiao, H.: A survey of methods and strategies for high-precision robotic grasping and assembly tasks—some new trends. IEEE/ASME Trans. Mechatron. **24**, 2718–2732 (2019)
17. Marsaglia, G., Bray, T.A.: A convenient method for generating normal variables. SIAM Rev. **6** (3), 260–264 (1964)
18. Melcher, D., Küster, B., Stonis, M. et al.: Optimierung von Fabrikplanungsprozessen durch Drohneneinsatz und automatisierte Layoutdigitalisierung. Wissenschaftliche Gesellschaft für Technische Logistik (2018)
19. Metzner, M., Leurer, S., Handwerker, A. et al. High-precision assembly of electronic devices with lightweight robots through sensor-guided insertion. Proc. CIRP (2020)
20. Nägele, F., Halt, L., Tenbrock, P. et al.: Composition and incremental refinement of skill models for robotic assembly tasks. In: The Third IEEE International Conference on Robotic Computing: IRC 2019: Proceedings: 25–27 February 2019, Naples, Italy. IEEE, Piscataway, NJ, pp. 177–182(2019)
21. Schnabel, R., Wahl, R., Klein, R.: Efficient RANSAC for point-cloud shape detection. Comput. Graph. Forum **26**, 214–226 (2007)
22. Schönberger, J.L., Frahm, J.-M.: Structure-from-motion revisited. In: Conference on Computer Vision and Pattern Recognition (CVPR) (2016)
23. Schönberger, J.L., Zheng, E., Pollefeys, M. et al.: Pixelwise view selection for unstructured multi-view stereo. In: European Conference on Computer Vision (ECCV) (2016)
24. Welch, B.: The generalisation of student's problems when several different population variances are involved. Biometrika **34**, 28–35 (1947)
25. Xu, J., Hou, Z., Liu, Z. et al. (2019) Compare Contact Model-based Control and Contact Model-free Learning: A Survey of Robotic Peg-in-hole Assembly Strategies
26. Zhang, K., Shi, M., Xu, J., et al.: Force control for a rigid dual peg-in-hole assembly. Assembly Automation **37**, 200–207 (2017)

Industry 4.0

Implementation of Innovative Manufacturing Technologies in Foundries for Large-Volume Components

Christian Klötzer, Martin-Christoph Wanner, Wilko Flügge and Lars Greitsch

Abstract

The development of new manufacturing technologies opens up new perspectives for the production of propellers (diameter < 5 m), especially since the use of the established sand casting process as a technology is only partially competitive in today's market. Therefore, different applications of generative manufacturing methods for the implementation into the production process were investigated. One approach is the mould production using additive manufacturing processes. Investigations showed that especially for large components with high wall thicknesses available systems and processes for sand casting mould production are cost-intensive and conditionally suitable. With our development of a large-format FDM printer, however, the direct production of large-format positive moulds for, for example, yacht propellers up to 4 m

C. Klötzer (✉) · M.-C. Wanner · W. Flügge
Fraunhofer Research Institution of Large Structures in Production Engineering IGP,
Albert-Einstein-Straße 30, 18059 Rostock, Germany
e-mail: christian.kloetzer@igp.fraunhofer.de

M.-C. Wanner
e-mail: martin-christoph.wanner@igp.fraunhofer.de

W. Flügge
e-mail: wilko.fluegge@igp.fraunhofer.de

W. Flügge
Chair of Production Technology, University of Rostock, Albert-Einstein-Straße 2, 18059 Rostock, Germany

L. Greitsch
Mecklenburger Metallguss GmbH—MMG, Teterower Straße 1, 17192 Waren (Müritz), Germany
e-mail: greitsch@mmg-propeller.de

© The Author(s) 2022
T. Schüppstuhl et al. (eds.), *Annals of Scientific Society for Assembly, Handling and Industrial Robotics 2021,*
https://doi.org/10.1007/978-3-030-74032-0_19

in diameter is possible. Due to the comparatively low accuracy requirements for the mould, the focus is on the durability of the drive system and the rigidity of this FDM printer. Equipped with simple linear technology in portal design and cubic design of the frame structure with rigid heated print bed, the aim is to achieve maximum material extrusion via the print head. The production of plastic models not only facilitates handling during the moulding process, but also allows considerable time and cost savings to be made during the running process. A further step in our development is the direct production of the components using WAAM. A possible concept for robot-supported build-up welding for the production of new innovative propeller geometries is presented using the example of a hollow turbine blade for a tidal power plant.

Keywords

Additive manufacturing · Propeller · Fused deposition modeling · Wire arc additive manufacturing

1 Introduction

Additive manufacturing technologies (AM) have not only developed rapidly due to their ability to produce near-net-shape components with complex geometry, but also offer various advantages over conventional processes in the area of individual component production. In addition to the geometric and design freedom, production times, material consumption and, as a result, costs can be reduced enormously for small batch sizes. Thus, the development of new manufacturing technologies opens up new perspectives for the production of components of maritime systems [3].

Due to the complexity of the components, the sand casting process has been established especially for propellers and components of the propulsion train. Figure 1 shows the conventional production process of maritime components using the example of a fixed pitch propeller with focus on the casting process.

Based on the technical casting adaptation of the CAD model through additional machining allowance (consideration of shrinkage during the casting process), an external production of a wooden model takes place. The positive molds, each consisting of a wing and hub segment, have very long production and delivery times. Provided with an additional coating is the manual molding of the wooden model with cement molding compound separately for top and bottom. The repeated rotation of the model around the hub axis creates the shape of the ship's propeller. After each rotation, alignment of the model on the propeller rotation axis is required. The exact positioning and alignment of the wing models during molding requires a great deal of effort. Handling the solid wood models weighing more than several 100 kg is only possible using crane systems.

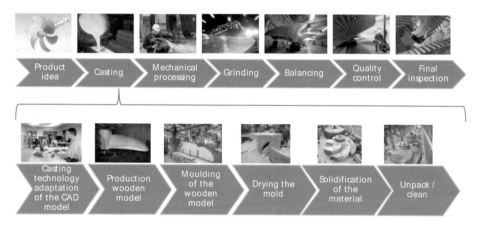

Fig. 1 Conventional production process using the example of a fixed pitch propeller

Afterwards, the mold must dry out depending on the surfaces present in the mould cavity. Casting and cooling are followed by measurement, on which the generation of the functional surfaces of the conical transverse press-fit and the finishing of propeller blade geometry by grinding to final dimensions are based. With a total duration of approx. 3 months, the production process is characterized by long service lives (drying and cooling times) and frequent machine changes.

Particularly in the production of small diameter propellers (D < 5 m), the established sand casting process can only partially be used as a competitive technology by European industry in today's market. In addition, the production of small propellers can be more complex than that of large propellers. Between the blades, there is a limited space for the mould construction and the later final contour machining. Therefore, the use of partially high-grade molding material and more complex molds is currently necessary. This leads to a further increase in manufacturing costs.

2   State of the Art and Related Work

Additive manufacturing is in use along the entire value chain, starting with the production of sample parts (rapid prototyping) and the manufacture of tools (rapid tooling) through to the production of fully functional components (rapid manufacturing). With the focus on a traditional application area, the shipbuilding industry, where most actions are based on experience and tradition, selected studies and applications are presented below. Taşdemir and Nohut [8] lists a variety of applications and Ludwig et al. [4] presents the need for AM processes and steps for implementation in the maritime industry.

Investigations of additively manufactured molding sand models of a propeller cap show the disadvantages for a cost-effective and timesaving alternative to traditional model making, especially for large-format castings such as propeller molds with wall thicknesses of up to 500 mm. To comply with strict environmental and worker protection regulations, inorganic binder systems such as water-based alkali silicate binders are increasingly in use. Inorganic binder systems require additional post-curing at high temperatures. The preferred microwave drying is possible for sand molds with a maximum wall thickness of approx. 40 mm. This leads in particular to a low storability of the moulds, a strong tendency to cracking and brittleness of the moulding material. Additional process fluctuations in the additive production of large, thick-walled molds and its post-treatment increased the scrap rate.

Another possibility is positive mold production from thermoplastics by fused deposition modeling. For example, patterns made of polylactide (PLA) and acrylonitrile butadiene styrene (ABS) for mold making serve for the production of miniature propellers or rudder segments [6]. The production of samples of newly developed customer-specific components within 24 h significantly shortens the overall product development time and ensures the fastest possible product changeover. This process is currently only used to produce sample components for mold making or to validate individual product properties of small series and is limited to small component dimensions due to the available systems.

For final products, the suitability of buildup welding for additive manufacturing of nickel-aluminum-bronze components is significant in the maritime sector. In Ding et al. [2], the authors investigated buildup welding of thin-walled structures with a focus on material behavior. In another joint research project [7] the optimization of the Wire-Arc-Additive-Manufacturing (WAAM) process for the production of large-format components for maritime applications was investigated. In Damen Shipyards Gorinchem [1], an additively manufactured ship's propeller was produced in 298 layers and is a welded solid propeller with subsequent complex finishing by manual grinding. Another research project [5] shows a demonstrator for hollow stainless steel propeller blades weighing 300 kg. An evaluation of the demonstrator in terms of fatigue, corrosion, residual stresses, material properties, and geometry shape accuracy is not available in Damen Shipyards Gorinchem [1] and NAVAL Group [5].

3   3D-Printed Mouldings for Sand Casting

A further approach for the integration of generative manufacturing processes into the existing production process is direct positive mould manufacturing by means of Fused Deposition Modeling (FDM). The first step is to develop an FDM printer for the additive manufacturing of large-format components. Subsequently, the existing production process must be adapted to implement the new manufacturing option. Finally, validation of the process and the printer is required using selected components.

3.1 Development of a Large Format FDM-Printing System

Based on the positively evaluated results from preliminary investigations of scaled propeller blades, Fraunhofer IGP developed a large-volume 3D printer with a working space of eight m^3 in cooperation with Mecklenburger Metallguss GmbH, shown in Fig. 2. The FDM printer additively manufactures ship propeller models with a diameter of up to four meters in addition to propeller hoods and special designs.

The casting production process achieves accuracies of between 100 and 1000 µm in the manufacture of ship propellers. In comparison, the achievable accuracies for milling are 10–100 µm and for sintering even 2–10 µm. The comparatively low accuracy requirements justify the elimination of cost-intensive precision linear technology in the development of the large-format FDM printer. The focus is on the durability of the drive system as well as the rigidity and dimensional accuracy of the entire system. Accuracy requirements limit standard linear guides in terms of maximum length. Due to the long travel of the print head within one axis, the deflection of the linear guides is a major challenge. By means of FEM simulations and tolerance analyses of the individual components and its adjacent construction, the permissible deformations and natural frequencies of the overall system were optimized. The frame structure of the printer is cube-shaped and is equipped with highly rigid aluminum profiles to withstand the stress of sudden changes in extruder direction. Additional elements in the corner points of the frame structure support the rigidity of the overall system. In the center of the system, there is a non-moving printing bed. Individual setscrews align the four precision-milled aluminum plates on the frame substructure to within tenths of a millimeter. Silicone heating mats installed on the underside heat the printing bed to a maximum of 110 °C. An external control for the heating elements monitors the even heating of the printing bed. The X and Y-axes in the Cartesian gantry system move the filament extruder above the print bed. Due to the very rigid basic structure and the desired parallelism of the axes, it is possible to operate the X and Y-axes with only one motor each. Closed belts transmit forces and motion from the stepper motors to the roller carriages on rails. Despite the

Fig. 2 Development of a large format 3D printer

travel distances of 2000 mm in X and Y direction, the deviations of the nominal to the actual contour of the component to be printed are only a few tenths of a millimeter. The movement of the X/Y portal and the extruder in the build-up direction is implemented by ball screws. Stepper motors control four ball screws on two sides of the frame structure. Additional guide rails in the frame structure support the dimensionally accurate movement of the extruder. Due to the minimal movement in the direction of assembly, motors are additionally equipped with holding brakes. Filament spools wound on the top frame ensure easy unwinding and insertion into the extruder. The system is equipped with a typhoon® filament extruder for the processing of thermoplastic material. The heating power of 400 watts and maximum printing temperature of 500 °C implement maximum material extrusion of 200 mm^3/s (0.9 kg/h), one of the highest flow rates for 2.85 mm filament. Usable with different die sizes up to max. 2.5 mm, the extruder can melt many times more material than standard 3D printers can with standard 0.4–1.2 mm dies. This is an important basic requirement for the additive production of large-format components. In order to determine the optimum process parameters, a defined test specimen was additively manufactured from different geometric elements (cuboid, sphere, inclined walls). In the statistical experimental design, the printing speed, layer height, printing temperature and infill are the factors to be investigated. The final visual inspection and verification of quality characteristics (dimensions, defects, overhangs) of the test specimens resulted in the following parameters in Table 1.

In order to check the positioning accuracy of the extruder and thus the relative dimensional accuracy of the printed object, the comparison of nominal and actual position is one way to evaluate. A large selection (n = 50) of randomly determined positions in the workspace of the 3D printer are used to check pose and repeat accuracy. A laser tracker Leica AT960LR measures the current position of the extruder and compares it to the specified mapped position in three-dimensional space. The evaluations result in an accuracy of ±1.5 mm. Causes for the deviations can be step angle errors and deviations of the step errors of the motors due to friction of the running rails or due to the belt drives. With regard to the component and process tolerances, this accuracy is sufficient.

Table 1 Process parameters for large format printing

Parameter (units)	Value
Nozzle diameter (mm)	2.5
Layer height (mm)	0.8–1.0
First layer height (mm)	0.6–0.7
Printing speed (mm/s)	90
Infill (%)	7–10
Extruder temperature (°C)	210–230
Perimeters	3

3.2 Implementation in the Production Process

A typical application for the additive production of a positive casting model at MMG is the MMG-escap®. This is a propeller cap to protect the shaft lock nut from seawater behind the propeller. The fin design, specifically adapted to the propeller, untwists the hub angle and reduces torque loss. The optimized flow behavior reduces cavitation at the rudder and hull, reduces wear and tear and increases both overall efficiency and operating time.

Figure 3 shows the adapted production process for a hybrid mould production using the example of a selected propeller cap. Fixed basic bodies exist for certain cap sizes, but for each propeller there are fins with different geometries. In the combined model setup, the basic bodies and cores are made of wood and the fin inserts of plastic. Starting with the CAD data from the propeller design, the geometries are adapted to production requirements in the pre-process. On the one hand, casting technology adjustments are necessary, mostly an additional machining allowance to take into account shrinkage during the casting process. On the other hand, transport holes must be defined and lifters must be specified for handling the mould. The user defines essential parameters for the fin bodies in the FDM process while the external pattern maker manufactures the base body. The Simplify3D software simulates the printing process for each component and generates the machine-specific code to start the fully automatic printing process. Once all printed parts are finished, the post process starts. In the first step, the operator removes any support structures, deburrs the component edges and fills any defects with filler. Alternating sanding and coating with a 2-component coating creates a smooth over surface. Prior to assembly, the employee assembles lifters and drawing belts. A final optical measurement of the model or later castings verifies compliance with the tolerance specifications.

Fig. 3 Left: propeller and propeller cap; right: customized process chain

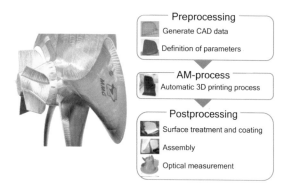

3.3   Validation and Results

The model production of fixed pitch propellers and propeller caps primarily uses the process. Process recordings, calculations and cost estimates document important parameters in the production process. Based on the collected production data, a comparison between conventional model making and mold making with additively manufactured components is possible.

The Fig. 4 shows the detailed costs and expenses for fin inserts of a propeller cap. Quotation prices and delivery times represent conventional model production by an external model maker. Printing time and post-processing processes are the main times in additive manufacturing. Additional non-productive times, e.g. for loading and unloading the jigs and fixtures, filament changes, transport times are included in the time recordings. The costs for additive model production include material and manufacturing costs. The manufacturing costs include energy costs, room costs, imputed depreciation and interest as well as production wages.

In addition to cost reduction (from 1500 € to 1120 €) and production time reduction (67%), the focus is on weight reduction of up to 60%. Plastic models of 20–25 kg are much easier to handle than comparable wooden models with a weight of approx. 60 kg. Above all, the assembly of the propeller canopy is more user-friendly. The data refer to the production of one fin insert. Since the propeller caps have several fin inserts, the cost, manufacturing time and weight savings for the entire model are many times greater. Thin-walled plastic models with optimized infill for fixed-pitch propellers reduce not only costs but also above all weight compared with solid wood models. The weight reduction makes it easier to handle the models during the moulding process.

The GOM Atos Triple Scan fringe light projection system then measures the components, see Fig. 5 the measurement of propeller cap PH00459 after casting. With a measuring field of 2000×2000 mm^2, the measuring accuracy is a few tenths of a

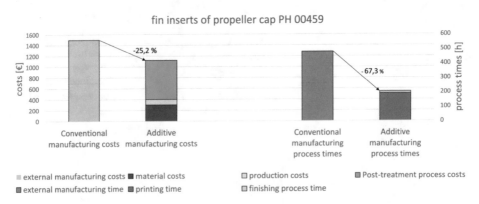

Fig. 4 Comparison of costs and process times between conventional and additive manufacturing for fin inserts of the propeller cap PH00459

millimeter. The best-fit process uses a CAD model to virtually align the resulting model in a defined configuration. Especially in the area of the fins, good dimensional accuracy and small deviations from the target geometry are given. Allowances of up to max. 5 mm allow an economical mechanical finishing of the freeform surfaces.

4 Robot-Supported Machining for the Additive Manufacturing of Maritime Components with Hollow Structure

In order to continue the path towards an effective manufacturing of small series of smaller propellers the direct component production using additive manufacturing processes has been investigated as an alternative to the casting process. Due to the relatively high deposition rate, small installation space restrictions as well as low investment and operating costs, arc-welding processes, especially WAAM, are predestined for the additive manufacturing of large structures. However, the main challenges are the large component dimensions in conjunction with the high accuracy requirements and the guarantee of freedom from defects as well as homogeneous mechanical-technological properties over the entire component. Figure 6 shows an experimental setup for conducting preliminary tests. An additional rotating and positioning device extends the working range of the 6-axis jointed-arm robot. The turning device can fix prefabricated cast work pieces so that buildup welding onto the intended base material is possible. The system also includes Fronius welding equipment, which is used in the standard arc process to build up the component layer by layer. The welding torch is located on the TCP of the robot, which connects wire feed hoses and connection hose packages with the TPSi 600 welding power source and the wire feed unit.

After the successful execution of initial joining tests and the production of test specimens for subsequent material tests, a model of a tidal power plant turbine blade was produced by build-up welding as a hollow blade. Current production possibilities for

Propeller cap PH00459

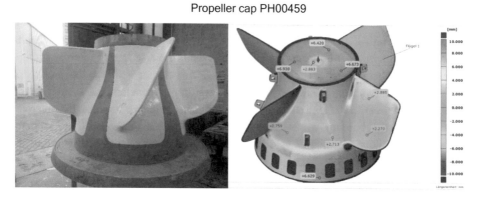

Fig. 5 Measurement results to illustrate the deviations of the casting surface from the planned one

Fig. 6 Left: test setup for the build-up welding, right: welded demonstrator blade with mechanical finishing

marine tidal turbines and hollow propellers are limited to casting processes or the construction of fiber composites. The use of heavy and stiff turbine blades in cast design requires high flow velocities and reduces the number of possible application locations. The use of fiber-reinforced plastics such as carbon fiber reinforced polymer (CFRP) has the disadvantage that they are not resistant to cavitation and are not yet fully recyclable. To reduce these disadvantages, the additive manufactured demonstrator is made of a copper–aluminum alloy, as good corrosion resistance to seawater, good weldability and high mechanical parameters characterizes it. The wing is part of a 3-blade turbine with a diameter of 4 m. With the aid of topology optimization and consideration of the manufacturing restrictions, the hollow demonstrator is 50% lighter than a solid casting (150 kg) with the same external shape.

The following investigations focus on hybrid machining strategies. A tool changing system on the robot provides the appropriate tools for the alternating additive and subtractive manufacturing steps. For hollow structures in particular, it is important to machine the inner surfaces in order to ensure that the components are also free of internal notches. Additional peripherals such as a line-cut sensor or arc camera monitor the production process. Together with the documentation of the welding parameters, the system detects weld seam irregularities. Continuous temperature measurements monitor the interpass temperature of the buildup welds. Due to the high thermal loads resulting from the layer-by-layer buildup, optimized welding strategies are essential for distortion-optimized path planning.

5 Conclusion and Future Work

The investigations and developments made so far show the possibilities of introducing additive manufacturing methods in the casting process. Especially the application of complex geometries of ship propellers and energy saving caps with their fin systems show a high potential for reducing process costs as well as processing time. Beside these quantitative advantages, the resulting weight reduction in case of 3D-printed moulding patterns gives further benefit in terms of assembly and handling these components.

The use of direct printing methods for sand moulding does so far not satisfy the requirements for large-scale moulds. Here further work must be established in enhancing the process towards increased wall thickness and moulding size. Furthermore, the cost side of this possible process needs reduction measures to ensure an applicability under commercial considerations.

However, the direct metal printing approach based on the WAAM process appears more promising especially to be applied for medium size propellers and similar components. The main advantage here is in the complete skipping of any model and mould making process steps. Based on this omission of related costs an overall cost benefit can be gained. A clear antagonist for this possible overall cost reduction is the cost factor for the material. The difference between raw material costs as used for the casting process and the wire costs for the WAAM process strongly triggers the economical maximum size of components. Here further works is needed to push the limits towards larger component sizes as e.g. the development of manufacturing hollow components. Beside the positive influence on the material usage, additional functionality can be realised with benefit for the product.

References

1. CJR Propulsion Ltd. https://www.cjrprop.com/products/cjr-propellers/. Accessed 19 Mar 2019
2. Damen Shipyards Gorinchem. https://www.da-men.com/en/news/2017/11/worlds_first_class_approved_3d_printed_ships_propeller_unveiled. Accessed 10 Apr 2019
3. Ding, D., Pan, Z., van Duin, S., Li, H., Shen, C.: Fabricating superior NiAl bronze components through wire arc additive manufacturing. Materials **9**(8) (2016). https://doi.org/10.3390/ma9080652.
4. Greitsch, L., Klötzer, C.: Neue Fertigungstechnologien in der maritimen Produktion, 8. Zukunftskonferenz: Wind & Maritim, Rostock (2019)
5. Ludwig, I., Loock, J., Kosubek, T., Steinmeier, O., Franke, C.: Bedarfsermittlung von additiven Fertigungsmethoden mit Fokus auf die maritime Wirtschaft in der erweiterten Metropolregion Hamburg, im Auftrag des Maritimen Clusters Norddeutschland e.V., Hamburg (2019)
6. NAVAL Group. https://www.naval-group.com/en/news/the-worlds-first-hollow-propeller-blade/. Accessed 17 Apr 2019
7. Queguineur, A., Rückert, G., Cortial, F., Hascoët, J.Y.: Evaluation of wire arc additive manufacturing for large-sized components in naval applications. Weld World **62**(2), 259–266 (2018). https://doi.org/10.1007/s40194-017-0536-8
8. Taşdemir, A., Nohut, S.: An overview of wire arc additive manufacturing (WAAM) in shipbuilding industry. Ships Offshore Struct. (2020). https://doi.org/10.1080/17445302.2020.1786232

The Digital Twin as a Mediator for the Digitalization and Conservation of Expert Knowledge

Dominik Hüsener, Michael Schluse, Dorit Kaufmann and Jürgen Roßmann

Abstract

A Digital Twin is a virtual representation of a physical asset. It reflects the current state of that machine through a model and the data as observed by sensors in the real machine; and enables effective and efficient interaction with the machine, i.e. for monitoring and control purposes. The Digital Twin facilitates the collection of data, as well as its analysis and visualization through its user interfaces, i.e. GUIs such as screens or Mixed Reality that provide intuitive access to the data and facilitates its manipulation. Embedded in Virtual Testbeds the Digital Twin becomes an "Experimentable Digital Twin" (EDT), in which experiments can be performed and the different outcomes can be compared or evaluated. The intuitive representation of the assets allows the experts to interact with the twin, without highly detailed knowledge in computer science. The digital twin observes, records, and benchmarks experiments performed by the operator. This way the operator's knowledge becomes digitized and thus preserved as an abstract representation of data, formulas, and models inside the digital twin. By introducing the Digital Twin into the processes carried out by different operators (not only the initially observed expert), formerly intuitive decision-making processes of the operators are enhanced based on empirical data. As a result, the Digital Twin serves as an assistance system that can guide future operators and the outcomes

D. Hüsener (✉) · M. Schluse · D. Kaufmann · J. Roßmann
Institute for Man-Machine Interaction, RWTH Aachen University, Ahornstr. 55, 52074 Aachen, Germany
e-mail: huesener@mmi.rwth-aachen.de

© The Author(s) 2022
T. Schüppstuhl et al. (eds.), *Annals of Scientific Society for Assembly, Handling and Industrial Robotics 2021*,
https://doi.org/10.1007/978-3-030-74032-0_20

of the experiments become reproducible. The specific representations of interactions and outcomes also facilitate collaboration between the machine operators and other stakeholders by providing different operators a common "perspective".

Keywords

Digital twin · Decision making · Knowledge digitalization

1 Introduction

With the advance of Industry 4.0, there is a shift from mass production to innovative, specialized products with small batch sizes. Therefore, machine parameters need to be changed more frequently. In a traditional production setting, this leads to higher personnel, time, and material requirements and consequently higher production costs [1]. In some cases, the new settings need to be tested before the actual production can begin further increasing time and cost requirements. Even if the test pieces are within tolerances, the final product can be faulty. On older machines, which are still in use in some industry sectors, machine parameters need to be adjusted manually for each product, thus limiting the possibility of batch size 1 and driving costs up for small batch sizes.

Especially in some disciplines belonging to the traditional mechanical engineering sector (as textile mechanical engineering), the success of the outcome is depending on the skills of the operator and his experience in setting special machine parameters. Thus, effective production is only possible with experienced, skilled personnel. This knowledge is not measurable and thus not transformable. This problem is enhanced by a lack of young employees, trainees, and skilled workers, and the changes in demographics [10].

The knowledge however is not conserved but is in the heads of the employees and depends on word of mouth to be passed along. This is because knowledge is empirical and not quantized. When an employee leaves the company, the knowledge is gone. Due to the limited use of digitalization in some industry sectors, the training takes long, and even skilled workers need time to learn to use an individual machine independently.

The rest of the paper is organized as follows. In Sect. 2 the Digital Twin concept is presented with the integration of simulation in a so-called Experimentable Digital Twin and mentioning its role as a mediator. In Sect. 3, a brief overview of knowledge management is given and combined with the Digital Twin approach as a means of knowledge management. In Sect. 4, this concept is transferred to a use case of the textile industry.

2 What is a Digital Twin?

The Digital Twin (DT) was first presented by Grieves [3] as a virtual model of a physical asset that receives data from the asset and virtually represents the state of the asset. In 2003 the "Digital Twin" concept was first presented. In 2014 Grieves published a white paper on the Digital Twin. A Digital Twin is a virtual representation of an asset that is fed with data from the real machine and also sends data to the real twin. The DT can be obtained by combining the data of the machine (the Digital Shadow) with a model of the machine (see Fig. 1).

The combination of the physical asset and the Digital Twin with its corresponding communication infrastructure is called a Cyber-Physical System (CPS). The physical asset can itself be a CPS. In Industry 4.0 CPS become increasingly important. Different machines have their own twin, while the twins can communicate with each other. The Digital Twin is used in health monitoring, also production planning (PLM), and the design of new products. In comparison to conventional planning software, what is new is that all data related to one product is stored together in one place (the Digital Twin) whereas before machines were able to store data but it was not combined into one unified presentation. This presentation however has brought a massive increase in value. Although it is only a simplified representation, it tries to model the behavior of the real twin accurately.

In Kritzinger et al. [5] literature on the Digital Twin in the manufacturing field is reviewed and classified as either Digital Model, Digital Shadow, or Digital Twin.

Fig. 1 The context of the Experimentable Digital Twin (EDT) and the Cyber-Physical System (CPS) [7, 8]

According to their definition, the difference with these terms is the communication between the digital and physical assets. Whereas the Digital Model is independent of the real asset, the Digital Shadow only receives data from the real asset and only the Digital Twin allows bi-directional communication between digital and physical assets. While the term "Digital Twin" is widely used in literature, it is often used for Digital Models and Shadows as well. Therefore, the need for a common definition of the Digital Twin is emphasized. The literature on actual Digital Twins according to their proposed definition however is scarce. Especially there is a lack of case studies on a higher level of integration.

A more recent literature review was performed in Tao et al. [11] with a focus on applications of the Digital Twin in industry. After the white paper of Grieves, the number of papers published is increasing steadily. It is also mentioned, that the definition of the Digital Twin is varying in different papers. The most relevant theoretical foundation of the DT is listed as DT modeling, simulation, verification, validation, and accreditation (VV&A); data fusion; interaction and collaboration; and service. The main applications currently where the DT is used are production and health management.

2.1 What is an Experimentable Digital Twin?

The experimentable Digital Twin brings together modern simulation technologies with the Digital Twin concept [9]. Whereas before several simulations (FEM, fluid dynamics, multi-body dynamics...) were performed individually, this can lead to wrong results if there is no exchange between the results. The Digital Twin enabled by co-simulation brings together different simulation techniques and stores the relevant data in the Digital Twin such that is exchanged between the various simulations (see Fig. 2).

With simulation, it is not only possible that a Digital Twin supervises a system and has a virtual representation but also a view into the future becomes feasible. The Digital Twin is not only able to collect the machine data but also to generate new data and through this improve the functionality of the physical asset. In manual production processes, it is difficult to use artificial intelligence to optimize the process since data of the machine and the produced object are not stored together. The Digital Twin however also opens the door to artificial intelligence techniques that can optimize the behavior.

2.2 Digital Twin as a Mediator

The Digital Twin displays its current status and can suggest actions to be taken through the human–machine interface. The machine sends its data which is processed by the Digital Twin to display valuable information to the user in an understandable way. The operator sees the data, can understand its meaning and can act accordingly (see Fig. 3).

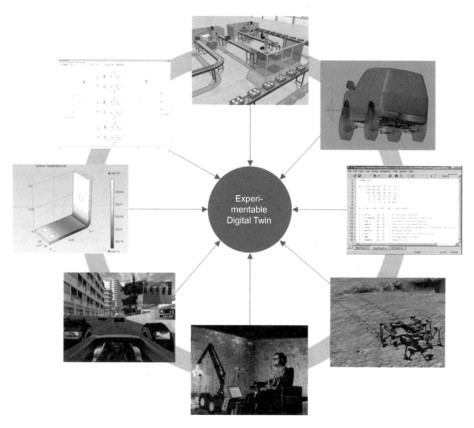

Fig. 2 "Experimentable Digital Twin approach". The EDT combines several simulation algorithms [9]

Fig. 3 Schematic concept of the digital twin as a mediator between the operator and the real twin

In his whitepaper Grieves [3] describes, how the virtual representation is beneficial to conceptualize, compare and collaborate in production processes since they provide a more intuitive perspective compared to 2D sketches or data in tables or regular graphs. The realistic visual view is valuable since sight is the most important human sense. Different users can see a representation that easily allows them to classify the current state of the machine which is also consistent for different users [4].

The user can simulate the behavior of the machine and is supported to find suitable settings for a given task. The simulation can be displayed in 3D and information that is otherwise not visible to the operator can be visualized through the Digital Twin. Hence, the Digital Twin can also be regarded as a mediator between the machine operator and the machine itself. In Cichon and Rossmann [2] a concept to facilitate the interaction of humans and machines is presented. Among others, joysticks, screens and AR/VR allow direct interaction with the digital twin. The user is able to interact with the virtual machine similar to the real machine, and the (real) machine informs the user of its current status through the visualization of the Digital Twin.

3   What is Knowledge Management?

Knowledge management is about conserving the (procedural) knowledge that is needed to perform a certain task. As mentioned earlier, the aging workforce requires that their knowledge is somehow stored in order not to lose the knowledge which is an essential asset for the company to remain productive. This is traditionally performed manually through observational studies and questionnaires [6]. Questionnaires however require the person conducting the study to have some basic knowledge of the process to be considered in order to ask the relevant questions. Otherwise, some necessary or useful information might be left out. Also, questionnaires require additional work to provide a benefit. They are only useful if the results are presented in a manner such that they are comprehensible to trainees. A textual description for example can be useful for experienced workers that need to learn how to operate a new machine but might be difficult to grasp for students that are not yet familiar with the functioning of the machine.

Observational studies are better in this way; however, it becomes costly very quickly if there is a high amount of different work tasks to do and infeasible to record tasks that only occur seldomly. Also, this requires a setup and is not simply integrated into the regular working process.

In Mohammad and Al Saiyd [6] a process for manually acquiring expert data in the context of artificial intelligence is presented. The required activities are the identification of the domain knowledge, finding experts that can provide that knowledge, finding a representation for modeling of that knowledge, and the construction of a knowledge base that can then be used in an inference engine (see Fig. 4).

Fig. 4 Activities of knowledge engineering and the integration of the EDT [6]

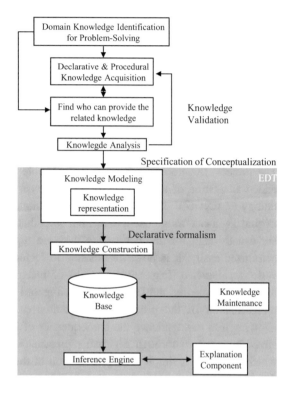

3.1 How to Conserve Knowledge with a Digital Twin

If the knowledge of the workforce is to be conserved, it can be digitized in some way. Digitization has many advantages and can increase the value of the captured information when it is combined with other information. Considering the activities in Mohammad and Al Saiyd [6], we propose to use the Digital Twin as a means of knowledge representation and to store the knowledge base. The Digital Twin also receives real-time data from the machine—as a result, there is one central data storage for all information related to that machine ("Single Source of Truth"). The Digital Twin can store data like a conventional database, but it also has the advantage of being able to visualize that data more intuitively. A further advantage of the Digital Twin is that at the end of the process a virtual model is generated that can be verified by the experts. Also, the virtual representation can help trainees as processes can be shown on a 3D model that—other than a video—can also be manipulated and is easier to grasp than a textual description. Through the use of augmented reality, it could even be possible to see a virtual representation of the machine along with the real machine that can guide the worker through the process of adjusting the machine parameters or maintenance with the advantage of having their hands free.

However, the data of the interaction is stored by the machine and transferred into the database of the Digital Twin, it becomes feasible to record processes on the go and understandably conserve them. Also, not very frequent tasks are conserved this way. The digital twin itself is subject to an ageing process, hence it needs to adapt to new production environments in order to sustain its relevance.

4 Use Case

The project 'Development of an experimentable digital twin for the analysis and auto-mated adaptation of textile manufacturing processes using the example of tufting tech-nology (T-EXDIZ)' aims to show in the example setting of a textile machine how the Digital Twin can bring benefit to manual tasks that are difficult to automize. The machines are usually kept for decades before they are replaced, and older machines cannot be automized easily. It is however possible to include sensors into the machine, such that various parameters (e.g. angle, length) of machine parts or settings can be read. To improve automation, the Digital Twin can record these settings and thus create a virtual representation of the process. The operator can experiment with the Digital Twin, allowing him to experience the consequences of his activities. These steps can be visu-alized on a screen through a virtual representation of the Digital Twin (i.e. rendering) which is usually a simplified representation of the real twin. Hence, through utilization, the settings are kept for further use. It is then possible to test these new settings using the simulation methodology without having to waste material. If this product has to be pro-duced again the exact values are stored and can be compared to the current values such that reproducibility is conserved. Besides, this also makes it possible to teach the process to new trainees. The Digital Twin can also simulate its behavior to provide training opportunities. Also, there is an opportunity to test new configurations without the costly procedure of testing each setting since the simulation can perform this testing in a fracture of the time. It can thus help trainees gain confidence and reduce the amount of training needed to operate the machine.

If a worker leaves the company, his procedural knowledge can be stored in an understandable form such that it is available to new trainees. Also, for experienced workers, it becomes simpler to check whether they have completed steps correctly. The human–machine interface (HMI)—a window in the virtual world where the digital twin lives—can help to visualize what steps must be taken. The spread by word-and-mouth also relies on a common understanding, the Digital Twin however has one unified rep-resentation for everyone. In training, it will be more easily possible to test multiple configurations.

This procedure is shown in the use case of a textile machine. In the textile industry automation is low and batch size one production is often not feasible. The computer has a model of the machine that can be used to simulate the machine. It receives data from the operation of the machine. The real data can be used to update simulation parameters and it

Fig. 5 The DT concept applied
to the T-EXDIZ use case

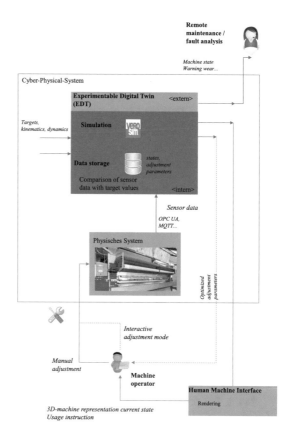

can also be used to detect irregularities and advise the worker to repair machine parts or to adjust settings. Instead of waiting for the finished product and noticing that it does not fulfill its specifications, the Digital Twin can simulate further behavior based on the current state. Thus, faults can be detected early and waste products, and thus cost is decreased.

The flow of data and information between the different components is shown in Fig. 5.

Currently, a DT model of the kinematics of a tufting machine is created. In the next step, this model can be refined with the input of experts. Later an automated mode will be integrated that can optimize machine parameters through simulation transforming the DT to an experimentable Digital Twin.

5 Conclusion

Due to an aging workforce, some personnel is leaving the production sites in the coming years. However, the knowledge of the workers is an important asset for the company to remain competitive, especially in sectors where manual labor is still widely present.

The Digital Twin is a virtual model of an asset that can be used to store and visualize machine data. As was shown in this paper, the Digital Twin can also be used to conserve the knowledge of the workforce and make it measurable in the first place. This helps in reducing training times for new employees and knowledge loss. The knowledge that comes from questionnaires, as well as observations, cannot only increase the value of the Digital Twin itself but the DT is also a useful tool to have a meaningful representation of that knowledge.

Acknowledgements The writing of this chapter was enabled within the context of the IGF project 'Development of an experimentable digital twin for the analysis and automated adaptation of textile manufacturing processes using the example of tufting technology (T-EXDIZ)' 21166 N/2 of the Forschungsvereinigung Forschungskuratorium Textil e.V.. It is funded via the AiF within the framework of the program for the promotion of joint industrial research and development (IGF) by the Federal Ministry of Economics and Energy based on a resolution of the German Bundestag.

References

1. Allahverdi, A., Soroush, H.M.: The significance of reducing setup times/setup costs. Eur. J. Oper. Res. **187**, 978–984 (2008). https://doi.org/10.1016/j.ejor.2006.09.010
2. Cichon, T., Rossmann, J.: Digital twins: assisting and supporting cooperation in human-robot teams. In: Proceedings of the 15 th International Conference on Control, Automation, Robotics and Vision (ICARCV 2018), November 18–21, 2018, Singapore, pp. 1–6 (2019)
3. Grieves, M.: Digital twin: manufacturing excellence through virtual factory replication. White Pap. **1**, 1–7 (2014)
4. Grieves, M., Vickers, J.: Digital twin: mitigating unpredictable, undesirable emergent behavior in complex systems. In: Transdisciplinary Perspectives on Complex Systems, pp. 85–113. Springer, Cham (2017). https://doi.org/10.1007/978-3-319-38756-7_4
5. Kritzinger, W., Karner, M., Traar, G., Henjes, J., Sihn, W.: Digital twin in manufacturing: a categorical literature review and classification. IFAC-PapersOnLine. **51**, 1016–1022 (2018). https://doi.org/10.1016/j.ifacol.2018.08.474
6. Mohammad, A., Al Saiyd, N.: A framework for expert knowledge acquisition. IJCSNS **10**, 145 (2010)
7. Reinhart, G.: Digital Twin—Synchronizing Reality and Virtuality (2015)
8. Roßmann, J., Schluse, M.: Experimentierbare Digitale Zwillinge im Lebenszyklus technischer Systeme. In: Handbuch Industrie 4.0: Recht, Technik, Gesellschaft, pp. 837–859. Springer, Heidelberg (2020). https://doi.org/10.1007/978-3-662-58474-3_43
9. Schluse, M., Rossmann, J.: From simulation to experimentable digital twins. IEEE Int. Symp. Syst. Eng. 1–6 (2016)
10. Schnitger, M., Windelband, L.: Shortage of skilled workers in the manufacturing sector in Germany: results from the sector analysis (2008)
11. Tao, F., Zhang, H., Liu, A., Nee, A.Y.C.: Digital twin in industry: state-of-the-art. IEEE Trans. Ind. Informat. **15**, 2405–2415 (2019). https://doi.org/10.1109/TII.2018.2873186

Web Service for Point Cloud Supported Robot Programming Using Machine Learning

Kristina Enes

Abstract

In industrial automation, the use of robots is already standard. But there is still a lot of room for further automation. One such place where improvements can be made is in the adjustment of a production system to new and unknown products. Currently, this task includes the reprogramming of the robot and a readjustment of the image processing algorithms if sensors are involved. This takes time, effort, and a specialist, something especially small and middle-sized companies shy away from. We propose to represent a physical production line with a digital twin, using the simulated production system to generate labeled data to be used for training in a deep learning component. An artificial neural network will be trained to both recognize and localize the observed products. This allows the production line to handle both known and unknown products more flexible. The deep learning component itself is located in a cloud and can be accessed through a web service, allowing any member of the staff to initiate the training, regardless of their programming skills. In summary, our approach addresses not only further automation in manufacturing but also the use of synthesized data for deep learning.

Keywords

Machine learning · Digital twin · Structured-light 3D scanner

K. Enes (✉)
RIF Institut Für Forschung Und Transfer E.V, Joseph-von-Fraunhofer Str. 20, 44227 Dortmund, Germany
e-mail: Kristina.Enes@rt.rif-ev.de

1   Introduction

This paper addresses three issues regarding machine learning in the industrial field.

Firstly, many problems can arise when applying machine learning in a production process. Artificial neural networks need a large amount of data to train on and while there is a lot available containing day to day situations, like indoor and outdoor scenes, this is not the case for industrial environments. Suitable training data need to be collected and labeled manually, which can be difficult and tedious. Another problem is that the environment can influence the quality of the data, like distortions due to vibrations or due to changes in temperature.

Secondly, while the production of large lot sizes is mostly automated, the demand for small lot sizes increases, as well as the variety of the products. Also, the number of times the production process needs to be adjusted to new products increases, too. Since both the reprogramming of the robots, that are included in the process, as well as the image recognition algorithms, are tailored to the environment and the products, the incorporation of a new product to the production process is costly. An expert with programming skills is required. But such experts are not always available.

And thirdly, the acquisition of suitable training data for a specific task. Though there are many datasets with large amounts of training data they do not cover every situation. There are many environments in which training data is nearly impossible to obtain, e.g. the deep sea or space. It is also hard to gather data on situations that rarely occur naturally. Creating your own dataset of real-world data will take time and effort, since besides gathering the data, they need to be labeled manually, too.

Our approach relates to all three challenges. We propose creating a digital twin mirroring a physical production line in the industry to generate synthesized training data to train an artificial neural network, accessible through a web service, to learn to recognize and locate products given observed point clouds of the scene.

In a simulation, we can generate an unlimited amount of training data and the labeling of the data is automated. Also, a simulation can easily show extreme situations that rarely occur. With this we can generate a wide range of training data, giving a possible solution to the problem of insufficient real-world training data.

If an artificial neural network can successfully be trained with synthetic data with or without a smaller amount of real-world data to recognize and locate products, the flexibility of the production process increases, and the costs decrease. A new product can easily be incorporated by every member of the staff without the need of any programming skills and without the need to consult a specialist. With this, producing small lot sizes of a large variety will become more efficient.

Our solution is also one step in the direction of introducing machine learning into the industry, showing that the issues with insufficient data and error-prone sensor data can be approached by using simulations.

The following parts of this paper are structured as follows: in Sect. 2 we describe some previous work on the above topics. In Sect. 3 we introduce some works which will be used as a basis for our solution, VEROSIM, and Rüstflex. The architecture and components of our system are described in Sect. 4. We will discuss further plans for our project in Sect. 5.

2 Related Work

Rossano et al. [12] describe how specialized knowledge is needed to program industrial robots. They give an overview of approaches to help with the program structure and the creation of new motion paths. The solutions suggested are either graphic-based, CAD-based, or include manually moving the robot arm. But mostly those solutions have drawbacks. A current approach used by Drag & Bot [5] or ArtiMinds [3] includes the use of function blocks in the form of graphic program modules. In IntellAct [16] a robot learns by observing a human manually demonstrating certain tasks.

Sahbani et al. [15] describe two different basic research approaches for a robot to grip unknown objects. Firstly, the analytical approach, which is based on a mathematical model. Secondly, the empirical approach, which includes the imitation of human motions or planning a motion based on observations of an object. Bohg et al. [4] give an overview of the data-based planning of gripping an object. However, the approach of using deep learning in this context is relatively new.

Digital twins are researched since approximately 2010. Negri et al. [9] give a survey of current research regarding digital twins. They define digital twins, in the context of production systems, as a virtual representation of a production process that can be used for different types of simulations. Rossmann and Schluse [13] define experimental digital twins, which can be combined into complex simulation models, namely virtual testbeds. They represent all important parts of an application close to reality and can be used for interactive experiments.

Georgakis et al. [7] synthesize existing real-world training data of indoor scenes by superimposing images. Textured object models are placed into different background scenes at different locations and with different sizes. This is done by either image blending or by using depth and semantic information to make smart positioning. An object detector was trained using these images in addition to an existing dataset with good results. This approach allows the expansion of existing datasets, but in case of applications with no suitable dataset, a different solution is needed.

Tsai et al. [17] create a large set of computer-generated 3D hand images to train a convolutional neural network to identify different hand gestures. They discovered that adding about 0.09% of real-world images to the training process increases the accuracy from 37.5 to 77.08%. Lindgren et al. [8] generate a dataset of synthetic hand gestures by using modern gaming engines. The synthetic hand gestures are created by making variations to the kinematics. They train a classifier purely on those generated data. The results are accurate and can be used on real-world data.

Richter et al. [11] use modern computer games to generate labeled training data for semantic segmentation tasks. They add different amounts of synthetic data to two different semantic segmentation datasets and compare the results of the trained networks. They have shown that by including synthetic data to real-world data the performance of the network increases, reducing the amount of hand-labeled data needed to train a successful neural network. But in this case, the generated data is completely dependent on the game. Using this approach for a specific application might not be possible, since the influence, the user has on the resulting training data, is restricted and there might not be a game suitable for your application.

3 Groundwork

3.1 Rüstflex

Rüstflex [14] is a web application created by Vathos GmbH [18] which can efficiently retool industry robots. It can adjust a robot's movements, mostly those that can be executed without sensory aid. The application runs either in a cloud or in a local computer center and can be accessed through all mobile end devices. It provides a formula where all relevant information for the setup of an article is stored. We will use Rüstflex in our project for the parametrization of the demonstrator (see Sect. 4.5) to reprogram the robot.

3.2 VEROSIM

With our simulation framework VEROSIM [19] we can create digital twins and virtual testbeds. A virtual testbed provides us with a completely virtual environment that can be used for experimentation. It can be integrated into real systems and can provide intelligent sensors, actors, and robots. We have already created a wide range of applications ranging from those in the industry to natural and urban environments as well as space. Three example applications are shown in Fig. 1. Our framework also provides several sensor simulations like ToF cameras, laser scanners, or radars, which can be built upon. One of the goals of this project is to upgrade our sensor simulations.

We will use VEROSIM in our research project to create digital twins of a production line and all its components, as well as to generate synthetic and automatically labeled data to be used as training data for an artificial neural network.

Fig. 1 Digital twins and virtual testbeds in industry, forestry, and space

4 System Components

Our goal is to enable a robot to handle unknown objects using machine learning methods trained mainly on synthetic data. Our overall system consists of several parts. First, there is a physical production line and its digital twin. A digital twin simulates all parts of its real-world counterpart. It can consist of several other digital twins and simulates the communication streams between different components. A deep learning component located in a cloud can be accessed by the physical production system and the digital twin as well as a user through a web service. All these components will be combined in a demonstrator. The overall architecture of our system is shown in Fig. 2. Further descriptions of each component are given in the following chapters.

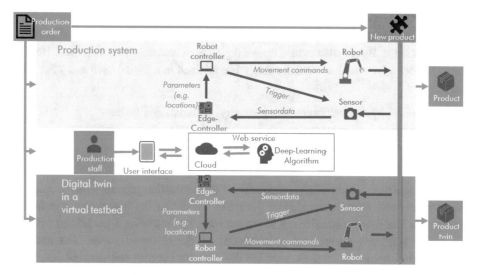

Fig. 2 System architecture

4.1 Production Line

We view a physical production process as the basis of our system. This process consists of a pick and place robot, a sensor, a gripper, a robot controller, an edge controller, and products of the production line. The robot controller dictates the robot's movements, opens, and closes the gripper, and triggers the sensor. The edge controller contains a copy of an artificial neural network trained in a cloud. The deep learning component is further described in Sect. 4.3.

The edge controller receives the data generated by the sensor and gives them as input to the copy of the neural network. In this case the edge controller returns the positions of the products to the robot controller.

The specific production line we chose is situated at aha! Albert Haag GmbH [1]. It is a deep drawing process where products are redrawn one or several times. Between each redrawing a robot from type Fanuc [6] takes and places the processed product from a machine to a palette with a piece of cardboard between each layer of products. Then a new product is taken and placed from a palette to the machine to be redrawn again.

There are many different variations of products that can be processed in this production line. They differ in size, material, geometry, and the number of required deep drawings. Also, each product is oiled. The sensor we use should be able to make accurate recordings of these products independent of their material or shininess. The observed data is than used to recognize and locate the products in the topmost layer of the current palette. The sensor we choose is a structured-light 3D scanner. It can produce both images and point clouds. The resulting point clouds are forwarded to the edge controller.

4.2 Digital Twin

Given the physical production line we build its digital twin using our simulation framework VEROSIM. But to what extend does a digital twin resemble its counterpart. Here are some, but not all, examples of what exactly is contained in a digital twin and what it can do:

- It can have physical attributes like geometry, material, and texture.
- It can manage working data, which is generated during its application.
- It can execute different functions and services.
- It contains interfaces for communication.

We create a digital twin of all components of the production process while keeping the above points and more in mind. For the robot we define the kinematics and inverse kinematics. The robot controller can move the robot by following a predefined path and the sensor generates synthesized point clouds of the observed scene. The difficulty regarding the simulation of the sensor are the sensor's internal and external errors that occur during the recording. Our sensor simulation framework should be able to simulate

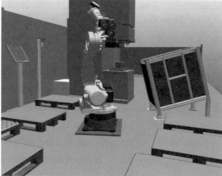

Fig. 3 Production process from aha! Albert Haag GmbH [1] and its digital twin

those errors, too, since our goal is to generate a simulation that is as close to reality as possible. But since the simulation contains all important information it is easy to automatically label all generated data.

In Fig. 3 you can see a picture of the physical production line on the left and its digital twin on the right.

4.3 Deep Learning

Both the physical production line as well as the simulated production line are capable of exchanging data through the physical and simulated edge controller with a deep learning component located in a cloud. The physical edge controller can load a copy of a trained neural network and provides a method for local inference. On the other hand, the simulated edge controller can send generated data to the cloud as training data or it can send unlabeled data for inference. In the letter case predicted parameters describing the class and the locations of the currently observed products are returned.

The deep learning component uses the training data, generated by the digital twin and the physical production system, to train a neural network to both recognize and locate new products. It detects objects in point cloud recordings of a scene. The training data we need to generate consists of three parts. Firstly, the point cloud of a scene. Secondly all visible bounding boxes in this scene and thirdly a mapping from each point in the point cloud to the center of the corresponding bounding box. The deep learning component and the webservice will be provided by Vathos GmbH.

4.4 Web Service

For easy access to the cloud and the deep learning component a web service is used. The web service allows a user in the production staff to upload the CAD data of new and unknown products to the cloud. He should also be able to initiate the start of the training process since this is the most expensive part of our system. The physical edge controller can synchronize with the cloud, loading a copy of the current neural network for a local usage. In this case the production system will keep working even during disturbances in the internet connection. Besides, the time needed for inference is shorter by using the edge controller in contrast to the time needed to access the cloud for inference. The simulated edge controller has a different duty than the physical one. Through the web service it can download the CAD data of the new products and replace old products with the new one in our simulation. After generating labeled training data, the edge controller can use the web service to upload them to the cloud where a neural net will use them to train.

Since Rüstflex also uses a web service for easy usage, it can be used as groundwork for the web service used in our project, which is currently in progress.

4.5   Demonstrator

The demonstrator is based on the production process from aha! Albert Haag GmbH. It combines all previously mentioned components. A simplified version of the described production line will be built by Arthur Bräuer GmbH & Co. KG [2], an integrator of robot arrangements. The goal of the demonstrator is to show that it is possible to easily adjust an automated process to new and unknown products using only data driven algorithms. We will use it to test our system in the applications from both aha! Albert Haag GmbH as well es Rewe Digital [10]. While the process from aha! Albert Haag GmbH is used to build our system on, a different process from Rewe Digital in logistics is considered to ensure the generalization of our system. For the parametrization of the robot, we will use Rüstflex. The results of the demonstrator will then be used to optimize the other components of our system.

5   Conclusions and Future Work

We are currently in the beginning stages of this project. Until now we have worked out the finer details of our system, specified the system architecture, which is partly shown in Fig. 2. We have specified the workflow of our chosen production line and built a first simulation model of said process. Vathos GmbH is developing the web service and the deep learning component. Currently we work on a digital twin of our chosen sensor and a component for labeling the generated data.

Further plans regarding this project contain finalizing our simulated sensor and training a network using the resulting data. At the end of our project, we hope to show that it is possible to improve a production process using our system and applying machine learning in industry using only synthetic data. Since a lot of clients wish for their production lines to be capable of dealing with small lot sizes, our approach would enable Bräuer GmbH &

Co. KG to offer their clients more flexible production systems. In addition, small and middle-sized companies like Rewe Digital and aha! Albert Haag GmbH could improve their current and future production systems. We hope for our results to help many other such companies to further automate their manufacturing processes.

And lastly, the success of our project will lead to the capability of industry robots to solve certain tasks more autonomously, making it possible to abstract the instructions given to the robot.

References

1. aha! Albert Haag GmbH (2020). https://www.aha-haag.de/. Accessed 10 Aug 2020
2. Arthur Bräuer GmbH & Co. KG (2020). https://www.braeuergmbh.de/. Accessed 10 Aug 2020
3. Artiminds (2020). https://www.artiminds.com. Accessed 14 Aug 2020
4. Bohg, J., Morales, A., Asfour, T., Kragic, D.: Data-driven grasp synthesis a survey. IEEE Trans. Rob. **30**(2), 289–309 (2014)
5. Drag & Bot (2020). https://www.dragandbot.com/product/. Accessed 14 Aug 2020
6. Fanuc Robots (2020). https://www.fanuc.eu/de/en. Accessed 10 Aug 2020
7. Georgakis, G., Mousavian, A., Berg, A.C., Kosecka, J.: Synthesizing Training Data for Object Detection in Indoor Scenes (2017)
8. Lindgren, K., Kalavakonda, N., Caballero, D.En., Huang, K., Hannaford, B.: Learned hand gesture classification through synthetically generated training samples. In: 2018 IEEE/RSJ International Conference on Intelligent Robots and Systems (IROS), pp. 3937–3942 (2018)
9. Negri, E., Fumagalli, L., Macchi, M.: A review of the roles of digital twin in CPS-based production systems. Proc. Manuf. **11**, 939–948 (2017)
10. Rewe Digital (2020). https://www.rewe-digital.com/. Accessed 12 Aug 2020
11. Richter, S.R., Vineet, V., Roth, S., Koltun, V.: Playing for data: ground truth from computer games. In: European Conference on Computer Vision (ECCV), vol. 9906, pp. 102–118. Springer (2016)
12. Rossano, G.F., Martinez, C., Hedelind, M., Murphy, S., Fuhlbrigge, T.A.: Easy robot programming concepts: an industrial perspective. In: 2013 IEEE International Conference on Automation Science and Engineering (CASE), pp. 1119–1126, Madison, Wisconsin, USA (2013)
13. Rossmann, J., Schluse, M.: Virtual robotic testbeds: a foundation for e-robotics in space, in industry—and in the woods. In: Developments in E-Systems Engineering, pp. 496–501, Dubai (2011)
14. Rüstflex (2020). https://uniktec.de/ruestflex.php. Accessed 10 Aug 2020
15. Sahbani, A., El-Khoury, S., Bidaud, P.: An overview of 3D object grasp synthesis algorithms. Robot. Auton. Syst. s **60**(3), 326–336 (2012)
16. Savarimuthu, T.R., Buch, A.G., Schlette, C., Wantia, N., Roßmann, J., Martinez, D., Alenyà, G., Torras, C., Ude, A., Nemec, B., Kramberger, A., Wörgötter, F., Aksoy, E.E., Papon, J., Haller, S., Piater, J., Krüger, N.: Teaching a robot the semantics of assembly tasks. IEEE Trans. Syst. Man Cybern. Syst. **48**(5), 670–692 (2018)
17. Tsai, C., Tsai, Y., Hsu, S., Wu, Y.: Synthetic training of deep CNN for 3D hand gesture identification. In: 2017 International Conference on Control, Artificial Intelligence, Robotics & Optimization (ICCAIRO), pp. 165–170, Prague (2017)
18. Vathos GmbH (2020). https://www.vathos-robotics.de/. Accessed 10 Aug 2020
19. Verosim Solutions (2020). https://www.verosim-solutions.com/. Accessed 10 Aug 2020

Usage of Augmented Reality for Improved Human-Machine Interaction and Real-Time Error Correction of Laboratory Units

Vadym Bilous, J. Philipp Städter, Marc Gebauer and Ulrich Berger

Abstract

For future innovations, complex Industry 4.0-technologies need to improve the interaction of humans and technology. Augmented Reality (**AR**) has a significant potential for this task by introducing more interactivity into modern technical assistance systems. However, AR systems are usually very expensive and thus unsuitable for small and medium-sized enterprises (SMEs). Furthermore, the machine's reliable data transfer to the AR applications and the user activity indication appear to be problematic. This work proposes a solution to these problems. A simple and scalable data transfer from industrial systems to Android applications has been developed.The suggested prototype demonstrates an AR application for troubleshooting and error correction in real-time, even on mobile or wearable devices, while working in a laboratory unit to simulate and solve various errors. The unit components (small garage doors) are equipped with sensors. The information about the state of the system is available in real-time at any given moment and is transmitted to a mobile or wearable

V. Bilous (✉) · J. P. Städter · M. Gebauer · U. Berger
Chair of Automation Technology, Brandenburg University of Technology Cottbus—Senftenberg, Siemens-Halske-Ring 14, 03036 Cottbus, Germany
e-mail: bilous@b-tu.de; fg-automatisierungstechnik@b-tu.de

J. P. Städter
e-mail: j.philipp.staedter@b-tu.de

M. Gebauer
e-mail: marc.gebauer@b-tu.de

U. Berger
e-mail: Ulrich.Berger@b-tu.de

© The Author(s) 2022
T. Schüppstuhl et al. (eds.), *Annals of Scientific Society for Assembly,*
Handling and Industrial Robotics 2021,
https://doi.org/10.1007/978-3-030-74032-0_22

263

device (tablet or smart glass) equipped with AR application. The operator is enabled to preview the required information in a graphical form (marks and cursors). Potential errors are shown and solved with an interactive manual. The system can be used for training purposes to achieve more efficient error correction and faster repairing.

Keywords

Augmented reality · Error correction · Human-machine interaction · Industry 4.0 · Technical assistance systems · Data transfer

1 Introduction

Augmented Reality (**AR**) is the computer-based extension of reality using smart glasses or other hardware devices. AR technology nowadays has a wide range of application areas [2, 15, 16]. By using AR, the reduction of error search and error correction is achieved. AR technologies have a great potential in commissioning [21] and maintenance [17], for assistance in training areas [10], assembly, repair, or automation [9]. AR applications (**AR apps**) serve as a useful tool in various technical and production fields (e.g., mechanical engineering and manufacturing automation).[1]

In the past years, AR technologies have been increasingly exploited in technical assistance systems, namely the visual ones [22]. AR is being used for the visualization of assembly tasks, machine operation, or repair processes [1, 6]. With the support of cameras (e.g., integrated into tablet or smartphone), the system automatically recognizes a malfunction or broken components, allocates the relevant information, and displays the exact assembly instructions directly in a real image. Thus, the automatic fade-in of the circuit diagram while working on an electrical system component is conceivable. For a technician, this simplifies recording a very complex system and the analysis of relevant measured values. The assistance via "instructions and feedback" [4] for technicians and other workers can even be originated from remote experts and executed in real-time [8].

Considering the current development stage of hardware and software in the field of AR technology support, writing user applications has become relatively simple and less time-consuming. Developers provide a vast amount of free software, which can be used for complex applications like error correction in mechanical problems [3].

---

[1]Usage of the created AR application is demonstrated here: https://www.b-tu.de/owncloud/s/ HwS8ZGFtd8ddToS.

2 State-Of-The-Art Applications and Motivation for the Study

As it was mentioned in the introduction, the implementation of AR in the fields of mechanical engineering and manufacturing promises significant perspectives to reduce the complexity and increase interactivity of the human-machine communications.

Despite numerous research and development projects in the field of AR technologies [5], it has not yet been widely used in mechanical engineering. The main reason is the hardware's high pricing and heavy weight. From the first experiments from 2003 to the increasing interest since 2008, the hardware available for AR apps shared the above-mentioned characteristics [4].

The recognition of real objects, which is based on artificial intelligence (AI), is essential for the further positioning of virtual objects in AR. However, the detection of 3D objects (e.g., machine parts) usually requires industrial hardware and expensive software. For example, the DIOTA company [7] offers common alternatives for hardware and software. Industrial tablets such as GETAC or head-mounted displays from HTC or ISAR Projection System are used. For the software, examples from DIOTA (DIOTA Player and DIOTA Connect) and future AR software from Siemens are in operation [7]. Most of the current solutions are mainly interesting for large industrial customers but not for SMEs.

Thus, the most important problem of the wide use of AR can be formulated as follows: many developers traditionally use rather expensive development tools and software products.

A project in the Brandenburg University of Technology Cottbus—Senftenberg (**BTU**) at the Chair of Automation technology [3] was initiated to solve this problem. The aim of the conducted development was to build an AR application for education and error correction. The software tools Unity3D and Vuforia, due to their availability to the project team, were used during the development process. Several new problems regarding the AR application development were identified and solved.

2.1 The Problem of Data Transfer

The data transfer between industrial systems (for example, programmable logic controllers, PLCs) and Android applications is a rather complicated task when using simple development tools. The basic approaches are primarily focused on transferring data from one single PLC to one or more Android applications, for example, SIEMENS PLC support the data transfer with TCP/IP protocol [18]. But this task is getting even more complex if the solution has to be scalable, i.e., effective not only for one unit, but also for several ones [12, 19, 20]. Overall, the problem of data transfer from industrial controllers (PLC) to AR hardware is currently not well developed. Thus, a new data transfer method has to be established.

2.2   The Problem of User Activity Indication

The data transfer between the programmable controller (PLC) of the plant or laboratory unit and the user is a part of the AR ergonomics problem, which is currently under development [14]. The user's Android application should not only receive data from the PLC but also transfer data back. The reason for the introduction of bidirectional information transfer is that the majority of the user's actions with the mechanical environment during the error correction or learning process cannot be detected by the environment's sensors. On the other hand, the importance of comfortable and easy communication between the user and the AR app is essential, which was explored in particular in [11]. Therefore, the user must be enabled to mark his actions with other tools, e.g., virtual keyboard, joystick, or application interface buttons (if any hand-held device as an AR hardware is used).

3   Method of Development

The following steps for the development of the project have been taken:

- Analysis and arrangement of the elements of an AR system;
- Creation of the "development logic"—instructions for the users' actions;
- Identification and selection of the open-source or low-cost AR software solutions;
- Combining the AR system elements and solutions into a functioning system;
- Developing of an AR application with "development logic" and tests.

This method of work with AR projects in BTU was formulated in the previous related work [3] and projects with AR HMI system for small laboratory plants). As for the application of the method, two important points required special attention. Firstly, the laboratory units needed a modular design to simplify possible modification.

Secondly, the supporting elements of the system were required. The complete scheme of the project released according to the method is shown in the Fig. 1.

The developed system simulated the operation of an automated garage doors facility and related maintenance processes.

3.1   Hardware Components

The laboratory unit was equipped with a Siemens PLC. AR information could be displayed on a tablet, smartphone, or HMD (Human-Machine Display). The data transfer from the door control hardware (PLC) to the user hardware (smartphone and cardboard) occurred via OPC UA. The latter is a universal interface to the plant network (control components of the industrial plants). The transmitted information constituted a list of

Fig. 1 Elements of the project

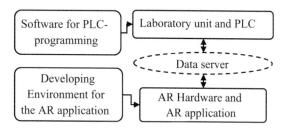

variables and their states. In this way, the information about the system components and the operating errors would be transferred to the AR application.

Reasons for the selection of OPC UA as the (data transfer) media were the following:

- Support of C# and Java as implementation programming languages;
- Scalability from embedded control software to operational or management information systems;
- Custom security implementation based on the latest standards;
- Configurability of timeouts for each service call.

In the first developing phase of the AR application, mobile hardware—an Android tablet—was used. In the final phase of the testing of the latest version of an AR app, a "wearable" device such as a smartphone and a cardboard box was used. A solution with real AR glasses, such as HoloLens from Microsoft, would be quite expensive for SMEs, therefore, such a configuration of an AR device was introduced to resemble the real future operating conditions. By using common and affordable devices such as tablets and smartphones, a distinctive balance of cost and capabilities may further be achieved.

3.2 Software Components

The software system used to develop the AR app consisted of the Vuforia Augmented Reality SDK in conjunction with the Unity3D physics engine.

The reason for the selection of Vuforia as a software solution was its broad functional capabilities. With Unity3D, complex dynamic models can be realized. The control of the models is possible, as well as communication with other programs. This development environment can be used for complex tasks and is free available for research and non-commercial projects. An alternative software solution to create an AR application could be another physical engine like, for example, Unreal Engine (**UE**). However, UE needs a high level of experience in C programming and is not free of charge. Nevertheless, the respective software is becoming more and more popular in large projects, as it offers significantly more options in the future.

Therefore, the main system components constituted Siemens PLC, OPC/UA interface and server, AR Hardware (Android smartphone and/or tablet, VR cardboard), Unity3D and Vuforia as the development environment.

3.3   Project Development Process

At the start of the project, a miniature laboratory facility (doors) was created and used. The experimentation unit contained a movable door, two user interfaces, and two tracker images. There were different tracker pictures for the positioning of AR elements and AR keyboard. The first prototype of the laboratory facility is demonstrated in Fig. 2.

The proposed system enabled the user to carry out advanced operations usually done by experts. The set of implemented standard errors and how they were solved showed the AR app opportunities for training and non-expert error correction.

Three errors could be simulated and indicated with this laboratory unit:

- "Short" door blocking. When the optical sensor detected an object while the door was moving down, the mechanism would be stopped. The user was guided to remove the object.
- "Long" door blocking. In case the optical sensor was able to detect the object for longer than 5 s, the facility's door motor was stopped. Then the manual restart by the technical expert through a specific interface (see Fig. 2) was done.
- Using the emergency button. The facility was stopped and the technical expert had to do the manual restart in the same way.

For the first prototype of the laboratory unit, a virtual keyboard (**SVC**) with two buttons was developed (Fig. 3.). This solution was created to solve a problem of the user activity indication (see Sect. 2.2). The user was required to switch between different instructions when fixing an error. The button "NEXT" could be activated to go to the next step of the error correction. The button "BACK" enabled returning to the previous step if checking off the correction action was needed.

The system under consideration was presented at the Hannover Expo in 2017 and 2018 as a part of the exhibition stand of the IHK (Industrie- und Han-delskammer) Berlin-Brandenburg. The overall number of the attendants of the exposition who tested the prototype amounted to about 100 people. Each tester was informed about the presence of simulated malfunctions in the laboratory unit, but was not informed about the possible course of actions explicitly. It was the job of the AR app to provide the users with

Fig. 2 Laboratory unit (small door, first prototype)

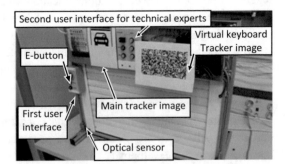

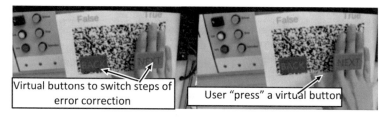

Fig. 3 Using virtual buttons ("NEXT" or "BACK") for step switching in the first prototype of the project

step-by-step instructions for problem resolving. Mobile hardware, namely a tablet with an Android system, was used. Such practical use cases have shown that SVC increased the user's error handling capabilities by freeing the hands from the joystick or any other remote control device. However, SVC seemed to require an extra tracker picture. It additionally reduced the user's mobility if subsequent movement around the unit was required.

The analysis of the application of the first prototype of the project allowed us to formulate a solution to the two-way communication.

3.4 The Solution to the User Activity Indication Problem

A universal solution to problem of the user activity monitoring and two-way communication was not possible by software methods alone. Essentially, it required the tracking and identification of the possible actions. The latter was only feasible if the laboratory plant was equipped with additional sensors capturing each action of the user. The transfer of subsequently appearing data was also required.

Thus, the second prototype of the system was developed. The laboratory unit has been modified by introducing a new console for technicians. The SVC tracker was no longer needed. The new sensors allowed to control the user actions and the steps due to laboratory tasks.

The operator of the system could use wearable AR hardware (smartphone with an AR application and cardboard). He acquired the related status information on the door, as well as a guided manual. The outline of the user interface and the prototype overview are presented in Fig. 4.

There were three control buttons on the new user console box, which allowed for extended interaction patterns:

"**Start**". If this button was pressed, the garage door would open. If nothing appeared to be passing through the door, the latter waited for 12 s, and then closed.

"**Stop**". If the Stop button was pressed, the door stopped moving.

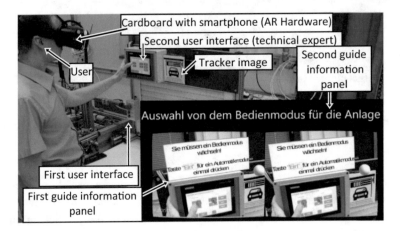

Fig. 4 Laboratory unit and all AR elements (small garage doors, second prototype)

"**E-Button**" (emergency stop button). If this button was pressed, the door was stopped, the movements of the system components were blocked, and a restart with a technical console box was required.

3.5 The Solution to the Data Transfer Problem

The primary idea for the development was a direct data transfer system. As previously described (see Sect. 2.1), the flexible and scalable data transfer between industrial solutions (e.g., OPC/UA) and Android applications appears to be a major challenge. Direct data transfer from the server to the Android system is also possible, but requires extensive knowledge of software development. This contradicted the philosophy of the project by increasing the complexity for the end-user and for the developer.

Therefore, a support web server was introduced. It received data from the OPC/UA. The Unity3D Web Socket (integrated into the AR app) collected the data from the web server (Fig. 5).

Such a communication structure had a feedback time of approx. 0.3 s, but was simple in development, reliable, and offered easy maintenance and scalability (numerous AR hardware units could take information from the support web server).

Furthermore, our solution appeared to be more universal, as it made OPC-UA available for every platform that supported Web Sockets. We thus enabled OPC-UA for platforms where no ad-hoc OPC UA support library existed. The programming language used in AR app did not play a significant role (C#, JavaScript, Unity, Python, etc.). It should be mentioned that we used JSON-RPC (JavaScript Object Notation Remote Procedure Call) via Web Socket, starting the calls on the web server as a result. In other words, our solution brought OPC-UA into the browser.

Fig. 5 The structure of the data transmitting between the machine and the AR application

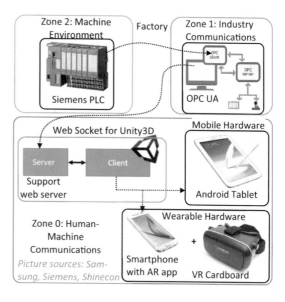

3.6 Resulting System Testing

To test the second prototype, we enlisted the help of laboratory visitors and university staff members. In total, the number of the participants of the experiment amounted to about 30 persons in 2019 and 2020. The same testing methodology as with the first system prototype was used.

The experimental results showed that 100% of testers were able to fulfil the service requirements in the laboratory unit for the first time in an error situation by following the AR app guidance. However, more than 60% of testers reported the problem of a the loss of position of virtual objects during sharp movements of the head. The issue was eliminated by modifying the tracking algorithms from Vuforia and using the extended tracking. With the modification under consideration, the system could work even when there were no trackers in the camera's field of view. Nevertheless, at the start of the program, the tracker image was still needed to initialize the object positions.

4   Technical Maintenance Simulation (Error Correction)

To solve the introduced error with the "short" door blocking, the user received a message to clear the door's path. (Fig. 6a).

In the case of "long" door blocking, the user had to clear the door's path at first. Then the user needed to follow the AR instructions to activate the manual control mode (Fig. 6b) and close the door manually with the "Down" button (Fig. 6c).

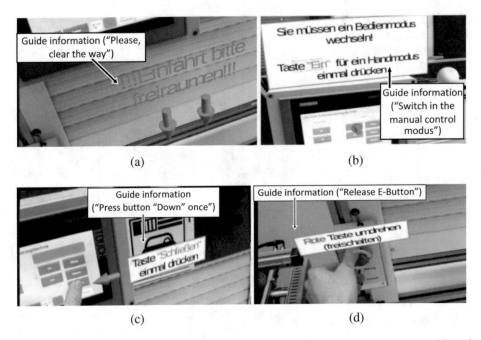

Fig. 6 **a** The first error ("short" door blocking). **b** Switching in the manual modus. **c** Manual control of the door movement. **d** Release of E-button

As previously described (see Sect. 3.6), the system could detect errors and control every relevant step of the user and error correction. For this reason, there was automatic switching between AR instructions for individual steps.

If a door was blocked with an E-Button, the user was informed about further steps. He had to release an emergency stop button (Fig. 6d) and then (as explained before) switch the system to the manual operation before closing the door and restarting (as on the Fig. 6a and b).

5   Results

The AR application developed in this project can be used as an assistance system to solve problems and errors with a laboratory unit. Such interactive facility replaces the traditional study of maintenance manuals. The visual feedback informs the user about errors in the system and whether they have been rectified.

Based on the recognized tracker pictures, AR objects can be displayed on the needed position. After the first recognition of the picture, the system can work when there are no more trackers in the camera's field of view. These AR objects reveal the error's exact location and give instructions for user guidance to resume the system's proper operation. The user is enabled to quickly and successfully follow the necessary steps of error correction. The use of the system for training and education purposes is foreseen, and this system has positive effects on reduced downtime.

6　Conclusion and Future Work

This work presents a practical application of Augmented Reality for machine environment maintenance. It is shown that communication between the current hardware and software components (OPC/UA and Android applications) and, therefore, technical assistance system realization, is possible without additional expenses. Moreover, the project demonstrates the possibility of developing and using an AR application with physics engine software (Vuforia Augmented Reality SDK in conjunction with the game engine Unity3D).

In the project presented in this article, a new scalable method of data transfer has been developed and tested. The achieved solution is portable and available for every platform that supports Web Sockets. The problem of user activity indication is solved with the implementation of new hardware elements in the mechanical environment.

The authors would like to further emphasize the importance of the equipment of modern generations of systems and plants with an excessive number of sensors [13] to precisely detect each user's action and the order of these actions for the further successful implementation of AR assistance systems. The next step may be providing the support of more machine operations. Further development in this direction might include the binding of additional AR visual effects libraries, as well as the introduction of new user-machine interactions. It might enable the use of a single OPC/UA solution for collecting data from even more units and plants in the laboratory. The collected data could finally be used for more AR hardware and operators.

References

1. Aottiwerch, N., Kokaew, U.: Design computer-assisted learning in an online augmented reality environment based on shneiderman's eight Golden Rules. In: 14th International Joint Conference (2017)
2. Arnoldy, S., Bautz, S., Bruns, L., et al.: Augmented Reality. Welche Branchen können in Zukunft profitieren? (2016). https://www.pwc.de/de/technologie-medien-und-telekommunikation/assets/tmt-studie-augmented-reality.pdf
3. Berger, U., Bilous, V., et al.: Anwendung von der Augmented Reality für die Mensch-Roboter Interaktion bei der Fehlerbeseitigung und bei der Maschinenbedienung. In: INNTERACT 2016 Conference, Chemnitz, p. 37 (2016)
4. Blattgerste, J., Renner, P., Pfeiffer, T.: Augmented reality action assistance and learning for cognitively impaired people: a systematic literature review. In: Proceedings of the 12th ACM International Conference on Pervasive Technologies Related to Assistive Environments, pp. 270–279 (2019)
5. Chen, J., Wang, Y., Guo J.: Augmented reality registration algorithm based on nature feature recognition. Springer, Berlin (2010)
6. Dini, G., Mura, M.: Application of augmented reality techniques in through-life engineering services. In: The Fourth International Conference on Through-life Engineering Services (2015)
7. DIOTA (2021). https://diota.com/en/home. Accessed 02 Feb 2021
8. Fang, D., et al.: An augmented reality-based method for remote collaborative real-time assistance: from a system perspective. Mob. Netw. Appl. **25**(2), 412–425 (2020)

9. Juhász, T., Schmucker, U.: From engineering CAD to a Modelica Model: Structural Manipulation throughout a Translation Process. IFF Fraunhofer, Magdeburg (2008)
10. Jung, T., et al.: Augmented Reality and Virtual Reality. Springer (2018)
11. Lee, J., Seo, J., et al.: End-Users' Augmented Reality Utilization for Architectural Design Review. The Hong Kong Polytechnic University, Hong Kong, China (2020)
12. Madasamy, N., Aishvarya, B., Kumar, K., Vinith, K.: Android application for accessing bosch rexroth PLC. In: International Journal of Research in Engineering, Science and Management, vol. 3, issue 4 (2020)
13. Mattes A., Gericke T.: Operating assistance system for machine tools. In: Nunes, I. (ed.) Advances in Human Factors and Systems Interaction. AHFE 2020. Advances in Intelligent Systems and Computing, vol. 1207. Springer, Cham (2020)
14. Munoz, L.: Ergonomics in the industry 4.0: virtual and augmented reality. J. Ergonom. **08** (2018)
15. Muñoz-Saavedra, L., Miró-Amarante, L., Domínguez-Morales, M.: Augmented and virtual reality evolution and future tendency. In: Architecture and Computer Technology Department (Universidad de Sevilla), E.T.S Ingeniería Informática (2019)
16. Palmarini, R., et al.: A systematic review of augmented reality applications in maintenance. In: Robotics and Computer-Integrated Manufacturing, pp. 215–228 (2018)
17. Quandt, M., et al.: User-centered evaluation of an augmented reality-based assistance system for maintenance. Proc. CIRP **93**, 921–926 (2020)
18. Siemes: TCP/IP Communication via Industrial Ethernet (2005). https://cache.industry.siemens.com/dl/files/612/22146612/att_113921/v1/t-bausteine_e.pdf
19. Softing Industrial Automation GmbH: OPC UA Kompetenz (2019). https://data-intelligence.softing.com/fileadmin/media/products/applications/OPC_Kompetenz_DE_Softing_Data_Intelligence_pdf
20. Troci, J.: Model for evaluation of OPC-UA for industry-4.0-compliant communication in wireless sensor networks. Master Thesis. Technische Universität Ilmenau (2019)
21. Yang, Z., et al.: Influences of augmented reality assistance on performance and cognitive loads in different stages of assembly task. Front. Psychol. **10**, 1703 (2019)
22. Yang, X., Plewe, D.A.: Assistance systems in manufacturing: a systematic review. In: Schlick, C., Trzcieliński, S. (eds) Advances in Ergonomics of Manufacturing: Managing the Enterprise of the Future, vol. 490. Springer, Cham (2016)

Scalability of Assembly Line Automation Based on the Integrated Product Development Approach

Florian Hoffmann⊙, Vanessa Wesskamp, Raphael Bleck
and Jochen Deuse⊙

Abstract

Product life cycles change, market developments and quantities are increasingly difficult to predict, as is the case in the production of charging stations. For these reasons, scalable assembly concepts with an adaptable degree of automation are becoming increasingly important. Currently, charging stations are still manufactured manually. With increasing quantities, however, manual production is no longer economical. New technologies such as lightweight robotics offer a great potential for making production more flexible in terms of quantity. At the same time, new challenges arise because these requirements must be taken into account from the very beginning of product development and process planning. Currently, there are no planning approaches and recommendations for action that take this into consideration. Therefore, the research project "Simultaneous product and process development of a charging station outlet module suitable for automation" (SUPPLy) develops an integrated, digital and simultaneous product and process development of a modular charging station suitable for automation. The aim of the project is to develop an assembly process which enables an economic production of charging stations in case of

F. Hoffmann (✉) · V. Wesskamp · R. Bleck · J. Deuse
Institute of Production Systems, TU Dortmund University, Leonhard-Euler-Straße 5, 44227 Dortmund, Germany
e-mail: florian.hoffmann@ips.tu-dortmund.de

J. Deuse
Advanced Manufacturing, School of Mechanical and Mechatronic Engineering, University of Technology Sydney, Sydney, Australia

T. Schüppstuhl et al. (eds.), *Annals of Scientific Society for Assembly, Handling and Industrial Robotics 2021*,
https://doi.org/10.1007/978-3-030-74032-0_23

fluctuating sales figures. The focus is not only on changes in the production process but also on a product design that is suitable for automation. The paper presents the ideas on a conceptual level.

Keywords

Integrated product development · Assembly · Scalability · Simultaneous engineering · Industry 4.0

1 Introduction

In order to maintain the competitiveness of the German automotive and supplier industry and to survive in global competition, Germany is to become the lead market for electric mobility [1]. To achieve this challenging goal, one million electric cars are to be in use on the roads by the end of 2022 [2]. The government's goal, and with it the establishment of electromobility as part of a sustainable transport system, will only be achieved if charging infrastructure solutions are available across the board that allow electric vehicles to be charged at any time and as needed. The development in this area is currently still in a premarket phase. Accordingly, demand for such products is still at a comparatively low level, meaning that extensive investments in plant and automation technology have not yet been economically viable for manufacturers. Due to low market demand and lack of economies of scale, the components and the manufacturing processes are very cost intensive, which in turn leads to high prices for the charging infrastructure. According to various forecasts, however, there will be a rapid and sustained increase in demand in the coming years. A nationwide charging infrastructure can only be set up if the corresponding charging stations can be built economically. So that a progressive industrialization of products and the corresponding production processes will become indispensable [2–4].

This paper presents a new approach for a scalable line automation. Chapter 2 gives an overview of existing planning approaches followed by a specification of scalability in charging station production, including the importance of human–robot-interaction (HRI) in Chap. 3. The developed concept is outlined in Chap. 4 including a theoretical application scenario. The paper concludes with an outlook.

2 Planning Approaches

Planning systems represent an important thematic basis for the development of scalable assembly concepts. They serve the analysis, planning, design and improvement of socio-technical work systems and can lead to more efficient processes through clever application. One of the conventional planning methods is the planning systematic according to REFA, a german association for work design, business organization and business development [5]. This contains general instructions for planning complex manufacturing and assembly systems. A central feature of the system developed by Rother and Harris [6] is the possible reduction of the degree of automation in order to enable increased flexibility and a continuous flow of employees. Dietz's [7] approach addresses the high and dynamically changing variant diversity in production systems over time and provides ideas for product structure dependent manufacturing system design. In contrary, the MABA-MABA (men are better at—machines are better at) planning approaches focus on skills-oriented process design between man and machine in the context of increasing automation [8].

Furthermore, production systems as a whole are increasingly exposed to time dynamic influencing factors, so that the planning of changeable and adaptable systems is also of increased importance for the planned research project [9].

Production oriented approaches are particularly suitable for planning manufacturing and assembly systems with high volumes and low product variance. In many cases, these methods are geared towards more capital intensive systems in order to produce the required quantities as economically as possible by exploiting economies of scale. In order to enable the production of an economically viable product even in the case of fluctuations in demand, the pure consideration of production oriented approaches is insufficient. These approaches must be combined with measures to optimize product development in order to achieve a holistic method.

In this context Hengstebeck [10] provides an insight into the different areas that are affected by the planning of semi automated work systems (e.g. product development, production planning, maintenance). Deuse identifies a great potential in the realization of cyber-physical production systems (CPPS), in which human capabilities are enhanced through the intelligent use of information and communication technologies. For this purpose, new collaborative forms of work must be developed in which humans can optimise processes and act as active decision makers [11].

Simultaneous engineering enables a shorter time-to-market through a close integration of product development and production planning, which can lead to competitive advantages. In literature there are promising approaches that show possible process models. Erlenspiel [12] describes a method that focuses on development and construction and addresses simultaneous planning across phases. Similarly, Bullinger contributes to simultaneous engineering, focusing on efficiency and cost aspects [13]. Overall, it can be stated that none of the planning approaches presented above sufficiently meets the requirements for designing a scalable production system. The core objective of SUPPLy is to develop a scalable assembly system for the production of charging stations that ensures cost-effective operation at all times during production. For this purpose, it is necessary to continuously adapt the production capacity to the respective situation on the sales market [14]. The basis for this in a high wage country, such as Germany, is a flexible adaptation of the degree of automation with the help of HRI. Product design is being rethought to maximise flexibility and aligned with the needs of HRI technology. As outlined in the next chapter, several approaches have to be combined to meet the challenges of the market.

3   Quantity Dependent Charging Station Production

Although the German government has set concrete targets for the number of electric cars on the road for the next few years, it is very difficult to predict how sales will develop. Political measures, such as targets for fleet consumption, purchase premiums, or an adjustment of the motor vehicle tax to match emissions, lead to fluctuating demand in the market. [15]

Due to the prevailing market situation, specific requirements for the work system planning arise. Short planning cycles as well as a good and economic scalability are the core challenges in the planning process. Conventional sequential planning of assembly systems is not able to meet these market dynamics. In high wages countries a scalability using only personnel deployment is not economic. As explained above, an adaption of the degree of automation is a frequently used instrument. However specialised semi-automated or automated systems cannot react to short-term fluctuations. Taking increasingly shorter product lifecycles due to a fast technological progress into account, these systems are not economical either. Lightweight robots offer a new possibility for an adaptable degree of automation. Accordingly, the planning concept must combine HRI-compliant and modular product design with scalable assembly system development [16, 17].

Comparatively low investment costs and the possibility to use them with small adaption in different production systems as needed and the option to realise HRI, promises a high potential to increase flexibility. HRI offers many possibilities for automating manual assembly processes making optimal use of the strength and weaknesses of humans and robots [18]. By exploiting these synergies, the output volume can be increased by using a robot to partially automate previously manual tasks [10].

When planning and realizing a flexible automatable assembly system, greater attention must be paid to the expandability of the assembly system and the reusability of components. A modular system structure is recommended in order to minimize the resulting costs for rebuilding measures or for the restart [19]. The lightweight robots can be integrated with only minor changes to manual work systems and are able to produce especially small and medium quantities efficiently and economically [20, 21].

To realise this potential, the product must be suitable for the handling and assembly by robots as well as manual handling. The geometry of a product significantly influences the way it can be picked up by the robot end effector. Sharp edges or protruding parts make handling just as difficult as fragile plug-in connections. In many cases, bendable parts are not suitable for automated handling. In this context, design alternatives can already be considered during product development. With regard to the payload, lightweight robots capable of interaction are usually limited to single-digit kilogram ranges due to safety requirements. Due to these framework conditions, the production use of HRI is significantly determined by the product design [22, 23].

For this reason, when planning a scalable work system with a variable degree of automation, it is very important to involve the construction and design department from the outset in order to minimize scaling costs during production [24]. In this context, Nyhuis defines the five (product) characteristics universality, scalability, modularity, mobility and compatibility as so called transformation enablers. With the help of these characteristics, the requirements for a flexible assembly system and its components can be described [25].

External disturbances influence the assembly system during ongoing production operations and require constant adjustments [26]. In the present case of charging station assembly, significant fluctuations in demand must be taken into account for the reasons mentioned previously. For the first concept of a scalable charging station assembly line, scalability and modularity are particularly important in order to create an optimal basis for further flexibility-increasing measures. In the context of the project, modularity relates in particular to product development, which is at the beginning of the development process and has a direct influence on the subsequent scalability of production.

Production currently involves the assembly of numerous individual components, including their cabling. A promising approach in this context is the development of a **modular product design** that reduces assembly complexity and favours HRI. This in turn enables the necessary **scalability of production** depending on market needs. Due to the need to consider assembly possibilities and HRI suitability already in the product design, the SUPPLy project, as shown in Fig. 1 focuses on **simultaneous engineering** in relation to the integrated product development approach.

By linking modularity and scalability, an assembly concept is created that offers maximum adaptability depending on the quantity requirements while addressing the need for short planning cycles. For subsequent development steps, compatibility, universality and mobility must be taken into account with the aim of further increasing flexibility. By

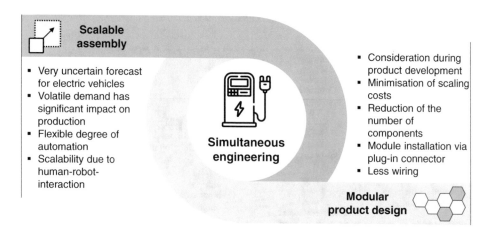

Fig. 1 Integrated product development approach in the SUPPLy project

taking these characteristics into account during product and process design, the future expandability and changeability of the overall system is guaranteed.

4 Application Concept

In the following, an application concept for the integrated product development ap-proach is described. As shown in Fig. 1, the simultaneous engineering concept is based on two approaches: The **modularisation of the product structure** and the **scalability of assembly line automation**. The two solution approaches are considered using the example of the manufacture of public AC (alternating current) charging stations. Many providers of charging stations are currently advertising with a modular design of their products. On closer inspection, however, it becomes clear that the modular approach is only from the customer's perspective. Similar to the vehicle configuration, the customer has the option to customize the station housing and can add or remove various technical components. However, a modular product structure from a design perspective is not yet applied.

A modern AC charging station for public spaces consists of a large number of individual parts, which are usually purchased separately. The assembly process for a charging station is currently very time-consuming and labour-intensive and is usually only carried out manually due to the extensive cabling effort. With an exemplary total assembly time of six hours, up to three hours must be allowed for the wiring of the components with subsequent functional testing. For this reason, the modularisation of the interior of a charging station is a promising approach. A large part of the components used is recurring standard parts, which are a fundamental part of every charging station regardless of the variant. Figure 2 shows standard components that can be combined in the course of a possible modularisation.

As part of the development of a modular product structure, an early concept for an outlet module is being developed that combines various basic components and can therefore be installed into any product variant. When developing this integral design, it is important to consider the subsequent usability in hybrid work systems according to the

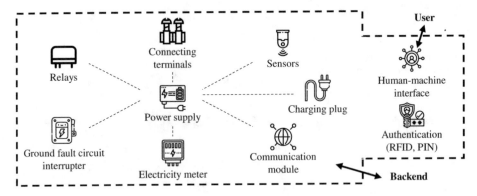

Fig. 2 Possible modularisation of standard components of a charging station

framework conditions for the use of HRI explained in Chap. 3. The aim is to reduce assembly time in order to increase quantities and to be able to adjust the degree of automation depending on market demand.

One of the main challenges is to reduce the amount of cabling by using different technologies. It is conceivable to design a motherboard that enables assembly by means of plug connections, as is the case with desktop computers, for example. In this way, optional interfaces could be created that enable the installation of customised modules (e. g. a user interface) without affecting the installation of the standard components.

By reducing or possibly even eliminating the manual cabling effort, the assembly time can be additionally reduced, which in turn favours high quantity requirements. In order to take into account, the payload of the robot, the module size can be adapted accordingly depending on the function combination, so that different handling and assembly processes can be automated. The integration of several standard electrical components, might make an in-house production economically feasible for certain parts. This would create the possibility of integrating the respective components firmly into the module instead of continuing to use the parts individually. In this way, additional influence can be exerted on the nature of the geometry and the surface, for example, to reduce the number of necessary gripper changes and eliminate associated set-up losses.

Due to the lack of product development, only basic framework conditions can currently be defined for the consideration horizon of the planning approach. It can be assumed that a large part of the wiring effort can be reduced by redesigning the individual electrotechnical components into functional modules, but that this is not completely eliminated. In order to enable very high piece count scenarios with simultaneous economic efficiency, flow production is used as the basic set-up principle already during the creation of a first planning draft. Figure 3 shows an exemplary layout for a corresponding assembly line.

Due to the prevailing framework conditions in charging station production, production must be designed for a high degree of flexibility in terms of unit numbers right from the start of planning. The assembly line has four workstations where the modular socket module is assembled manually. The work planning should take into account a possible parallelisation of the assembly so that, if necessary, the cycle time of the workstations can

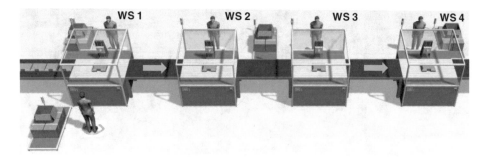

Fig. 3 Layout of an assembly line with manual operations

be reduced due to a necessary adjustment of the line cycle by means of additional employees (e. g. workstation 1) and thus the output quantity increased.

This is conceivable, for example, through an individual assembly sequence, which must be taken into account in the product design. In the event that economic production is no longer possible due to manual assembly activities, the next stage can be applied by using HRI. Figure 4 shows an example of a possible line adaptation.

Workstation three and four are converted into HRI workstations with the aim of further reducing the cycle time of the line. A skill-oriented division of labour between humans and robots must be taken into account. Necessary cabling, for example, will continue to be carried out by humans. Appropriate analyses can be used to determine the HRI potential of individual workstations [27].

For this purpose, an HRI preassembly could be integrated, as shown in Fig. 4 for the fourth workstation, in which the work content is divided accordingly. The tasks could be divided in such a way that one employee carries out the necessary wiring at the pre-assembly workplace. The wired module is then brought into the assembly line by a robot and mounted there in collaboration with another employee. Depending on the degree of decoupling from a necessary assembly sequence, any line layouts are conceivable. Appropriate line layouts are to be worked out depending on the quantity requirements and the work contents are to be distributed to the respective workstations. In addition to the sensible division of labour between humans and robots, the assembly sequence also plays an important role.

The concept presented in Chap. 4, which follows the approach of an integrated product development in order to maximise the production flexibility, represents the first stage of a reference planning process for charging station production. The planned procedure must be substantiated by initial product development approaches and market requirements. In this context, the use of simulation software would be useful to analyse and economically evaluate different scenarios. In addition to market development, other factors must be taken into account that influence the choice of a production system technology. These

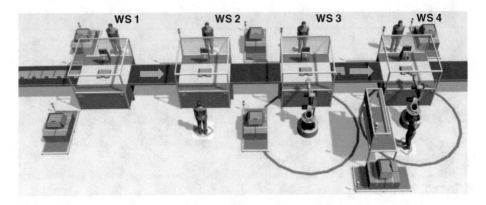

Fig. 4 Layout of an assembly line with scalable degree of automation

include, for example, the duration of the product life cycle, political deregulation or general technological progress [28]. Since electromobility is a comparatively young technology, it can be expected that various technical innovations will be developed, especially at the beginning. Due to the many external influences and uncertainties, it can also be assumed that a very high degree of flexibility in the number of units is required to enable an economic production of charging stations in the high-wage country Germany. The following planning stages must therefore further favour and promote the flexibility of production.

5 Conclusion and Outlook

At present, assembly systems planning processes do not adequately address the prob-lem of volatile sales. The current situation on the electromobility market contributes to the relevance of the research approach. The integration of interacting human–robot systems is a promising approach for making production more flexible in terms of quantity. However, the requirements for a flexible degree of automation must be already taken into account during product development. For these reasons, the project goal was the integrated, digital and simultaneous product and process development for modular, automation capable charging stations. The current working status of these two approaches is still at concept level. After reviewing and evaluating the relevant literature, uniform goals and the further procedure were defined as part of a joint kick-off event. The project partners Institute for Factory Automation and Production Systems (FAPS) and Compleo Charging Solutions are currently working on the technical aspects of product development, which will be used to define the framework for a scalable production process. On this basis, the production planning is carried out. The research project SUPPLy started in January 2020 and has a funding period of 36 months. In addition to the new development of a modular outlet module, the objectives include the conceptual design and commissioning of an assembly line developed especially for this purpose.

Acknowledgements The research and development project SUPPLy is funded under Grant No. 01MV19001B within the scope of the German "Elektro-Mobil" technology program run by the Federal Ministry for Economic Affairs and Energy and is managed by the German Aerospace Center in Cologne. We would like to thank our project partners Compleo Charging Solutions GmbH and FAPS for their valuable contributions.

References

1. German Federal Government: National Development Plan for Electric Mobility. https://www.bmu.de/en/download/the-federal-governments-national-development-plan-for-electric-mobility/ (2009). Accessed 11 Dec 2020
2. German National Platform for Electric Mobility (NPE): Progress Report 2018 - Market ramp-up phase. http://nationale-plattform-elektromobilitaet.de/en/the-npe/publications/ (2018). Accessed 11 Dec 2020
3. Federal Ministry of Transport and Digital Infrastructure (BMVI): The future of mobility is electric. https://www.bmvi.de/SharedDocs/EN/Dossier/Electric-Mobility-Sector/electric-mobility-sector.html (2019). Accessed 11 Dec 2020
4. Federal Ministry of Economic Affairs and Energy (BMWI): Electric mobility in Germany. https://www.bmwi.de/Redaktion/EN/Dossier/electric-mobility.html (2020). Accessed 11 Dec 2020
5. REFA: Grundlagen der Arbeitsgestaltung. Hanser, München (1993)
6. Rother, M., Harris, R.: Creating continuous flow. The Lean Enterprise Institute, Brookline, USA (2001)
7. Dietz, C., Richter, R., Deuse, J.: Variantenmanagement im Simultaneous Engineering. Ein Vorgehen zum Abgleich von Produktstruktur und Fertigungssystem. Zeitschrift für wirtschaftlichen Fabrikbetrieb (ZWF) **108**, 325–329 (2013)
8. Dekker, S.W.A., Woods, D.D.: MABA-MABA or Abracadabra? Progress on Human-Automation Co-ordination. Cogn. Technol. Work **4**, 240–244 (2002)
9. Baudzus, B., Krebs, M., Deuse, J.: Design of manual assembly systems focusing on required changeability. In: Dimitrov, D., Schutte, C. (eds.) Proceedings of International Conference on Competitive Manufacturing (COMA), Stellenbosch, South Africa, 30.01.-01.02.2013, pp. 269–275 (2013)
10. Hengstebeck, A., Weisner, K., Deuse, J., Roßmann, J., Kuhlenkötter, B.: Betriebliche Auswirkungen industrieller Servicerobotik am Beispiel der Kleinteilemontage. In: Wischmann, S. (ed.) Zukunft der Arbeit - Eine praxisnahe Betrachtung. Springer, Berlin (2017)
11. Deuse, J., Weisner, K., Hengstebeck, A., Busch, F.: Gestaltung von Produktionssystemen im Kontext von Industrie 4.0. In: Botthof, A., Hartmann, E.A. (eds.) Zukunft der Arbeit in Industrie 4.0, pp. 99–109. Springer, Berlin, Heidelberg (2015)
12. Ehrlenspiel, K.: Integrierte Produktentwicklung. Carl Hanser Verlag GmbH & Co. KG, München (2009)
13. Bullinger, H.-J. (ed.): Systematische Montageplanung. Hanser, München, Wien (1986)
14. Dove, R.: Agile Production: Design Principles for Highly Adaptable Systems. In: Maynard, H. B., Zandin, K.B. (eds.) Maynard's industrial engineering handbook. McGraw-Hill standard handbooks, 5th edn. McGraw-Hill, New York, USA (2001)
15. Federal Ministry of Economic Affairs and Energy (BMWI): Regulatory environment and incentives for using electric vehicles and developing a charging infrastructure. https://www.bmwi.de/Redaktion/EN/Artikel/Industry/regulatory-environment-and-incentives-for-using-electric-vehicles.html (2020). Accessed 11 Dec 2020

16. Krüger, J., Lien, T.K., Verl, A.: Cooperation of human and machines in assembly lines. CIRP Ann. **58**, 628–646 (2009)
17. Kock, S., Vittor, T., Matthias, B., Jerregard, H., Kallman, M., Lundberg, I., Mellander, R., Hedelind, M.: Robot concept for scalable, flexible assembly automation. In: IEEE International Symposium on Assembly and Manufacturing (ISAM), pp. 1–5, Tampere, Finland (2011)
18. Haddadin, S., Suppa, M., Fuchs, S., Bodenmüller, T., Albu-Schäffer, A., Hirzinger, G.: Towards the robotic co-worker. In: Siciliano, B., Khatib, O., Groen, F., Pradalier, C., Siegwart, R., Hirzinger, G. (eds.) Robotics Research, vol. 70, pp. 261–282. Springer, Berlin, Heidelberg (2011)
19. Mehrabi, M.G., Ulsoy, A.G., Koren, Y.: Reconfigurable manufacturing systems: Key to future manufacturing. J. Intell. Manuf. **11**, 403–419 (2000)
20. Wu, P.S., Tam, H.Y., Venuvinod, P.K.: Hybrid assembly: A strategy for expanding the role of "advanced" assembly technology. Comput. Electr. Eng. **22**, 109–122 (1996)
21. Hägele, M., Schaaf, W., Helms, E.: Robot Assistants at Manual Workplaces. Effective Co-operation and Safety Aspects, Stockholm, Sweden, 199–204 (2002)
22. Selevsek, N., Köhler, C.: Angepasste Planungssystematik für MRK-Systeme. ZWF **113**, 55–58 (2018)
23. Boothroyd, G.: Assembly automation and product design, 2nd edn. Manufacturing engineering and materials processing, vol. 66. Taylor & Francis, Boca Raton, USA (2005)
24. Lanza, G., Stähr, T., Sapin, S.: Planung einer Montagelinie mit skalierbarem Automatisierungsgrad. Zeitschrift für wirtschaftlichen Fabrikbetrieb (ZWF) **111**, 614–617 (2016)
25. Nyhuis, P., Heinen, T., Brieke, M.: Adequate and economic factory transformability and the effects on logistical performance. Int. J. Flex. Manuf. Syst. **19**, 286–307 (2007)
26. Wiendahl, H.-P., Hernández, R.: The transformable factory—strategies, methods and examples. In: Dashchenko, A.I. (ed.) Reconfigurable Manufacturing Systems and Transformable Factories, pp. 383–393. Springer, Berlin, Heidelberg (2006)
27. Ermer, A.-K., Seckelmann, T., Barthelmey, A., Lemmerz, K., Glogowski, P., Kuhlenkötter, B., Deuse, J.: A Quick-Check to Evaluate Assembly Systems' HRI Potential. In: Schüppstuhl, T., Tracht, K., Roßmann, J. (eds.) Tagungsband des 4. Kongresses Montage Handhabung Industrieroboter (MHI). 4. Fachkolloquium der Wissenschaftlichen Gesellschaft für Montage, Handhabung und Industrierobotik, Dortmund, 26. - 27.02.2019, 128 - 137. Springer Vieweg, Berlin (2019)
28. Cisek, R., Habicht, C., Neise, P.: Gestaltung wandlungsfähiger Produktionssysteme. Zeitschrift für wirtschaftlichen Fabrikbetrieb (ZWF) **97**, 441–445 (2002)

Machine Vision

Robot-Based Creation of Complete 3D Workpiece Models

David Singer, Dorian Rohner and Dominik Henrich

Abstract

A complete object database containing a model (representing geometric and texture information) of every possible workpiece is a common necessity e.g. for different object recognition or task planning approaches. The generation of these models is often a tedious process. In this paper we present a fully automated approach to tackle this problem by generating complete workpiece models using a robotic manipulator. A workpiece is recorded by a depth sensor from multiple views for one side, then turned, and captured from the other side. The resulting point clouds are merged into one complete model. Additionally, we represent the information provided by the object's texture using keypoints. We present a proof of concept and evaluate the precision of the final models. In the end we conclude the usefulness of our approach showing a precision of around 1 mm for the resulting models.

Keywords

3D models · Camera vision · Perception for grasping and manipulation · Robot-object interaction · Object recognition

D. Singer (✉) · D. Rohner · D. Henrich
Lehrstuhl für Angewandte Informatik III (Robotik und Eingebettete Systeme), Universität Bayreuth, 95440 Bayreuth, Germany
E-mail: david.harrer@uni-bayreuth.de
URL: https://robotics.uni-bayreuth.de

1   Introduction

In the last years, the usage of robotic manipulators expanded to small and medium sized enterprises (SME) as well as service environments. The tasks for the robot in these cases often depend on their surroundings (e.g. position of workpieces). Therefore, it is necessary for the robot to recognize its environment. Several object recognition methods use an object database, which stores models for all potential workpieces [1–3]. Depending on the underlying recognition method, different types of models are stored, e.g. coloured point clouds or CAD data. The necessary models are not always available in household or SME settings, due to the specialized or rare nature of the objects. Generating these models manually is often a tedious process or requires expert knowledge. Therefore, our goal is to generate these workpiece models in an automated process.

In order to utilize models for an object database, they usually have to fulfil the requirement of completeness, meaning every geometric feature (e.g. faces, edges, vertices or point based representations) and colour feature should be represented. Additionally, our goal is to develop and evaluate a fully automated concept, meaning no human interaction should be necessary in generating the models. As an additional assumption, we only consider rigid objects. To enable easy handling, we assume that the objects have at least one side, on which they can be placed for a stable stand on a horizontal planar surface.

In the following sections, we discuss the current state of the art and identify possible improvements. Based on this, we develop our overall approach and select a suitable model type for the object database. For a single object, we will discuss how the object is recorded from multiple views in different poses and how to generate the final model. We validate our approach and measure the quality of the models regarding their size compared to ground truth data. Finally, we discuss the contribution of this paper and give an outlook.

2   State of the Art

Existing work on this topic can be classified into two groups: On the one hand there are techniques that use additional and specialized hardware to create object models similar to a 3D-scanner, while on the other hand there are robot-based methods that do not use additional hardware except for the robot and sensors.

In the first group, there are approaches that are semi-autonomic but require dedicated hardware. The object is commonly placed on a rotary plate allowing the camera to record all sides of the object. Depending on the concept, multiple cameras on a rigid frame [4] or a single camera on a slide [5] are used to capture the object. Some systems use a robot mounted camera to capture images of the object on a rotary plate [6]. Techniques from this group do not meet our requirements. Firstly, the approaches are not fully automated, since the objects must be presented to the system one by one. Secondly, the generated models are not complete, since the bottom side of the object is not included in the models.

The second group of approaches for the generation of object models make use of robot-based systems. Unlike the techniques of the first group, no additional hardware beside a robot and a camera is necessary. To capture images from multiple views, the object is moved by the robot itself instead of a rotary plate. The object can also be moved indirectly, by placing it on a robot-mounted plate, which is then moved in front of the camera [7]. In another approach, the robot may grasp the object directly and show it to the camera in different poses [8, 9]. Finally, the object can be rotated in place by pushing it with the gripper of the robot, while a stationary sensor generates the model [10].

Some of the techniques based on the robot manipulating the object have strategies to pick up the workpiece by themselves and solve the first requirement regarding the full automation. This can be done if the robot is able to pick objects up by itself [8] and extract them from cluttered scenes [10]. However, approaches from this group still share the drawback, that the object models are not complete since either the bottom side of the object is missing [7, 10] or some parts of the workpiece are occluded by the gripper of the robot [8]. Approaches exist to handle this occlusion by detecting the gripper, meshing the models and performing hole closing algorithms on the meshes [9].

Overall we can conclude, that no approach exists, that generates complete workpiece models only using a robotic manipulator with a mounted depth camera. In this work, we propose a approach to generate complete 3D object models completely automated and without the usage of specialized hardware and occlusions based on the manipulator grasping the object. This allows us to keep the original point cloud based representation of the model.

3   Our Approach

The problem treated in this work can be described as follows: Our input is an unknown number of objects with unknown pose and shape, which we assume to stand on a known table surface. We use a lightweight robot with an eye-in-hand RGBD-camera to create an object database for object recognition and for retrieval of additional information about those objects.

3.1   Basic Approach

Our approach to the aforementioned problem is structured in several parts (see Fig. 1): In a first step, the camera is moved to a predefined position and captures an initial view on the scene. The captured point cloud is then segmented into pixels, representing the background and objects respectively groups of objects by utilizing the assumption of a known table surface. Individual objects respectively groups of objects can then be identified by applying connected components labeling [11] on the segmented point cloud. Following this, one of the objects is extracted from the cluttered scene and transferred to an examination area. To

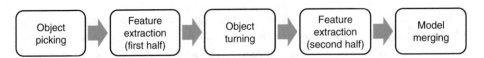

Fig. 1 Flowchart of our complete approach for one isolated object. The boxes depict the relevant parts of the software framework

extract the objects we use a technique loosely based on the one described in [12]. The basic idea is to try a grasp calculated on the segmented point cloud. If the grasp is successful, we transfer the object to the examination area. Otherwise, the manipulator is used to push the group of objects towards the centroid of the corresponding pile. Afterwards, we repeat the process of grasping and pushing, until at least one object is isolated from the pile. At the examination area, RGB-D images are taken from multiple fixed views to include all sides of the object. Those images are then combined into a so called *half-model* of the object. Afterwards, the object is picked up and turned around to allow for inspection of the previous, not visible bottom side. A second half-model is then generated from images of the turned object. The two half-models are then merged to a complete model of the object, including all of the objects sides.

As a model representation we chose a point cloud representation because with that the input data of the depth camera, which provides a point cloud representation of the scene, doesn't have to be converted. Additionally, by using the coloured point cloud as representation we do not loose any data from the measurement. This allow later creation of more abstract representations, e.g. a polygon-model via meshing or a CAD-model [13] with higher precision. The geometric information of the point cloud is extended by texture information. To reduce data we do not keep the images itself but use well known methods [1] to extract keypoints and their corresponding feature vectors.

3.2   Sensing

The main goal of this work is to generate complete object models in order to be able to recognize the modelled object from every possible direction. To generate such models we must record point clouds from different views to include all sides of the object. These point clouds have to be merged into one single object model. Due to insufficient calibration precision, noise, and imprecise pose estimation of the robot-mounted depth sensor there is an error in the point cloud poses that prevents us from simply merging the different point clouds based on the pose information acquired through the robots kinematic. A common solution to tackle this problem is the usage of the ICP-algorithm [14]. This algorithm geometrically fits the point clouds together to balance out those errors. The main advantage of this algorithm compared to e.g. keypoint based approaches is, that it uses all points from the overlap area and not just some significant ones, which benefits the accuracy since the data basis is much larger.

Fig. 2 An object with many symmetries and few edges and corners is placed on the calibration object (left). Multiple point clouds taken from different views can still be merged with the ICP algorithm due to the additional imposed edges and corners by the calibration object (right)

The drawback of the algorithm is that besides a starting guess also sufficient unambiguous edges and corners are needed, to function properly. To ensure the availability of such edges and corners even with round or symmetrical objects like the salt shaker in Fig. 2, we place the object on a special asymmetrical calibration body. This calibration body is designed to have many clearly identifiable edges and corners but no symmetry axes. Together with a starting guess from the camera pose the ICP algorithm seems suitable for this task. After the transformation is applied to all point clouds, we remove the known calibration object from the half model as it is not needed anymore.

3.3   Grasping and Turning

To be complete, the object model must include all sides of the object. The half-model generated in the previous section however is missing at least the underside of the object. To include the missing side the object has to be turned around. The idea is to grasp the object sideways, lift it up, turn it 180° and place it back on the calibration body to include the underside in the model. The problem with this approach is, that with relatively flat objects, grasping exactly horizontally is not possible due to the gripper geometry as seen in Fig. 3 (left side). As seen in Fig. 3 (right side) the object can not be placed evenly if it was not grasped horizontally. The solution here is to place the object at the edge of the table to allow for horizontal grasping. The resulting procedure is as follows: In a first step the object is transferred to the edge of the table surface. At this position a grasp is planned and executed under the constraint to grasp the object at vertical edges. The object is then lifted, turned by 180 degrees around the A-axis of the NSA-coordinate-system and placed back on the table. After successfully turning, the object is transferred back to the calibration body.

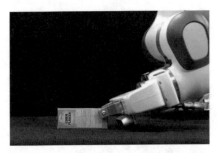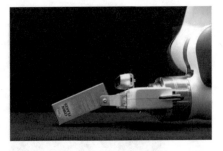

Fig. 3 The geometry of the robots hand (gripper and wrist) does not allow for horizontal grasping of objects lying flat on the tables surface (left). As a consequence the object can not be placed correctly after turning it around (right)

3.4   Merging of the Half-Clouds

After creation of the first half model and turning the object, we repeat the sensing process for the other half. We now have two partial models of the object which need to be fitted together. The geometrical approach used to create the half model from the different point clouds has two major drawbacks here: Firstly, there is no good starting guess available in this case. The object is turned by 180 degrees but we cannot be sure that it stays rotated by this angle. On the one hand, the object will move a bit while grasping and turning. On the other hand, the object may topple based on its shape. While a box-shaped object will likely stay as placed, a pyramid-shaped object will almost certainly topple in one direction. While the toppling of the object is not a problem for the whole process since the previously occluded side is still visible, we can not estimate a meaningful starting guess for the geometrical approach. Secondly, the two half models need not have common corners or edges to allow the ICP-algorithm to run. While this problem could be fixed in the aforementioned case, here it is not possible to use such a calibration object, since this would have to be rotated exactly as the object and would be required to topple in the same way.

An alternative to the geometrical approach are texture based methods such as e.g. SIFT [1], or SURF [2]. These methods are specifically suitable, since we have relatively big overlapping areas at the sides of most objects. The drawbacks with these methods is that they are vulnerable to symmetries and that they depend on overlapping sides. Thus it is not possible to process largely symmetrical and very flat objects like a ruler. In practice we used the SIFT keypoint detection algorithm to extract keypoints and to find correspondences and then used geometric consistency grouping [15] to calculate the correct transformation based on these keypoint correspondences. An example of the merging process can be seen in Fig. 4. Finally, we can apply the calculated transformation to obtain the resulting workpiece model. In the complete process no human interaction is necessary. The models should be complete, due to multiple views for each side, as well as the turning of the object to capture every part.

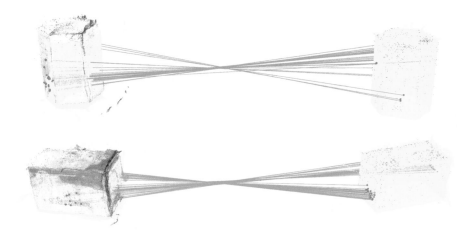

Fig. 4 Merging of two half-models of a pen box (top) and a tea box (bottom). One half-model is shown in original colours, while the other one as well as the resulting transformed half-model are shown in yellow. The keypoints are depicted in blue and correspondences as green lines, with their respective keypoints in red

4   Experimental Results

The following section describes the validation, where we present a proof-of-concept. Afterwards, we will evaluate the precision of the generated models in regard to the deviation from ground truth data. For our evaluation we used a Franka Emika Panda with a mounted Intel® Realsense™ D435.

4.1   Validation

To validate our approach we first generate a complete model as described above. The results of this process for one workpiece model can be seen in Fig. 5. The workpiece (a) is recorded in two different poses. For each pose multiple point clouds are merged using an inital pose estimation together with the ICP algorithm to provide reasonable half-clouds ((b) and (c)). These are merged afterwards into one complete model (d) using the keypoints to determine the transformation. In an optional step, this point cloud is transformed into a planar Boundary Representation model (BRep) [16] (e). This last step can also be utilized to eliminate noise (black voxels) from the used depth sensor (b–d).

We utilize these models in two scenarios as a proof-of-concept: On the one hand, we record a new point cloud and utilize SIFT to recognize the model in the point cloud. The SIFT keypoints are used to determine correspondences and with the underlying 3D information the 6 DOF transformation from the model to the point cloud is calculated. On the other hand,

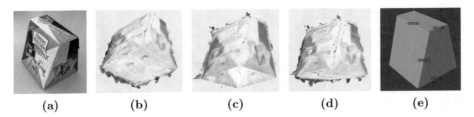

| (a)   (b)   (c)   (d)   (e) |

Fig. 5 The complete process of one recording: starting with a workpiece (**a**), we generate two point clouds from multiple views (**b**) + (**c**) and merge them into the complete model (**d**). Additionally, a BRep (**e**) can be generated

Fig. 6 The results of the proof of concept: the learned model is recognized correctly within a cluttered scene (left). Also, the generated models were transformed into a BRep, added to an object database, and recognized in a reconstructed scene (right)

we apply a surfaced based object recognition method [3] based on boundary representation models (BReps). Our new model is transformed into a BRep by utilizing [13] and added to the object database. Then we capture a scene as a BRep [13] with a Kuka LWR 4 and an Ensenso N10 and use the method from [3] for object recognition. In Fig. 6 the results regarding the object recognition are visible. In both scenarios the objects are classified correctly, which indicates, that the used colour and surface-based features are extracted correctly during the generation of the model. Furthermore, it shows the independence of the models from different depth sensors.

4.2   Evaluation

For our evaluation we generate models from multiple objects and transform them into planar BReps and measure their characteristic lengths (all edges to uniquely describe an objects geometry). For non-planar objects we determine these lengths by segmenting the recorded point cloud into single areas and measuring the length of them. Additionally, we determine the corresponding ground truth lengths, calculate the difference and determine the mean error. To generate these models, we use 20 views overall and turn the object once. This is

Table 1 The results of the evaluation regarding the geometry. Row (a) shows the test objects with their relevant edges in green, row (b) shows the **measured size of the model** and the ground truth. The following row (c) is the mean error. All values are given in [mm]

a)				
b)	**112.7** - 115.7 **48.8** - 48.8 **118.0** - 116.6 **116.2** - 113.4	**73.3** - 75.2 **159.7** - 157.0 **38.7** - 37.1 **37.0** - 36.4	**120.7** - 121.0 **48.0** - 46.6 **64.0** - 64.8	**120.2** - 120.3 **64.1** - 64.0 **76.0** - 76.6
c)	1.8	1.7	0.8	0.3

a)				
b)	**102.7** - 105.3 **101.5** - 102.7 **65.0** - 65.3	**100.0** - 101.0 **154.0** - 156.2	**62.8** - 63.8 **143.2** - 143.5	**54.5** - 55.1 **76.5** - 77.4
c)	1.4	1.6	0.7	0.8

used to evaluate the geometric correctness of the generated models. This measure is relevant due to the fact, that multiple objects have similar form and size (see Table 1). For a correct classification it is therefore necessary to have a precise workpiece model.

The results regarding the size can be seen in Table 1. For each object (first and fourth row) we measured the length of multiple edges (first five objects) or the height and the diameter (last three objects), as shown in the second and fifth row, left values for each object. The other value (right) is the ground truth length. The last row is the mean error. These values indicate multiple results: Overall, the error is around 1 mm with some variance depending on the object and the concrete length inspected. The maximum difference is 3 mm. The minimum is 0 mm, due to imprecision caused by rounding. Nevertheless, a difference lower than 1 mm occurs several times, which shows the precision of our approach. This result may be improved furthermore by using a depth-camera more suitable for close up views or with a higher resolution. Furthermore, the workpiece models are complete with the exception of minor holes in some of the models, which may be closed automatically by our work [17].

Overall we conclude the usefulness of our approach: The final models are precise enough to distinguish similar sized objects. The necessary colour and surface information is extracted and can be used to recognize the objects successfully in two different scenarios. The concept of the generation works autonomously without human interaction and the models are generally complete.

5   Conclusion

In this paper we presented an automated approach to generate complete workpiece models. We developed an overall process with the goal of creating complete models without any human interaction. To do this, a given pile of objects is separated into isolated workpieces by a robot manipulator. One half of each of these objects is recorded while the object is placed on a calibration object. The generated point clouds are merged by using the ICP algorithm with an initial pose estimation. Afterwards, the object is grasped and turned around. The other half is recorded as well. Both half-models are merged using SIFT keypoints. In a validation we showed the functionality of our approach and evaluated its precision.

Future work may include more sophisticated methods regarding the separation of objects, another feature representation or the consistent merging of keypoints. To reduce the amount of required memory it is possible to downsample the resulting models. Furthermore, the amount of necessary views for each half-model to generate a complete model should be evaluated. Additionally the view directions could be calculated by active vision methods and the trajectories should be optimized.

Acknowledgements This work has partly been supported by the Deutsche Forschungsgemeinschaft (DFG) under grant agreement He2696/21 SeLaVi.

References

1. Lowe, D.G.: Distinctive image features from scale-invariant keypoints. Int. J. Comput. Vis (2004)
2. Bay, H. et al.: SURF: speeded up robust features. Comput. Vis. – ECCV (2006)
3. Rohner, D., Henrich, D.: Object recognition for robotics based on planar reconstructed B-rep models. In: IEEE International Conference on Robotic Computing (2019)
4. Singh, A., et al.: Bigbird: a large-scale 3d database of object instances. In: IEEE International Conference on Robotics and Automation (2014)
5. Kasper, A., et al.: The KIT object models database: an object model database for object recognition, localization and manipulation in service robotics. Int. J. Robot. Res. (2012)
6. Banerjee, D., et al.: Robotic arm based 3D reconstruction test automation. IEEE Access (2018)
7. Wang, W., et al.: Textured/textureless object recognition and pose estimation using RGB-D image. J. R.-Time Image Process. (2013)
8. Venkataraman, A., et al.: Kinematically-informed interactive perception: robot-generated 3d models for classification (2019). arXiv:1901.05580
9. Krainin, M., et al.: Manipulator and object tracking for in-hand 3D object modeling. Int. J. Robot. Res. (2011)
10. Bevec, R., Ude, A.: Pushing and grasping for autonomous learning of object models with foveated vision. In: IEEE International Conference on Advanced Robotics (2015)

11. Wu, K., et al.: Optimizing two-pass connected-component labeling algorithms. Pattern Anal. Appl. (2009)
12. Katz. D., et al.: Clearing a pile of unknown objects using interactive perception. In: IEEE International Conference on Robotics and Automation (2013)
13. Sand, M., Henrich, D.: Incremental reconstruction of planar B-rep models from multiple point clouds. Vis. Comput. (2016)
14. Arun, K., et al.: Least-square fitting of two 3-D point sets. IEEE Trans. Pattern Anal. Mach. Intell. (1987)
15. Chen, H., Bhanu, B.: 3D free-form object recognition in range images using local surface patches. Pattern Recognit. Lett. (2004)
16. Sand, M., Henrich, D.: Matching and pose estimation of noisy. Partial and planar b-rep models. In: Proceedings of the Computer Graphics International (2017)
17. Rohner, D., et al.: User guidance and automatic completion for generating planar b-rep models. Annals of Scientific Society for Assembly, Handling and Industrial Robotics (2020)

Classification of Assembly Operations Using Recurrent Neural Networks

Björn Papenberg, Patrick Rückert and Kirsten Tracht ⓘ

Abstract

Visual sensor data of manual assembly operations offers rich information that can be extracted in order to analyze and digitalize the assembly. The worker's interaction with tools and objects, as well as the spatial–temporal nature of assembly operations, makes the recognition and classification of assembly operations a complex task. Therefore, classical methods of computer vision do not provide a sufficient solution. This paper presents a recurrent neural network for the classification of manual assembly operations using visual sensor data and addresses the question as to what extent such a solution is feasible in terms of robustness and reliability. Since complex assembly operations are a combination of basic movements, four main assembly operations of the Methods Time-Measurement base operations are classified using a machine learning approach. A dataset of these four assembly operations, reach, grasp, move and release, containing RGB-, infrared-, and depth-data is used. A Convolutional Neural Network—Long Short Term Memory architecture is investigated regarding its applicability due to the spatial–temporal nature of the data.

Keywords

Machine learning · Manual assembly · Image processing

B. Papenberg (✉) · P. Rückert · K. Tracht
Bremen Institute for Mechanical Engineering (Bime), University of Bremen, Badgasteiner Str. 1, 28359 Bremen, Germany
e-mail: papenberg@bime.de

T. Schüppstuhl et al. (eds.), *Annals of Scientific Society for Assembly, Handling and Industrial Robotics 2021*,
https://doi.org/10.1007/978-3-030-74032-0_25

1   Introduction

Globally acting, manufacturing companies must have efficient and versatile production structures in order to be able to meet customer demands for individually tailored products of high quality. These changing demands and shortened product lifecycles force modern production to be highly flexible [1]. This flexibility can be provided by human robot collaboration, by assigning exhausting and non-ergonomic tasks to the robot [2].

In order to enable such collaborations, it is of utmost importance to implement a visual sensor system that allows the robot to classify and analyze the assembly steps conducted by the human in real time. Industrial collaborative robots are able to detect collisions with humans or other objects and execute an emergency stop as soon as a collision is detected. While this safety mechanism forms the baseline for a safe collaboration it still requires the human to observe the robot at all times to avoid collisions, therefore burdening the human with an additional assignment. Thus, true collaboration along with a decrease in stress levels for humans can only be achieve by transferring that burden to the robot.

In this context, the presented paper addresses the question if and how assembly operations can be classified using methods of machine learning based on visual sensor data. Existing solutions generally do not classify assembly operations directly. Instead, these methods track activities in certain regions of the assembly area to draw conclusions regarding the assembly process [3, 4]. A key challenge in the classification of assembly operations using methods of machine learning is the interaction with tools and objects. Due to this interaction, conventional implementations fail, since this interaction is not represented in the used datasets [5]. In order to classify manual assembly operations efficiently, it is explored how they can be delimited and defined to enable a meaningful classification.

The assembly station and dataset proposed by Rückert et al. [6] is used for the implementation of the algorithm. Finally, the results are evaluated and interpreted in order to clarify the implementations strengths and weaknesses.

This paper contributes an outlook on the possibilities of classifying manual assembly operations using recurrent neural networks and an extension of the algorithm proposed by Rückert et al. [6].

2   Classification of Manual Assembly Operations

Existing implementations often rely on the assumption that the product is being assembled when the hands of the worker reside inside a predefined area in the work space [3, 4]. Other approaches draw bounding boxes around hands and objects and assume that the objects were interacted with, when the bounding boxes intersect [7].

While these implementations do not classify actual assembly operations, they offer insights on the challenges of the detection and classification of manual assembly operations. Therefore, it is evident that manual assembly operations need to be analyzed in detail.

The two most widely used methods to analyze manual assembly operations in such detail are REFA and Methods-Time Measurement (MTM) [8]. The time data collection according to REFA measures the actual time needed to perform certain assembly tasks and calculates the target times based on this information Verband and für Arbeitsstudien und Betriebsorganisation e.V.: Methodenlehre des Arbeitsstudiums: Teil 2Datenermittlung, Hanser, MünchenVerband and für Arbeitsstudien und Betriebsorganisation e.V.: Methodenlehre des Arbeitsstudiums: Teil 2Datenermittlung, Hanser, München [9]. Method-Time Measurement divides every manual assembly operation into several basic movements. For the finger-, hard-, and arm-system, these movements are reach, grasp, move, position and release [8]. These basic movements are defined beforehand an can be combined to represent 85% of all manual assembly operations [11].

Reaching is the movement of the fingers or hands to a specific place. Grasping involves bringing an object under control by closing the fingers. Moving an object from its initial position to its new destination describes the basic operation move. The operation position involves bringing an object into its final position at the end of its transport path. Releasing is defined as lifting control of an object by opening the fingers [10].

During the basic operations move and reach, the hand and arm are in motion. This is not the case with grasp and release, since these assembly operations primarily concern the motion of the fingers. Therefore, it is to be expected that the classification algorithm will mainly misclassify the motions move and reach as well as grasp and release, which results in a chess-like pattern of the confusion matrix. The key difference between reaching and bringing is the presence of an object in the hand. Unlike grasping, releasing is characterized by removing the fingers from an object.

Since MTM enables the detection of different motion sequences by identifying its typical characteristics and offers the prospect of deriving weaknesses in the work processes, it is more suited for the underlying approach than REFA Deuse et al. [8].

3 Machine Learning Algorithms for Activity Recognition

The recognition of hand gestures made significant breakthroughs in the last years due to methods of machine learning. While static gestures can be easily classified without machine learning, dynamic gestures prove to be more challenging, since they contain spatial–temporal information.

Recurrent Neural Networks (RNN), due to their ability to save states and thus handle spatial–temporal information, are suited for tasks where this property is needed [12]. This is enabled by the special neural connections of RNN architectures. Compared to normal feed-forward networks, where one neuron can only be connected to neurons in the next layer, a recurrent neuron can be connected to itself, neurons in the same layer and neurons in the next and previous layer, thus enabling it to store states, providing a vast number of trainable parameters [13]. Vanilla RNN-architectures are prone to the vanishing or exploding gradient problems, which are caused by the temporal links between the different time steps [13]. Long Short-Term Memory (LSTM) is a RNN architecture that solves the vanishing and exploding gradient problem and is used in applications where spatial–temporal information has to be processed [12].

Molchanov et al. proposed a sequential architecture for the dynamic classification of hand gestures, where short video sequences were fed into a 3-Dimensional-Convolutional Neural Network (3D-CNN), extracting spatial temporal features. Afterwards, features were fed into an RNN where the spatial temporal information was processed. Multiple modalities were used to feed the network individually and combined, so that the results could be compared. With a precision of 80.3%, the depth-data provided the best individual result, while the combination of all modalities results in a precision of 83%. The robustness of the depth-data towards changing illumination might be the reason for the superior results, as suggested by the authors [14].

Lai et al. fused a CNN and RNN architecture on different architecture levels and reported an overall accuracy of 85.46% using skeleton-, and depth-data [15].

Since the recognition of hand gestures is a much more common problem than the recognizing and classifying assembly operations, considerably more datasets are available on this topic. While gesture recognition and assembly recognition are closely related, the classification of assembly operations is much more challenging, since the interaction with objects and tools might occlude parts of the hand and vice versa. Additionally, video sensors for assembly operations have to be mounted above the worker as to not hinder the assembly process. Thus recording the image sequences from a bird view perspective is necessary, thereby hindering the effective use of skeleton recognition.

4   Activity Recognition in Manual Assembly

While papers regarding the field of the recognition of manual assembly operations often focus on detecting objects and hands in certain areas inside the workspace, they generally do not detect individual assembly operations. Nonetheless, these papers offer insights regarding the challenges of the recognition and classification of assembly operations.

One challenging aspect of the recognition of manual assembly operations is processing the classification in real time. Root et al. detect the hands using the You-Only-Look-Once-v3 algorithm and surround them by a bounding box and subsequently calculates their positions [5]. They reported an accuracy of up to 89% but could not reproduce this result in a real assembly scenario due to the interaction with tools and objects [5].

Liu et al. investigated the feasibility in classifying manual assembly operations in a real production environment. In this context, they elaborated three key points that need to be addressed. They concluded that individual characteristics of each worker hinder a precise classification of assembly operations and that the visual sensors must not interfere with the assembly process. Furthermore, they mention the financial costs of the sensors. The suggested algorithm distinguishes seven different movements found in assembly scenarios. The most important ones are classified as general assembly, reaching for assembly pieces and the recognition that no assembly takes place at specific moments in time. For the classification, a sequential architecture is proposed.

The hands are detected and their trajectory is fed into a 3D-CNN, which classifies the assembly operation with an accuracy of 89.1% Liu et al. [4].

Petruck et al. pursue a similar approach, using nine classes including general assembly and reaching for tools and objects. The movements during assembly are tracked using markers for a visual camera system, generating a numerical 27-dimensional vector, which contains the position and velocity and is fed into a CNN-architecture [3].

Andrainakos et al. propose a system that detects different objects and the hands of the assembler, drawing bounding boxes around them. Once the bounding box of an object and a hand intersect by a certain amount, the object is considered to be grabbed.

The water pump, which is used as an example, must be assembled from top to bottom. Whenever one detected object is correctly stacked on top of the other, the assembly is classified as correct [7].

5   Network Structure and Implementation

5.1   Dataset

For the experimental part of this paper, the dataset presented by Rückert et al. [6] was used. It consists of 2100 assembly operation sequences. The MTM basic operations reach, grasp, move and release are equally represented. The depth-data which was used for the training of the neural networks is stored in a resolution of 480×270 pixel [6]. While this dataset is small in comparison to other machine learning datasets, it is able to form the basis for a proof of concept.

5.2   Neural Network Architecture

During the training runs conducted in this paper, the depth-data was used since it highlights changes in the height of the hands and objects relative to the table. The images were scaled down to 50% of their original size in order to reduce the number or parameters in the neural network and to counteract overfitting. The initial learning rate of the Adaptive Moment Estimation (*adam*)-optimizer was set at 0.0001. Whenever the validation loss did not decrease during the training runs for five epochs, the learning rate was further decreased by a factor of five, until a minimum value of 10^{-6}. This step helps the loss converge.

The architecture of the neural network is shown in Table 1. The input layer contains no trainable parameters. As the name suggests its purpose is to bundle the input and forward it to the next layer. The input is a five-dimensional tensor. The first dimension represents

Table 1 Architecture of the neural network

layer (type)	Rows	Columns	Dimensions	Parameters
Input	106	128	32	0
ConvLSTM2D	106	128	32	40,448
Batch normalization	106	128	32	128
Max pooling 3D	53	128	32	20
ConvLSTM2D	53	64	8	11,552
Batch normalization	53	64	8	32
Dense	53	64	8	72
Flatten	1	27,136	27,136	0
Dense	1	1	4	108,548

the number of samples fed into the neural network during a discrete training step, also known as the batch size. The second dimension represents the number of individual images per sequence. Dimension three and four represent the height and width of the input frames. The last dimension represents the number of color channels. The Convolutional LSTM 2 Dimensions (ConvLSTM2D)-layer that was proposed by Shi et al. [16] extends a LSTM to have a convolutional structure, therefore enabling it to process the spatial temporal data of image sequences. The activation function of the second layer is hyperbolic tangent, while the output dimensionality is 64. In order to reduce overfitting, 50% of the neurons are dropped out at random during every epoch. A batch normalization is performed in the third layer. Thereafter a three dimensional max pooling operation is performed, reducing the height and width of the initial image in half, while the time dimension remains unchanged. Another ConvLSTM2D-layer, with a dropout of 50%, is applied after max pooling. This layer has eight output dimensions. It has to be noted that the output of this layer is a four-dimensional tensor, since the individual images are no longer represented at this point. Thereafter another batch normalization is performed. Afterwards follows the first densely connected neural layer with eight output dimensions. The second to last layer flattens the neural network to a series of individual neurons, which are then fed to the last densely connected layer. This last layer can be interpreted as a 4×1 vector, where each column represents one of the possible classes for classification. The activation function of the last layer is the SoftMax function. The neural network has 160.780 parameters in total.

5.3 Implementation

For the training of the neural network an NVIDIA RTX 2060 graphics card is used. The code was implemented in python, using the TensorFlow 2.3 framework [17].

To evaluate the model, a variant of a stratified tenfold cross validation was conducted. At first, the dataset was shuffled to randomly distribute the assembly operations. Afterwards, 20% of the dataset was reserved for the final evaluation. This test dataset was stratified so that the assembly operations were distributed evenly. The remaining 80% were divided into ten splits. Therefore, ten individual training runs were conducted. During these runs, nine splits were used for training and the remaining split was used for validation. This procedure was chosen so that a potential overfitting of the model can be evaluated. After each training run, another split was used for validation, until eventually all ten runs were completed. Each run was conducted for 30 epochs.

6 Results

The loss and accuracy of the training and validation data are depicted in Figs. 1 and 2. The training losses decay exponentially and converge at values from 0.25 to nearly zero. Because of this decrease, the training accuracy increases significantly in the first few epochs before converging near 100%. Up until the third epoch the accuracies and losses of the training and validation dataset rise and decay in unison.

The validation accuracy converges at about 60% after the fifth epoch. In the first few epochs, the validation losses generally decrease while still being very volatile. The validation loss of one of the runs even increases at this stage. While this particular run seems to be an outlier, the general phenomenon occurs due to the initial learning rate. At first, the learning rate is set at a value of 0.0001 in order to escape local minima of the loss function, which leads to volatile validation losses.

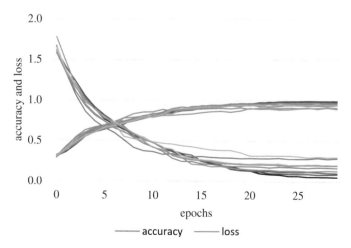

Fig. 1 Accuracy and loss of the training datasets converge to values near 1 and 0.25 to 0 for the ten different training runs

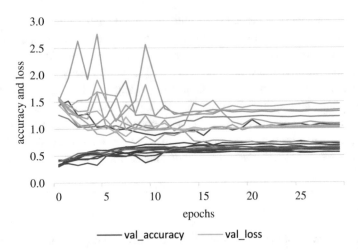

Fig. 2 Accuracy and loss of the validation datasets converge around 0.6 and 1 to 1.5 for the ten different training runs

The first reduction of the learning rate, which takes place between the tenth and eleventh epoch, significantly reduces the volatility of the validation losses. The validation loss converges at a value of 1 to 1.5 after 15 epochs. While the model does not overfit, it stops improving after the first 15 epochs.

The evaluation of the test dataset confirms the results of the validation dataset. The average accuracy is 59.6% with a standard deviation of 3.37% as shown in Table 2. The best training run has an accuracy of 64.3% while the worst run has an accuracy of 52.1%.

The confusion matrix of the average training run is shown in Fig. 3. The networks rarely predict reach/move to be grasp/release and vice versa, since those motions are fundamentally different from another. While reach/move has a dynamic course of movement, the assembly operations grasp/release show static characteristics. Although the distinction works most of the time in the case of reach/move, it does not work as well for grasp/release. Reach and move differ in the position of the tool or object after the assembly operation is finished. The difference between grasp and release is the inverted movement.

Since the algorithm is able to make a relatively clear distinction between the assembly operations reach and move, it can be assumed that it is able to implicitly detect objects. As

Table 2 Overview of the accuracy of the different training runs on the test dataset

No	1	2	3	4	5	6	7	8	9	10
Accuracy in %	64.3	61.0	61.0	56.7	59.5	63.3	52.1	59.0	61.7	57.4

Mean accuracy: 59.6%
Standard deviation: 3.37%

Fig. 3 Confusion matrix of the average training run. Correctly classified assembly operations are depicted on the main diagonal

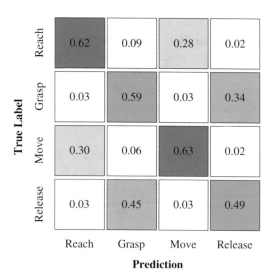

the range of motion differs greatly between reach/move and grasp/release, it is deducted that the recurrent part of the neural network is able to process the spatial–temporal information contained in the image sequences very well.

7 Conclusion and Outlook

The recognition of assembly operations in an industrial manufacturing scenario is a challenging task due to the interaction with objects and tools as well as the constraints of manual assembly work stations. Due to these challenges, many approaches do not track individual assembly operations. Instead, they define classes like "general assembly" to circumvent the issue of complexity. This paper aimed to directly classify the assembly operations reach, grasp, move and release of the MTM-1 basic operations.

The dataset used in this approach consists of 2100 image sequences, which were fed into a RNN to process spatial temporal information. The average accuracy of the ten runs that were performed during the tenfold stratified cross validation is 59.6%. Most of the wrongful predictions are reach/move and grasp/release.

In order to achieve significant improvements in the training results, the dataset used must be enlarged, which allows the use of more complex neural networks. In addition, more people need to be involved in data collection so that individual characteristics have a smaller influence on the dataset and the task of classifying assembly operations can be better generalized.

Assembly operations are an interaction between the mechanic's hands and the tools or objects. The recognition and classification of these open up great potential for improvement. By adding this feature, conclusions about the assembly process can be drawn based on the positions and movement patterns of the hands and objects, which can be directly processed by the neural network in order to significantly boost its performance. Therefore, this enhanced architecture is suited to decrease the wrong predictions for the case reach/move, since the position of the objects is detected by a sub-architecture of the neural network. Complementary to this, detecting the position of the hands as well as their orientation and trajectory results in a better performance regarding the case grasp/release, since more individual characteristics of the movements can be identified.

Acknowledgements The modular assembly system "Experimentelle Modulare Montageanlage (EMMA)", which was used in recording the data, was funded by the German Research Association (DFG).

References

1. Landherr, M.H.: Integrierte Produkt- und Montagekonfiguration für die variantenreiche Serienfertigung. Fraunhofer-Verlag, Stuttgart (2014)
2. Schröter, D.: Entwicklung einer Methodik zur Planung von Arbeitssystemen in Mensch-Roboter-Kooperation. Fraunhofer-Verlag, Stuttgart (2018)
3. Petruck, H., Mertens, A.: Using convolutional neural networks for assembly activity recognition in robot assisted manual production, In: M. Kurosu (eds), Human-Computer Interaction. Interaction in Context, LNCS, vol. 10902, pp. 381–397. Springer International Publishing, Cham (2018)
4. Liu, L., Liu, Y., Zhang, J.: Learning-based hand motion capture and understanding in assembly process. IEEE Trans. Industr. Electron. **66**(12), 9703–9712 (2019)
5. Root, M., Jauch, C.: Challenges of designing hand recognition for a manual assembly assistance system, In: Multimodal Sensing: Technologies and Applications, PROC SPIE, Munich (2019)
6. Rückert, P., Papenberg, B., Tracht, K.: Classification of assembly operations using machine learning algorithms based on visual censor data, In: 8th CIRP Conference of Assembly Technology and Systems, Procedia CIRP, Athens (2020)
7. Andrianakos, G., Dimitropoulos, N., Michalos, G., Makris, S.: An approach for monitoring the execution of human based assembly operations using machine learning. Procedia CIRP **86**, 198–203 (2019)
8. Deuse, J., Stankiewicz, L., Zwinkau, R., Weichert, F.: Automatic generation of methods-time measurement analyses for assembly tasks from motion capture data using convolutional neuronal networks—A proof of concept. In: Nunes, I.L. (ed.) Advances in Human Factors and Systems Interaction, pp. 141–150. Springer International Publishing, Cham (2020)
9. REFA Verband für Arbeitsstudien und Betriebsorganisation e.V.: Methodenlehre des Arbeitsstudiums: Teil 2 Datenermittlung, Hanser, München (1978)
10. Bokranz, R., Landau, K.: Handbuch industrial engineering: Produktivitätsmanagement mit MTM. Band 1: Konzept, 2nd ed., Schäffer-Poeschel, Stuttgart, (2012)
11. Deuse, J., Busch, F.: Zeitwirtschaft in der Montage, In: B. Lotter, H.-P. Wiendahl (Eds.): Montage in der industriellen Produktion, pp. 79–107. Springer Berlin Heidelberg (2012)
12. Hochreiter, S., Schmidhuber, J.: Long short-term memory. Neural Comput. **9**, 1735–1780 (1997)
13. Buduma, N., Locascio, N.: Fundamentals of deep learning: designing next-generation machine intelligence algorithms. O'Reilly, Sebastopol, CA (2017)

14. Molchanov, P., Yang, X., Gupta, S., Kim, K., Tyree, S., Kautz, J.: Online detection and classification of dynamic hand gestures with recurrent 3D convolutional neural networks, In: 2016 IEEE Conference on Computer Vision and Pattern Recognition (CVPR), pp. 4207–4215. Las Vegas, NV (2016)

15. Kenneth, L., Yanushkevich, S.N.: CNN+RNN depth and skeleton based dynamic hand gesture recognition, In: 24th International Conference on Pattern Recognition (ICPR), IEEE, pp. 3451–3456, Beijing (2018)

16. Shi, X., Chen, Z., Wang, H., Yeung, D.Y.: Convolutional LSTM network: a machine learning approach for precipitation nowcasting, In: 28th International Conference on Neural Information Processing Systems, vol. 1, (NIPS'15). MIT Press, pp. 802–810. Cambridge, MA (2015)

17. M. Abadi, P. Barham, J. Chen, Z. Chen, A. Davis, J. Dean, M. Devin et al.: TensorFlow: A system for large-scale machine learning. In: 12th USENIX Symposium on Operating Systems Design and Implementation, USENIX, pp. 265–283. Savannah, GA (2016)

Configuration and Enablement of Vision Sensor Solutions Through a Combined Simulation Based Process Chain

Johann Gierecker⬤, Daniel Schoepflin⬤, Ole Schmedemann⬤
and Thorsten Schüppstuhl⬤

Abstract

Machine vision solutions can perform within a wide range of applications and are commonly used to verify the operation of production systems. They offer the potential to automatically record assembly states and derive information, but simultaneously require a high effort of planning, configuration and implementation. This generally leads to an iterative, expert based implementation with long process times and sets major barriers for many companies. Furthermore the implementation is task specific and needs to be repeated with every variation of product, environment or process. Therefore a novel concept of a simulation-based process chain for both—configuration and enablement—of machine vision systems is presented in this paper. It combines related work of sensor planning algorithms with new methods of training data generation and detailed task specific analysis for assembly applications.

Keywords

Synthetic data • Machine vision • Sensor planning • Object recognition

J. Gierecker (✉) · D. Schoepflin · O. Schmedemann · T. Schüppstuhl
Hamburg University of Technology, Institute for Aircraft Production Technology, Denickestraße 17, 21073 Hamburg, Germany
E-mail: johann.gierecker@tuhh.de
URL: https://www.tuhh.de/ifpt

© The Author(s) 2022
T. Schüppstuhl et al. (eds.), *Annals of Scientific Society for Assembly,
Handling and Industrial Robotics 2021,*
https://doi.org/10.1007/978-3-030-74032-0_26

1   Introduction

Increasing complexity and variety of production processes enhances the demand for process control systems to reduce downtime, guarantee sufficient quality and avoid rejects. Where manufacturing processes are frequently automated and therefore already have a high density of information, aircraft assembly is mainly performed manually. For this reason, the feedback of information and assembly progresses is typically done by the worker. In this environment, optical sensor systems offer the possibility to automatically record assembly states and derive the required information without intervening the actual working process. The implementation of those vision sensor solutions can be a challenging and time consuming task. Influences from different fields, such as inspection task, hardware, image processing algorithms and optics, have to be considered which either makes it an iterative trial-and-error process or requires experienced engineers. Supporting or even automating this process through an appropriate software pipeline would provide a great advantage as it could reduce configuration and commissioning time and increase the use of machine vision systems.

Such sensor planning tools are well established for mechanical inspection procedures where inspection sequences are automatically generated from CAD models. This is different for the process of designing a machine vision system, as both—the mechanical setup and the generation of a machine vision program—change task-specifically and correlate with each other. Consequently there is a need for a task-oriented description of objects and environment to ensure a successful sensor planning process.

In a novel approach we aim to re-use the 3D models and semantic task descriptions acquired during this planning phase to generate synthetic AI training data and to enable the vision system application. Creating task-specific AI training data is often necessary as industrial applications feature highly individualistic objects and environments and can hardly be generalized. Since handcrafting data is widely considered a costly and tedious approach, generating synthetic data is becoming increasingly popular. Generating data, however does not alleviate the necessity for domain experts [4, 10] and in turn causes high efforts. However, the necessary analysis and semantic descriptions have high similarity to the ones used for the sensor planing. Re-using this process chain to enable the already planned sensor set-up can benefit the use of vision systems with AI applications.

We therefore aim to introduce a combined pipeline for planning and enabling of a vision system and application. Our conceptual work contributes the following to the fields of vision sensor planning and synthetic AI training data generation in the assembly domain:

- Derivation of a assembly feature and task analysis with respect to a possible visual process inspection
- Formulation of a semantic task description and scene grammar
- Introduction of a sensor planning framework for the calculation of viewpoints based on that task description

- Introduction of a data generation pipeline based on that grammar, for generation of training data with the goal of object recognition.

2   Related Work

This section presents related work in vision sensor planning with regard to industrial applications followed by the recent developments in the generation of training data.

2.1   Vision Sensor Planning

The topic of vision sensor planning is an ongoing research for many years in fields of surface inspection [8], active robot vision [19] or public surveillance [11]. Sensor Planning in industrial application is most commonly classified by the available knowledge about the scene. This divides the publications into *Scene Reconstruction*, *Model-Based Object Recognition* and last *Scene Coverage*, which includes our use case and requires detailed knowledge about objects, positions and environment [12]. Tarabanis [14] published a survey to categorize work in the field of sensor planning for *Scene Coverage Problems* in which even current research can still be classified.

Generate-and-test approaches [9] generate sensor configurations by equally dividing the solution space and evaluating the single configurations based on the task requirements. Cowan [3] shows with a *synthesis* approach that a configuration can be generated by an analytic description of inspection task, sensor parameters and several feature detectability constraints such as visibility, concealment, perspective, field-of-view, resolution or depth-of-field. *Expert systems* [2] describe databases which contain information about successful implemented viewing and illumination system, expressed in several rules to support the user while planning his configuration. The last category is *Sensor Simulation* where the scene is visualized within a framework to render sensor-real data based on configuration generated with either of the presented methods. All these different approaches try to find a set of viewpoints from which a maximized set of feature points can be detected. These feature points highly differ between every task. Where for a use case like object reconstruction the set can be a discrete description of the entire surface [9], the assembly inspection requires a task-specific analysis of the relevant features as it is not purposed to have a visibility of the complete object.

However most works regarding sensor planning assume, that the modeled sensor poses will exactly be executed. Manual influences or deviations in positioning systems often result in pose errors. Scott [18] introduces pose errors in the sensor planning process and suggests methods to minimize these. As we manually transfer the calculated sensor poses into our test setup, the pose error problem is relevant for our use case.

2.2   Training Data Generation

Following the planning phase, the vision application has to be implemented. As the use of AI based solution increase in this field, the need for appropriate training data has risen alike. Due to the time-consuming and expensive nature of manual data acquisition processes, use of synthetic data that is rendered out of 3D models has gained in popularity in recent years [5]. Successful training with synthetic training data was achieved in the fields of autonomous driving [7], picking [16] or identification of household objects [5]. Other industrial applications [15] have used CAD data to utilize Reinforcement Learning of a robotic grasping trajectory. Similar tasks were solved by [1, 6]. The insertion of pegs through object identification trained on synthetic data was shown by [17]. Synthetic Data enabled object recognition of assembly related objects like screws and the like was shown by [20]. These datasets are publicly available, yet not necessarily transferable to every industrial vision system and task, due to unknown environments and objects. Therefore, when designing a vision based application, often new data generation pipelines have to be created alongside.

The creation of such data generation pipelines is in need of defining what is to be displayed, a proper scene grammar [10], and implementation in a toolbox. Later can be provided by data synthesizing tools like NDDS[1] for Unreal Engine, SynthDet[2] for Unity and similar for Blender,[3] the semantic definition of a grammar however has to be done by the user for each problem variation he wants to train. Supporting the user in this process is not sufficiently addressed in recent approaches, but could be an important element in widespread use of AI based vision systems.

Where the presented work in sensor planning lacks a task individual feature definition of what has to be visible, the process of training data generation requires similar information about the object and environment. Consequently we display a simulation based process chain with a combined task semantic for both fields. Furthermore the introduced problem of pose errors can be handled with the use of an AI solution where the calculated pose is part of the variation parameters for training data generation.

3   System Overview

The aim of our work is to combine the presented fields of sensor planning and training data generation and to complete the process chain with a task analysis for the field of assembly inspection. Figure 1 shows the proposed system pipeline which individual steps will be explained in the following sections.

---

[1]https://github.com/NVIDIA/Dataset_Synthesizer, September 2020.
[2]https://github.com/Unity-Technologies/SynthDet, September 2020.
[3]https://github.com/921kiyo/3d-dl, September 2020.

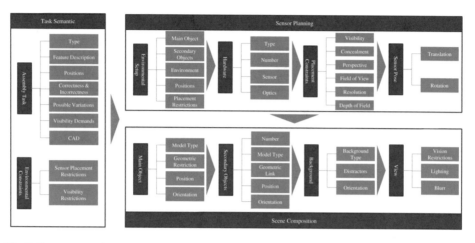

Fig. 1 System overview

3.1   Assembly Feature and Task Analysis

The inspection task of assembly verification is in this work proposed as a problem in object detection, where it is assumed that algorithms need to make decision based on a 2D-image of assembly joints. To verify a robust detection as a result of a automated system configuration, the task has to be analyzed before sensor planning process starts. The resulting task semantic (see Fig. 1) describes all information, which is required by the sensor planning and scene composition. Assembly tasks can differ from positioning of single or multiple objects over connecting them via screws or rivets to welding or soldering. Furthermore each category itself differs in its specifications depending on the geometry. Therefore we describe the parameters, which characterize a successful assembly task and convert them in geometrical features. Figure 2 shows exemplarily how the relevant features (marked red) differ in the category of a bolt connection. To detect a hexagon socket srew, features inside of the head have to be visible from the sensor view, whereas a hexagon cap has its characterizing features on the outside. Depending on the connection type there are additional parts (e.g. washer or nut) on the underside, which may not be visible from the same viewpoint as the features on the topside.

Beside those visibility demands there may also be visibility restrictions inside the environment. These are areas or objects (e.g. humans) which must not be visible within the sensor data. Sensor restrictions are positions where a mounting of sensors is restricted due to interference of the assembly process. Together with CAD-data the scene can be modeled within the sensor planning and scene composition processes.

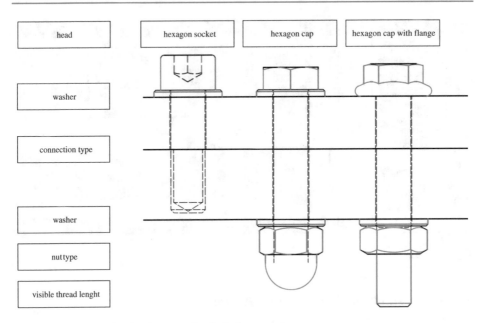

Fig. 2 Feature extraction by the example of a bolt connection

3.2   Calculation of Possible Viewpoints

Calculation of viewpoints and final pose selection as part of the sensor planning process uses mathematical descriptions of feature detectability constraints (see Sect. 2.1) together with sensor parameters and 3D-scene to generates possible poses which satisfy those constraints. Where relevant features for visibility demands are formulated within the task semantic, other optional features can be extracted from the STL-CAD-file. The description of surfaces as triangles allows us to simply generate cell center points and the relating surface normals all over the object. The relationship between the optical axis of the camera and those center points or normals is the base for most features (e.g. visibility, concealment and perspective) within the calculation process.

Where the visibility demands must be fulfilled, the results of the calculations with optional features are important for the rating of viewpoint candidates. The amount of features that satisfy the feature detectability constraints allow us to quantify the *degree of visibility* of the assembly joint. Start poses for sensor planning result from the positioning restrictions of the task semantic. After the calculation of every combination of sensor pose and feature, a final pose is selected based an the amount of features, which satisfy all constraints viewed from this pose. To avoid the expected pose error problems, this final viewpoint has to be slightly varied in the following process of training data generation.

3.3   Scene Composition for Training Data Generation

The rules and formulations describing the possible compositions of a scene can be referred to as the grammar of that scene. In order to implement a pipeline that creates scene variations according to such a grammar a parametrization has to be defined. We will first focus on the more generalized scene grammar and derive the parameter space in the later presented Use-Case.

We define the goal of the training data to enable an assembly process supervision application. Through object recognition this application should state whether the object in the sensor view is assembled correctly or not. However, this is not viable for every assembly type. We discard cases for which measurements e.g. for a slit have to be taken, to determine whether the assembly was done successfully. Our focus is the presence of certain objects e.g. screws or larger components in a view. Through the aforementioned sensor planing, appropriate view poses for detection of the objects within the view are defined.

We distinguish between the main object which is assembled and the secondary objects which are assembled to the main object, which is to be detected. The necessary parameters to define are shown in Fig. 1. The to be rendered scene is built around the main object, whereas the position and the orientation may change with respect to the sensors field of view. According to the type of assembly done, the secondary objects are placed.

4   Use Case

For reasons of confidentiality the suitable use cases from the project can not be used for the presentation. Therefore we select a tool for scarfing of CFRP structures in aircraft MRO [13] as our main object for the validation of the presented process chain. The relevant joint is a bolt connection, which consists of a hexagon socket srew (M12x35mm) with two washers and a nut. These parts represent the secondary objects. A sensor frame is placed around the object where the camera can be flexibly mounted. The available machine vision hardware is a IDS uEye camera with a 5MP sensor[4] and a Schneider-Kreuznach lens with a focal length of 12 mm.[5]

Correct assembly is considered when all objects and the correct type of srew is mounted. A correct srew can be expressed by the geometry of its head and its threat length which is visible underneath the nut. This results in visibility demands for the inner lines of the hexagon socket and the thread, which can be described as linear and cylindrical features. The smallest feature of interest is the thickness of the washer, which is has to be considered for the calculation of the resolution constraint.

---

[4]https://de.ids-imaging.com/store/ui-5280cp-rev-2.html, September 2020.
[5]https://schneiderkreuznach.com/en/industrial-optics/lenses-2-3-c-mount-3-5-mp,   September 2020.

 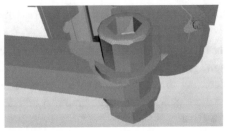

(a) View frustum (green lines) from the final viewpoint

(b) Field-of-view (blue areas) and visibility mapped on the objects

Fig. 3 Visualization of the final viewpoint

Table 1 Parametrization and variation of the synthetic training data generation

Object	Parameter	Variation
Main object	Position	10% Translation variation
	Orientation	Rotation z: 10°
	Colour/texture	As CAD model
Secondary object	Screw length	30–60 mm
	Position	Mounted, Thread showing
	Screw nut	Off–on
	2 x washer	Off–on
Background		13 variations
Underground		9 variations
Light	Type	Sun, spot, point, plane
	Intensity	2000–20000

The sensor frame is selected as a positioning restriction and 248 viewpoint candidates are generated equally over the frame. All STL-CAD-files of the four secondary objects add up to 3242 cell normals. Some are directed to the inside, which reduces the amount of cell normals for the calculation to 2590. Figure 3 shows the sensor simulation framework with a visualization of the view frustum from the final viewpoint (Fig. 3a) which supports the user to verify the calculated pose. The field-of-view is mapped on the objects which shows that the features to detect the correct type of srew are visible (Fig. 3b).

We utilize the semantic grammar and view-point definitions to derive a parametrization space for the synthetic training data variations which in turn is based on the parameters of the vision system. Our main object is the scarfing tool, mounted target objects are srew, washers and nut. The quantified values of the parameters as ruled by the grammar are shown in Table 1. This parameter space is then implemented Blender. We rendered 1600 combinations for correct assembly and 3500 for incorrect assembly. Examples are shown in Fig. 4a, b.

(a) True Class Image   (b) False Class Image   (c) Real Capture

Fig. 4 Examples of rendered training data compared with an real image. One Image represents the True Class with correct assembly, whereas the other image image displays an incorrect screw

4.1   Presentation

We trained a VGG-16 network with our dataset. A binary classification task was trained, with an added top layer of 1024 Dense units, 20% Dropout and a single unit output layer. This was trained for 10 epochs with adam. Afterwards the entire network was trained for 5 epochs with an SGD optimizer and a learning rate of 0.0001. To compensate for the lack of applied Domain Adaption techniques, we fine tune to the real domain with 15 real images of each class. Afterwards the network was tested against real world data, picturing the real scene in various lighting situations and with multiple distractors applied. In sum 164 images were gathered for testing. The confusion matrix of the results can be seen in Table 2. The classification accuracy results to 97%.

4.2   Discussion

With 97% classification accuracy, the aim of enabling a Deep-Learning task can be considered achieved. However, it is to be noted, that the task of object detection and localization

Table 2 Confusion matrix of test results. True class, corresponding to the correctly mounted assemblies and false class indicating the incorrect assemblies

	$True_{pred}$	$False_{pred}$
$True_{real}$	78	1
$False_{real}$	4	81

are only moderately challenging in this set-up and classification accuracy cannot be the lone indicator towards success of the synthetic training data generation. Our aim was to demonstrate how some of the information and parametrizations that are generated by the process of designing a vision system for a task can be reused to develop a data generation pipeline. This in turn can lead to a quick enablement of a image processing task. No additional analysis had to be done to identify suitable variations of the scene to obtain 3D models or to define the labels of a scene composition. Although the results of the demonstration are promising, additional testing with more complex environments can provide deeper insight into the applicability of the presented method. Suitable domain adaption techniques are to be applied, to improve the networks transfer capability towards the real application data. Further applications, e.g. measurement tasks can be developed. For this adaptions for the synthetic data generation pipeline have to be implemented, to include labeling of appropriate key-points.

5   Conclusion and Future Work

In this work an approach of a combined simulation based process chain for both—configuration and enablement—of a machine vision is presented. It states, that sensor planning and generation of a processing pipeline contribute from a common detailed task and object analysis. The resulting task semantic includes basic geometrical description of the object and environment as well as relevant features for the certain assembly process and environmental constraints which affect the sensor placement. This database is relevant for the sensor planning process and can simultaneously be used for the scene composition as part of AI training data generation. Using a sensor simulation and the example of a bolt connection we show that it is possible to set up a working machine vision solution for assembly verification from CAD data only. To improve the presented process chain and to extend the applicability of this concept, future work includes:

1. **A joint framework** for sensor planning and training data generation would use synergies between both fields and allow an evaluation of viewpoint candidates based on realistic sensor data renderings.
2. **Further visibility constraints**, such as illumination and overexposure, have to be integrated into the framework in order to increase the realism and thus the quality of calculations.

Acknowledgements Research was funded by the Luftfahrtforschungsprogramm LuFo V-3 "Hi-Digit Pro 4.0" and "DEPOT." The authors wish to thank Sönke Bahr and Sebastian Sauppe *(3D.aero GmbH)* for cooperation during the work on the simulation framework.

Gefördert durch:

Bundesministerium
für Wirtschaft
und Energie

aufgrund eines Beschlusses
des Deutschen Bundestages

References

1. Bousmalis, K., et al.: Using Simulation and Domain Adaptation to Improve Efficiency of Deep Robotic Grasping. In: 2018 IEEE International Conference on Robotics and Automation (ICRA), pp. 4243–4250 (2018)
2. Burla, A., et al. An assistance system for the selection of sensors in multiscale measurement systems. In: Furlong, C., Gorecki, C., Novak, E.L. (eds.) SPIE Optical Engineering + Applications. SPIE, 77910I (2010)
3. Cowan, C., Kovesi, P.: Automatic sensor placement from vision task requirements. IEEE Trans Pattern Anal Machine Intelligenz **10**, 407–416 (1988)
4. Dahmen, T., et al.: Digital reality: a model-based approach to supervised learning from synthetic data. In: AI Perspectives 1, Springer, Heidelberg, pp. 1–12 (2019)
5. Hinterstoisser, S., et al.: An Annotation Saved is an Annotation Earned: Using Fully Synthetic Training for Object Detection. In: 2019 IEEE/CVF International Conference on Computer Vision Workshop (ICCVW) (2019)
6. Fang, K.: Multi-task domain adaptation for deep learning of instance grasping from simulation. In: 2018 IEEE International Conference on Robotics and Automation (ICRA,) pp. 3516–3523 (2018)
7. Gaidon, Adrien: VirtualWorlds as Proxy for Multi-object Tracking Analysis. In: 2016 IEEE Conference on Computer Vision and Pattern Recognition (CVPR), pp. 4340–4349, (2016)
8. Gospodnetic, P., et al.: Flexible Surface Inspection Planning Pipeline. In: 6th International Conference on Control, Automation and Robotics, pp. 644–652 (2020)
9. Jing, W., et al.: (2016) Sampling-based view planning for 3D visual coverage task with Unmanned Aerial Vehicle. In: 2016 IEEE/RSJ International Conference on Intelligent Robots and Systems (IROS), pp. 1805–1815 (2016)
10. Kar, A., et al.: Meta-sim: Learning to generate synthetic datasets. Proceedings of the IEEE International **2019**, 4550–4559 (2019)
11. Liu, J., Sridharan, S., Fookes, C.: Recent Advances in Camera Planning for Large Area Surveillance. ACM Comput Surv **49**, 1–37 (2016)
12. Mittal, A., Davis, L.: A General Method for Sensor Planning in Multi-Sensor Systems: Extension to Random Occlusion. Int J Comput Vis **76**, 31–52 (2008)
13. Rodeck, R., Schüppstuhl, T.: Repair of composite structures with a novel human-machine system. In: Proceedings of ISR 2016: 47st International Symposium on Robotics (ISR), pp. 660–666 (2016)
14. Tarabanis, K., Allen, P., Tsai, R.: A survey of sensor planning in computer vision. IEEE Trans Robot Automation **11**, 86–104 (1995)
15. Thomas, G, et al.: Learning Robotic Assembly from CAD. In: 2018 IEEE International Conference on Robotics and Automation (ICRA), pp. 3524–3531 (2018)
16. Tobin, J., et al.: Domain randomization for transferring deep neural networks from simulation to the real world. In: 17 IEEE/RSJ International Conference on Intelligent Robots and Systems (IROS), Vancouver, BC, pp. 23–30 (2017)

17. Triyonoputro, J.: Quickly Inserting Pegs into Uncertain Holes using Multi-view Images and Deep Network Trained on Synthetic Data. In: IEEE/RSJ International Conference on Intelligent Robots and Systems (IROS), p. 5729–5779 (2019)
18. Scott, W.: Model-based view planning. Machine Vision and Applications **20**, 47–69 (2009)
19. Zeng, R., et al.: View planning in robot active vision: A survey of systems, algorithms, and applications. Comp Visual Media **6**, 225–245 (2020)
20. Židek, K., et al.: An Automated Training of Deep Learning Networks by 3D Virtual Models for Object Recognition. In: Symmetry 11, (2019)

Towards Synthetic AI Training Data for Image Classification in Intralogistic Settings

Daniel Schoepflin⑩, Karthik Iyer⑩, Martin Gomse⑩
and Thorsten Schüppstuhl⑩

Abstract

Obtaining annotated data for proper training of AI image classifiers remains a challenge for successful deployment in industrial settings. As a promising alternative to handcrafted annotations, synthetic training data generation has grown in popularity. However, in most cases the pipelines used to generate this data are not of universal nature and have to be redesigned for different domain applications. This requires a detailed formulation of the domain through a semantic scene grammar. We aim to present such a grammar that is based on domain knowledge for the production-supplying transport of components in intralogistic settings. We present a use-case analysis for the domain of production supplying logistics and derive a scene grammar, which can be used to formulate similar problem statements in the domain for the purpose of data generation. We demonstrate the use of this grammar to feed a scene generation pipeline and obtain training data for an AI based image classifier.

Keywords

Synthetic data • Training data generation • Image classification • Intralogistic • Production supplying logistic

D. Schoepflin (✉) · K. Iyer · M. Gomse · T. Schüppstuhl
Institute for Aircraft Production Technology, Hamburg University of Technology, Denickestraße 17, 21073 Hamburg, Germany
E-mail: Daniel.Schoepflin@tuhh.de
URL: http://www.tuhh.de/ifpt/

1   Introduction

The intralogistic transport of components on the plant-site of an aircraft manufacturer is comprised of multiple handling and repackaging processes [1]. Such components are often transported on material delivery units, and due to the manual handling, are subjected to error. This is mostly reflected in delivery units being loaded with the wrong components. For this reason identifier tags may be used to verify the loading. However, due to process and manufacturing requirements, tagging the components may be prohibited and a visual object identification system is needed. We consider the usage in a system setting as seen in Fig. 1, where components are placed in boxes in a load carrier and a camera achieves a top-view on those components. Evaluating this top-view raises the need for proper training of an AI image classifier. For the high variety of components in aircraft manufacturing, obtaining and labeling images manually is a tedious and costly process. Thus, synthetic training data is considered a viable alternative.

In recent years, different generation pipelines have been introduced with different fields of application [2–4]. As stated by [5], creating such virtual worlds is in need of domain experts and the formulation of a representative scene grammar. To the authors knowledge, this has not been done sufficiently for the intralogistic transport of components. We therefore contribute a use-case analysis for this domain and define a parametrization that is usable to derive a tool independent scene composition grammar that is transferable to intralogistic use-cases. We implement this in a simulation and rendering pipeline to obtain training data for an AI image classifier.

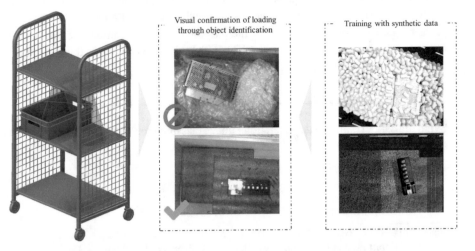

Fig. 1 Use-Case of intralogistic component transport with delivery units. Loading is confirmed with a top view visual object identification system, which is trained with synthetic data

2   Related Work

Various approaches for creation of synthetic driving scenes have lead to multiple synthetic data-sets of different applications like driving scenes [6], household objects [4, 7] to robotic picking [3, 8]. Sets like [8, 9] enable the community to research on industrially relevant box-picking tasks or object identification tasks. A synthetic data generation approach was undertaken by [10] to enable handling of gears and similar components in industrial settings. Due to the success in these domains, it can be inferred that training an object identifier in the highly variational intralogistic scenery with synthetic data is a viable approach. Common object data-sets [2, 4, 11] can be used by multiple users to enable AI services relating to universally common environments e.g. household objects. However, this universal applicability is not necessarily the case for industrial applications or data-sets, as they address non-common objects in specialised environments [12]. Thus, in order to enable broad applicability of synthetic training data approaches, we strive to contribute towards adaptability of generation pipelines with regards to user needs.

In most cases, pipelines place 3D models of the objects of interest in a scene and render this scene to obtain the training image. In general this creation and composition of a scene is varied for each rendering process e.g. changing position and orientation of objects within a pre-defined parameter space. The NVIDIA Deep Learning Data Synthesizer[1] provides a data creation tool based on the Unreal Engine 4, with which the household object datasets SIDOD and FAT were created. Such a tool can be used to create synthetic training data for different domains, when used with a modeled environment or scene creation grammar for that domain. Similar tools can be found for the Unity Game Engine[2] or the Open Source Tool Blender.[3] Although these tools provide easy use for modelled environments, they do not automatically provide semantics for creation of new 3D scenes. This semantic formulation of a scene is mostly provided by the user.

Approaches like [13] use a fully automated and randomized composition of front- and background, whereas approaches like [9] utilize context true scene compositions. In both cases a set of rules is created, on which the implementation of a composition algorithm is based. These accumulated rules and parametrizations are referred to as grammar [5] or model [12] and are vital for the creation of data generation pipelines. We provide such a generic grammar for the use-case of intralogistic transportation of material.

3   Process Analysis and Problem Statement

We first describe the intralogistic transport domain for production supplying logistic with delivery units. To further analyse this, a categorization with respect to the complexity of

---

[1]https://github.com/NVIDIA/Dataset_Synthesizer, September 2020.
[2]https://github.com/Unity-Technologies/SynthDet, September 2020.
[3]https://github.com/921kiyo/3d-dl, September 2020.

scenes is developed. This then leads to a generalized grammar for this domain. To validate this grammar, a parametrization for later implementation of a specific use-case is formulated.

3.1   Use-Case Description

As the generated data shall enable the training of an AI image classifier, the scope of this specific application is defined. With the considered use as shown in Fig. 1, the main application that a visual based AI image classifier may enable is the validation of commissioning. Through identification of the loaded component and comparison with an expected loading, the validity of loading can be concluded. This may be combined with a counting of the objects. However these tasks can not be realised for every component type. The components have to be visually detectable and thus, can not be fully wrapped in packaging or appear in great numbers with mutual concealment. Thus, the likely number of objects to identify in one scene is in the single to low double digit range. The components are often placed in boxes, with little position and pose constraints or on shadowboards with pre-defined positions. Further we assume that the identification task is applied to components that may not carry RFID or similar identification markers. This is mostly the case for assembly ready components and lesser for semi-finished products. Figure 2 visualizes these constraints of the possible scenarios which we aim to synthesize.

As seen in Fig. 1, the considered setting of the cameras on a delivery unit itself provides a top-view of the loaded components. However, in other cases an askew view may be provided, e.g. with a camera above a robotic effector. Thus it is neccessary, that the derived scene grammar will take such cases in consideration.

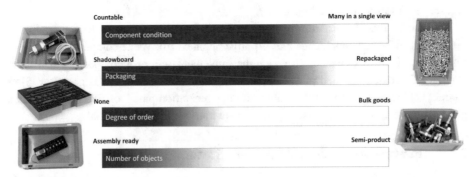

Fig. 2 Definition of the scope for the approach considered in this use-case. Objects appear in the visually countable numbers and are visually detectable

3.2   Categorization of Transportation Settings

As the sceneries to which the AI is applied may differ significantly, a categorization is undertaken. This enables a more detailed analysis of the domain and then leads to the description of the domain in a more generalized grammar formulation. These categories are defined regarding the complexity of parameter variations occuring in these settings:

1. **Simple settings**: objects and components are placed on a uniform background such as anti-slip mats. Besides material changes leading to differences in texture and color of the background, parameter variations occur with the placing of the object on such a mat. Objects can occur in different translatory positions as well as rotational placements. However, many components are limited in their contact points to the mat, to which they are forced by their axes of inertia and gravity. This leads to discrete number of stable resting states, with only one axis as rotational degree of freedom.
2. **Intermediate settings**: many components, in particular small components, are transported in boxes or cartonages. Those are also subject to the same placement restrictions as the simple cases above. Causing a more complex scenery is the variant lighting condition as well as the shadowing caused by the boxes.
3. **Complex settings**: some shock sensitive components may be transported in boxes filled with packaging flips or in bubble wrap. Some flat components are transported by placing them between struts of a support structure. Additional straps may be used to secure the components. Besides creating a complex scenery by adding a complex setting of other objects, they also may allow the components to be placed in different positions than the previous two settings. Additionally, lighting situations are more complex, due to local shadowing.

3.3   Formulation of Scene Grammar

It is now necessary, to define a scene composition grammar with respect to the parameters of three above presented categories. We first focus on the object composition and briefly introduce further variations like background, view and lighting.

3D Composition Semantic

Arrangement of real components on a delivery unit follows a set of semantic rules, that are mostly intuitively met by the persons handling the commissioning. In order to later simulate a loading scenario, relations between the objects to place have to be defined, which together forms a semantic 3D composition grammar. This is displayed in Fig. 3: We start with the type of unit in use. This defines the geometrical restrictions in which the boxes and components are placed. Further, this defines the location and type of the camera and further

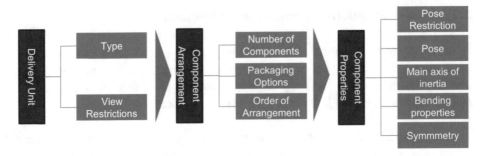

Fig. 3 Sequence of the semantic scene grammar parameters with respect to the object composition relation

view restrictions. Afterwards, the principal component arrangement has to be defined. The number of components to be modelled is defined and their relative arrangement set. This is done with the components properties as well as possible packaging options. Those packaging options refer to the arrangement of components in a box, packaging material like flips and possible distractors like transfer papers. Afterwards for each component the individual pose spectrum has to be defined. As many components have a preffered orientation, caused by its geometry, the pose spectrum can be restricted. This later limits the necessary variations of scene compositions required. Possible symmetry in shape can also help to reduced this parameter space. For possible physics-simulation the properties of the components with respect to flexibility have to be considered.

Viewing Properties

Depending on the vision system used for the task, multiple parameters have to be defined to generate training data. As some variational restrictions might arise from the 3D composition, a relational semantic has to be defined. In our use-case the type of delivery unit defines the placement of the camera. However, for different use-cases this might not be the case. E.g. for creation of training data out of the view of a robotic manipulator, more and inherently variant viewpoint variations have to be considered. Further as the visual set-up may differ between mono- and stereo-setups, field of view and similar parameters, these relations are defined in the scene grammar.

Background and Lighting Properties

Depending on the application, the background of the composed scene may be in the field of vision. In principle it is possible to use a parametrization and scene composition for the surrounding as well. Such might be needed for autonomous mobile robotics with variable camera handling, but lesser for static top-view settings. Therefore, randomized cluttered backgrounds similar to [4] may be utilized, if necessary.

As lighting is a highly variational and environmental dependent scene parameter, it can be emulated by randomized light sources of variant strengths. A relationship may occur if cameras are equipped with ring lights or similar.

3.4   Derivation of a Parameter Space

We now utilize the formulated grammar and derive a scene parameter space for the transportation of two components in boxes with varying packaging infill on a delivery unit with three levels. This setting is shown in Fig. 4: A pointer indicates the to be modelled tray setting. We assume camera settings on top of the box as indicated in the schematic. Each component is transported in its own box. Different box colours and fillings are predefined. For the filling packaging flips, bubble wrap and anti-slip mat are considered.

As one of the objects is equipped with a display and buttons on top, we assume a preferred orientation with the display upwards. For the variations with anti-slip mat and no in-fill, the pose of the objects are restricted to rotation around the vertical axis and translation in plane. For the fillings with packaging flips and bubble wrap, 6D poses are possible. However, to achieve randomized yet physic accurate 6D poses, a physics simulation is done. The objects are considered stiff, leading to rigid body simulation. The quantitative values of this parameter space are shown in Fig. 5a, b.

With this formulation of the use-case derived, a pipeline can be utilized to randomize the composition of scenes and render them.

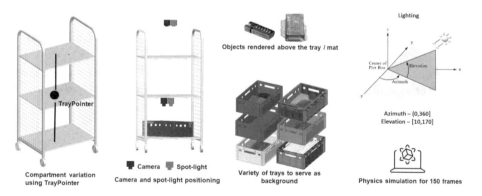

Fig. 4 Semantic interpretation and definition of variations according to the defined grammar

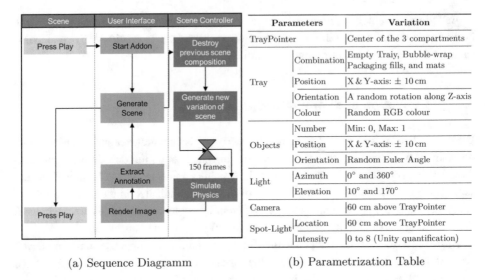

Parameters		Variation
TrayPointer		Center of the 3 compartments
Tray	Combination	Empty Traiy, Bubble-wrap Packaging fills, and mats
	Position	X & Y-axis: ± 10 cm
	Orientation	A random rotation along Z-axis
	Colour	Random RGB colour
Objects	Number	Min: 0, Max: 1
	Position	X & Y-axis: ± 10 cm
	Orientation	Random Euler Angle
Light	Azimuth	0° and 360°
	Elevation	10° and 170°
Camera		60 cm above TrayPointer
Spot-Light	Location	60 cm above TrayPointer
	Intensity	0 to 8 (Unity quantification)

(a) Sequence Diagramm   (b) Parametrization Table

Fig. 5 Sequence diagramm of the pipeline in (**a**), parametrization of the setting in (**b**)

4   Data Generation Pipeline

Implementation of the derived parameter space is done with the Unity Game Engine. We wrote an addon that generates scenes according to the defined parameter space. A sequence representation of that pipeline is shown in Fig. 5a. After generation and simulation of a scene variation, images are rendered and annotations extracted. In each loop, the parameters are changed according to the defined parameter space and Fig. 4. Quantitative formulation of the parameter space is shown in Table 5b.

Annotations are generated by our written Unity addon used for creation of the scene. Besides class of the picture, also the bounding box, and pose of the object is extracted and saved.

5   Validation

In order to validate that the scene grammar and implementation is capable of generating viable training data, we train a Deep-Learning image classifier with the synthetic images and test it against a hand-annotated real world data.

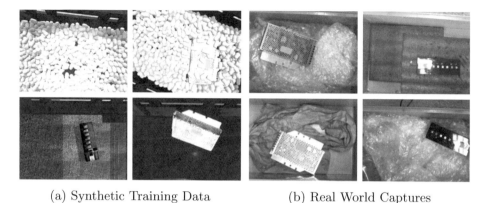

(a) Synthetic Training Data   (b) Real World Captures

Fig. 6 Examples of generated synthetic training data (**a**) and real world data (**b**)

5.1   Model and Training

We use a ResNet34 architecture. The final classification layer consists of three classes, representing the Object 1, Object 2 and None Class. The model is trained in two stages. First the top fully connecte layers are trained, afterwards the entire model is trained. One cycle policy was used, with learning rate maximum 0.001 and minimum 0.0002. In both stages 5 epochs were trained. The pipeline generated 1000 synthetic images.

5.2   Results

We evaluate the trained network on for the network previously unknown data sets captured from a real word scene. This test set consists of 745 images with 329 *Object 1* images, 361 *Object 2* images and 55 *None* class Images. The confusion matrix of the results is shown in Table 1: ten out of the 745 images were miss-classified with 6 *Object 1* images not being detected and three being classified as *Object 2*. One *Object 2* image was classified as *Object 1*. This translates into a classification accuracy of 98.50%.

Table 1 Confusion matrix of the the real world classification test-set

	Predicted		
	Object 1	Object 2	None
Object 1	328	3	6
Object 2	1	358	0
None	0	0	49

5.3   Discussion

With few missclassifications, mostly concerning the object 1 class, the general classification task can be regarded successful. However, it is to be expected that more classes and a more challenging test set would have a negative impact on the results. In general, when training with synthetic data and applying that AI, the domain adaptation problem poses to open world capability of the AI. Missmatches between the synthetically created content and the real world as well as the difference between a rendered image and a real sensor perception, may prevent a successful use of the AI. In our case, with strict modelling of realistic transportation settings, we achieved to narrow the content gap between synthetic and real world for our use-case. As seen from Fig. 6, the real images contain blurring and glares, that are not accounted for in the synthetic images. Closing such appearance gaps is focus of different approaches, which could be combined with our developed semantics and pipeline but was out of scope for this work.

With the task of enabling an image classifier for the presented use-case being successfully fulfilled, the derived grammar can be considered a viable contribution to industry ready synthetic training data. However, we'd like to point out that the actual implementation of the parametrization space is not necessarily universal to every use-case. For example, the developed pipeline does not handle multi-object detection tasks and is as such less viable for vision systems with a greater field of view. Generalizing this implementation to include most of the grammars problem statements is aim for future work.

6   Conclusion and Outlook

In this work we aimed to enable an image classifier network for usage in intralogistic transport scenarios. To achieve this, we formulated a scene grammar for such scenarios and derived a parameter space for a given use-case. We implemented this use-case in a pipeline and utilized this to generate training data. This data is then used to train a Deep-Learning image classifier and validated against real world data.

Future work will focus on further generalization of the pipeline, to reduce the necessary transfer effort between formulating a scene parameter space and implementing it in a pipeline. Also a native integration with state of the art domain adaptation may be necessary, when used for more challenging tasks.

Acknowledgements Research was funded by the German Federal Ministry for Economics and Energy under the Program LuFo V-3 DEPOT.

Gefördert durch:

Bundesministerium
für Wirtschaft
und Energie

aufgrund eines Beschlusses
des Deutschen Bundestages

References

1. Sliwinski, M., Raabe, C.M., et al.: Modulare Ladungsträger für den Kleinteiletransport. ZWF Zeitschrift für wirtschaftlichen Fabrikbetrieb **115**, 418–21 (2020)
2. Jalal M, Spjut J, et al. SIDOD: A synthetic image dataset for 3D object pose recognition with distractors. In: IEEE Computer Society Conference on Computer Vision and Pattern Recognition Workshops. Vol. 2019-June. 2019:475–7. https://doi.org/10.1109/CVPRW.2019.00063
3. Bousmalis K, Irpan A, et al. Using Simulation and Domain Adaptation to Improve Efficiency of Deep Robotic Grasping. In: 2018 IEEE International Conference on Robotics and Automation (ICRA):4243–50. https://doi.org/10.1109/ICRA.2018.8460875
4. Hinterstoisser S, Pauly O, et al. An Annotation Saved is an Annotation Earned: Using Fully Synthetic Training for Object Instance Detection. In: 2019 IEEE International Conference on Computer Vision Workshop (ICCVW)
5. Kar A, Prakash A, et al. Meta-sim: Learning to generate synthetic datasets. In: Proceedings of the IEEE International Conference on Computer Vision. 2019:4550–9. https://doi.org/10.1109/ICCV.2019.00465
6. Gaidon A, Wang Q, et al. VirtualWorlds as Proxy for Multi-object Tracking Analysis. In: 29th IEEE Conference on Computer Vision and Pattern Recognition. Piscataway, NJ, 2016:4340–9. https://doi.org/10.1109/CVPR.2016.470
7. Tremblay J, To T, et al. Deep Object Pose Estimation for Semantic Robotic Grasping of Household Objects. In: Conference on Robot Learning. 2018
8. Kleeberger K, Landgraf C, and Huber MF. Large-scale 6D Object Pose Estimation Dataset for Industrial Bin-Picking. In: 2019 IEEE/RSJ International Conference on Intelligent Robots and Systems (IROS). 2019:2573–8. https://doi.org/10.1109/IROS40897.2019.8967594
9. Brucker M, Durner M, et al. 6DoF Pose Estimation for Industrial Manipulation Based on Synthetic Data. In: Proceedings of the 2018 International Symposium on Experimental Robotics. Vol. 11. 2020:675–84. https://doi.org/10.1007/978-3-030-33950-058
10. Andulkar M, Hodapp J, et al. Training CNNs from Synthetic Data for Part Handling in Industrial Environments. In: IEEE International Conference on Automation Science and Engineering. Vol. 2018-August. 2018:624–9. https://doi.org/10.1109/COASE.2018.8560470
11. Lin TY, Maire M, et al. Microsoft COCO: Common Objects in Context. In: Lecture Notes in Computer Science book series. Vol. 8693:740–55. https://doi.org/10.1007/978-3-319-10602-1 48
12. Dahmen, T., Trampert, P., et al.: Digital reality: a model-based approach to supervised learning from synthetic data. AI Perspectives **1**, 1–12 (2019)
13. Tobin J, Fong R, et al. Domain Randomization for Transferring Deep Neural Networks from Simulation to the Real World. In: 2017 IEEE/RSJ International Conference on Intelligent Robots and Systems (IROS). 2017. https://doi.org/10.1109/IROS.2017.8202133

Evaluation of ML-Based Grasping Approaches in the Field of Automated Assembly

Oliver Petrovic, Philipp Blanke, Manuel Belke, Eike Wefelnberg, Simon Storms and Christian Brecher

Abstract

Current trends in the manufacturing industry lead to high competitive pressure and requirements regarding process autonomy and flexibility in the production environment. Especially in assembly, automation systems are confronted with a high number of variants. Robot-based processes are a powerful tool for addressing these challenges. For this purpose, robots must be made capable of grasping a variety of diverse components, which are often provided in unknown poses. In addition to existing analytical algorithms, empirical ML-based approaches have been developed, which offer great potentials in increasing flexibility. In this paper, the functionalities and potentials of these approaches will be presented and then compared to the requirements from production processes in order to analyze the status quo of ML-based grasping. Functional gaps are identified that still need to be overcome in order to enable the technology for the use in industrial assembly.

Keywords

Robotics • Grasping • Assembly • Computer vision • Machine learning

O. Petrovic (✉) · P. Blanke · M. Belke · E. Wefelnberg · S. Storms · C. Brecher
Laboratory for Machine Tools and Production Engineering (WZL) of RWTH Aachen University, Steinbachstr. 19, 52074 Aachen, Germany
E-mail: O.Petrovic@wzl.rwth-aachen.de
URL: http://www.wzl.rwth-aachen.de

337

1   Introduction and Motivation

In the manufacturing industry, a trend towards robot-based automation has been observed for years. From 2013 to 2018, the number of new robot installations has increased by an average of 19% per year [1]. Reasons for this are the rising quality standards and labor costs which lead to high competitive pressure. In countermove, economic automation is becoming more and more difficult as product life cycles become shorter and batch sizes smaller. This places great demands on the flexibility and autonomy of the technologies to handle this variety [2]. Especially in assembly, the degree of automation is often still very low, as the generation of variants is usually shifted as far back in the value chain as possible to final assembly in order to minimize its impact. Therefore, a key challenge is the flexible interaction of the robot with its environment. It must be able to handle a wide range of components, which are often fed in an unknown position and orientation, and this with an economical level of implementation effort. In recent years, the robotics and computer vision community has contributed a wide range of different approaches to solve the grasping problem. Analytical approaches consider kinematic and dynamic formulations in grasp synthesis [3]. However, these approaches are characterized by high computational complexity and cannot be generalized well to unknown objects, which is why the developed ML-based methods are promising approaches [4].

The aim of this paper is to provide an overview of these current approaches in research and to highlight the remaining challenges for their use in production, especially in assembly: In Sect. 2, state-of-the-art on ML-based grasping approaches in research is given. Section 3 analyzes the production requirements for the application, followed by the presentation of a derived integration approach of grasping into the digital process chain of assembly in Sect. 4. Finally, Sect. 5 identifies the gaps that need to be closed in order to implement an integration to meet the requirements.

2   State of the Art

Sensor-based perception of the environment are fundamental capabilities of a smart robot. In this regard, autonomous or partially autonomous grasping based on vision systems is one of the sub-disciplines of robotics that can contribute greatly towards improving the flexibility of robotic applications. According to Kumra et al. the vision-based grasping process can be seen as a sequence of three sub-steps: Grasp detection, trajectory planning and execution of the grasp [5]. This paper will mainly focus on the first step of this sequence, which again can be divided into three sub-problems shown in Fig. 1.

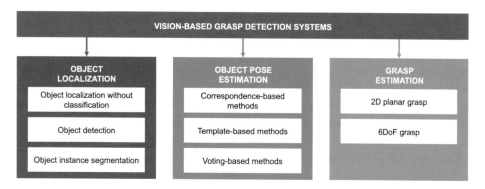

Fig. 1 Subtask classification of vision-based grasp detection systems [6]

2.1 Object Localization

The object localization task can be further divided into pure localization and localization including the detection of the class of the object. Since the manipulation requires spatial knowledge about the object, only the 3D localization methods will be described. To use these 3D localization methods, a RGB-D camera is used, that provides depth data in addition to the RGB image [6]. The pure localization task can applied to simple objects like cubes or cylinders but the method can be improved by combining shape primitives with the use of triangular meshes to be able to map various types of objects Rusu et al. [7]. The localization of objects with no restriction regarding its shape is called salient object detection. Many approaches take the pixels of an image as inputs. Instead, salient feature vectors can be used as inputs to a CNN to learn the combination of different salient features for the recognition of salient objects [8]. Outputs of the detection of objects are the 3D bounding box and the class label of the object, which can either be detected sub-sequentially as in ImVoteNet [9] or at once by using a regression method like 3DSSD [10]. To further refine the position of the object, instance segmentation can be used. The starting point is the bounding box of the object, within which the 3D position of the object is detected. OccuSeg uses the occupancy to cluster segments despite partial occlusions [11].

2.2 Object Pose Estimation

The second subproblem is the Object Pose Estimation, where the 6D pose of the localized part must be determined. The degrees of freedom to be determined can be reduced by a predefined part feeding. Du et al. cluster the existing methods into three categories [6]. Firstly, there are the correspondence-based methods, in which corresponding feature points between captured image information and the object to be grasped are searched for. It is possible to utilize deep learning algorithms and to work with 2D RGB images like HybridPose [12] as well as with

3D point clouds like 3DMatch [13]. These methods are suitable if the object has a rich texture and geometric details.

The second group of algorithms are the template-based methods. A multitude of templates are labeled with corresponding 6D poses for the object to be grasped. If 2D images are used as in [14], the 6D problem is reduced to an image retrieval problem, because the image is only compared with a known set of 2D images of the object. If, on the other hand, 3D template-based methods are used, the recorded point cloud is directly compared with the 3D model of the object as in MaskedFusion [15]. In general, template-based methods are suitable especially if the object has few distinctive textures and geometric details.

The third category are the voting-based methods. Here, not the whole image is analyzed at once, but every single 2D pixel respectively every 3D point is considered separately and contributes a vote to the estimation. If the objects to be grasped have a high degree of occlusion, then voting-based methods can be effective. On the one hand, there are indirect voting-based methods in which the image points first contribute a vote for higher-level features from which the 6D pose can be indirectly derived. This is shown in YOLOff Gonzalez et al. [16]. On the other hand, direct voting-methods can be used, where the pixels vote directly for the 6D pose of the object as in DenseFusion [17].

2.3   Grasp Estimation

The goal of Grasp Estimation is to find a robust grasp pose. According to Du et al. the algorithms for Grasp Estimation can be divided into 2D planar grasps and 6D grasps [6]. The 2D planar grasp has two fixed axes of rotation, so that only the height of the plane, the position in the plane and the rotation around the normal vector are determined. The developed algorithms of both categories refer to either analytical or ML-based approaches. Due to the dependence on assumptions to be made (friction, object stiffness, object complexity etc.) analytical approaches in practice do not generalize well over new objects [18].

One of the ML based grasp methods is the project Dex-Net presented by Mahler et al.. The input is a recorded point cloud, which is evaluated by a CNN regarding the grasp quality of all grasping candidates. Zeng et al. performs a pixel-wise evaluation of the grasp affordance for different grasping primitive actions and perform the end-effector position and orientation with the highest affordance. Furthermore, the project Form2Fit [21] not only deals with grasping new objects but also with placing them in the desired position. A trained fully convolutional network (FCN) detects correspondences between the object surface and the shape of the target position.

3   Production Requirements on Vision-Based Grasping

In order to evaluate the industrial applicability of today's algorithms for ML-based grasping, the production requirements for such a system must first be analyzed. In this chapter, these requirements are categorized into six categories (Fig. 2) to identify gaps in the usability and functionality of today's solutions.

First, the required performance of the system is derived directly from quality and productivity requirements, which can be translated into the required precision and speed of the grasp detection. The second category is the robustness of the system against external influences such as poor lighting conditions, humidity and a dynamic image background. Another important factor are the components to be grasped. On the one hand, the components themselves, i.e. their variance, dimensions, shape, transparency and surface, and on the other hand the way they are fed to the process, has to be considered. The feeding can vary in the level of order, the degree of occlusion and hooking as well as the distance between the components. The hardware is required to provide the necessary computing power for the execution of the algorithms in a cost-effective manner in order to enable a profitable operation of the system. The available interfaces of the software as well as the range of compatible hardware like robots, grippers and sensors like cameras have a great influence on the integratability and transferability of the solution. Finally, required data sets and programming efforts should be mentioned, which directly impact the implementation effort and the competence hurdle for the programmer. The number, scope and quality of compatible data sets for training the algorithms, on the other hand, have great influence on the performance of the system. Furthermore, to achieve good industrializability, it must be possible to integrate existing product and process data. Physical component data, functional

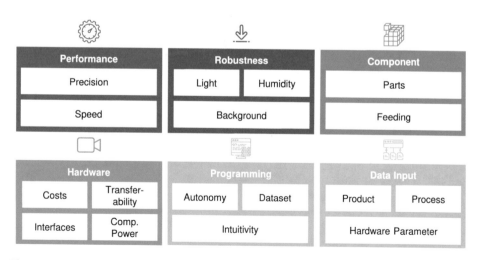

Fig. 2 Clustering of production requirements on vision-based grasping

surfaces and the requirements of subsequent process steps are some examples of important parameters when selecting a grasp. In addition, parameters of the equipment, such as force limits and workspaces of robots and grippers, must also be taken into account.

4 Integration of ML-based Grasping in Assembly Processes

To meet these requirements, the three presented steps of a grasping system need to be embedded into a novel end-to-end system and closely linked to the digital process chain and its corresponding product lifecycle. Such a concept is proposed in Fig. 3. The product lifecycle can be sub-divided into engineering, production planning, production, usage and recycling. During the engineering phase, the product is designed and can be disassembled into product specifications, drawings, CAD models of each single part. In the production planning, the data of the engineering phase is used to plan the production and especially the process and assembly sequence. The grasping system is implemented in this phase. In the subsequent production, the actual gripping process is carried out. Throughout the entire life cycle, product and process data must be made available in accessible formats via a

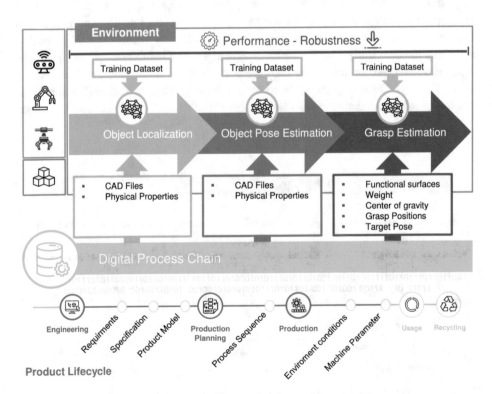

Fig. 3 Integration of ML-based grasping approaches in the digital process chain

central digital process chain which is an important enabler for the seamless integration of engineering data into the robot-based assembly process.

Before the individual components are selected, the overall performance and robustness of the system required to fulfill the task at hand have to be defined. The precision and speed required to assemble the components is determined by the product while the light and background as robustness parameters are given by the environment. This narrows down the suitable algorithms to perform the tasks. Another important factor for selecting the algorithms is defined by the programming requirements. In order to make the system versatile, it should be intuitively operable and have enough autonomy to make the supervision by a human operator redundant.

The object localization task requires RGB and depth images belonging to a CAD model. With the ImVoteNet architecture for example, an object locater is trained on both RGB and depth data to efficiently detect the 3D bounding boxes of the objects as well as the class [9]. As batch sizes of products continue to shrink, multiple object classes are placed at the assembly station at once. The classification is therefore a crucial step during the object localization to be able to choose the right object which must be assembled next. Depending on the algorithm, the hardware is chosen based on the required interfaces, the transferability of the system and economic aspects. This is closely connected to the data input coming from the digital process chain. The latter serves as the connection between the product lifecycle and the grasping process and has to deliver the product data, process data and hardware parameters in the format processable by the algorithms.

Based on this first classification and the calculated bounding box, the pose estimation of the object follows. Most objects which are assembled in the production do not have rich texture which make the correspondence-based methods unsuitable in many cases. For weak texture and geometric detail, the template-based methods perform good, while for occlusion which is common in the production, voting-based methods are a good choice. The DenseFusion algorithm uses both RGB images and depth data for the pose estimation of objects, which are fed into the process via the digital process chain [17]. Before estimating the pose of the object, DenseFusion does an object segmentation on the RGB image to detect the pixels belonging to a specific object. After this step, both the RGB and depth data are fused accurately to predict the 6D pose of the desired object. Each pixel of the RGB image votes for a 6D pose which results in a good estimation even if parts of the object are occluded.

The last step is the selection of grasps based on the object pose. With the ML approach of DexNet, object localization and pose estimation do not have to be done but possible grasps are generated directly based on depth data [19]. The biggest drawback of this approach is the lack of object specific data for grasp generation. In the production the exact grasping location is highly relevant. The functional surfaces, the weight, center of gravity and the position where the object has to be placed in the assembly are known from the engineering phase. These factors are combined in the component requirement consisting of parts and feeding. Using these factors, grasp positions are generated. During the production, after the

6D pose of the object was detected, one grasp is selected. The selection process takes into account the pose of the object, the position of the assembly, the used robotic hardware and the environment to avoid collisions but at the same time optimize the time used to assemble the object. To make use of DexNet's good grasp selection and at the same time use object specific data, we used DexNet's grasp selection as a starting point. With this selection, we estimated the 6D pose of the object with an iterative closest point algorithm. The advantage of this approach is the good quality of the pre-selection by DexNet followed by an exact pose estimation of the object.

To train the ML-algorithms, training datasets generated either synthetically or by physical experiments are necessary. The advantage of synthetical data is the cheap generation and the possibility to include unexpected scenarios, but the different physical conditions and parameters have to be considered nonetheless. This makes the transfer of the algorithms from the simulation to reality a challenging task. The conduction of physical experiments to collect the data is more expensive and time consuming, but the data is closer to reality and can thus lead to more robust solutions [19].

5   Current Challenges

The previous section highlighted examples of how an intelligent combination of information from the product life cycle with existing ML approaches can sustainably improve the robustness and performance of gripping systems and thus find more use in assembly. However, it also becomes clear, that it is difficult to compare the existing algorithms on a common ground. This is partly because they sometimes focus on individual steps or combine several steps, and partly because they are tested with different data sets. This makes it difficult to find the optimal combination for the individual application. In order to make this possible, a test framework is required in which the models or a combination of algorithms can be tested against each other in a defined setting, as shown in Fig. 3. In such a model, the constraints of the environment are set. Since the environmental conditions and specific hardware properties can only be modeled to a limited extent, there must be a defined input stream that, in addition to the input data, also provides reference data for evaluating the result. Based on this, the individual models can then be exchanged or arranged differently until the intended requirements are met. Via defined interfaces, the algorithms can also access information from the product life cycle to improve the overall result.

Beside that, the robustness and safety of such systems must be further improved. While robustness to different lighting conditions can be achieved by training with a heterogeneous data set, the problem of reflective surfaces remains even when using stereo camera systems. Strategies must also be developed to continue operating efficiently in the event of a system failure. The system should be able to overcome such errors by having an alternative solution especially in safety critical processes and to learn from its mistakes for a continuous optimization of the solution.

Moreover, it is important to incorporate significantly more process and product knowledge into the decision-making processes of the algorithms. Therefore, there is a demand for research regarding the incorporation of domain knowledge in the training process, in transfer learning as well as on methods of data augmentation for those data types that are particularly relevant for industrial use. On the one hand, the algorithms have to offer appropriate interfaces and on the other hand, the corresponding data has to be converted into compatible and standardized formats. In general, ML approaches must be considered more in the overall tool and value chain of the process in which they are to be integrated.

Finally, the acceptance and transparency of ML solutions must also be addressed. It is important that ML based systems shift from current black box models into comprehensible systems. Explainable AI is an important keyword here, without which the broad industrial use of the algorithms is difficult to implement.

6   Conclusion and Outlook

In the context of this paper it became obvious that there are still some challenges to be solved in order to enable ML-based gripping for broad industrial use in assembly. It was shown that the requirements from production are very complex and multilayered. In particular, the parameters influence each other very strongly, so that a generalization is only possible to a limited extent or only for individual domains. On the other hand, it became clear that the described approaches offer advantages over classical, analytical approaches. For example, flexibility was derived from the assembly perspective as a central requirement, which can be achieved much more easily through ML.

However, it also became apparent that a chaining of different modules with an underlying end-to-end data process chain is absolutely necessary to achieve the higher-level objectives. For the daily use in production, the whole tool chain should be considered in a holistic approach and it should be clarified how the individual modules can be linked together in an effective way and how robustness and precision can be increased by use of underlying data. To do this, a test framework is needed to benchmark the existing models and approaches against each other in a defined environment. Therefore, it is planed to develop such a framework to enable the user to decide which approach fits to his requirement and to reveal remaining potentials for further research.

Acknowledgements The IGF-projekt 20922 N (FlexARob[2]) of the research association FVP was supported via the AiF within the funding program "Industrielle Gemeinschaftsforschung und – entwicklung (IGF)" by the Federal Ministry of Economic Affairs and Technology (BMWi) due to a decision of the German Parliament.

References

1. International Federation of Robotics. Executive Summary World Robotics 2019 Industrial Robots
2. Martin, C., Leurent, H.: Technology and Innovation for the Future of Production: Accelerating Value Creation. In collaboration with A.T. Kearney, Geneva (2017)
3. Sahbani, A., El Khoury, S., Bidaud, P.: An overview of 3D object grasp synthesis algorithms. Robotics and Autonomous Systems **60**(3), 326–336 (2012). https://doi.org/10.1016/j.robot.2011.07.016
4. Mahler, J., Goldberg K.: Learning Deep Policies for Robot Bin Picking by Simulating Robust Grasping Sequences. In: Proceedings of the 1st Annual Conference on Robot Learning, pp. 515–524. Proceedings of Machine Learning Research (2017)
5. Kumra, S., Kanan, C.: Robotic grasp detection using deep convolutional neural networks. In: IEEE/RSJ International Conference on Intelligent Robots and Systems (IROS), pp. 769–776. IEEE, Canada, Vancouver (2017). https://doi.org/10.1109/IROS.2017.8202237
6. Du, G., Wang, K., Lian, S., Zhao, K.: Vision-based Robotic Grasping From Object Localization, Object Pose Estimation to Grasp Estimation for Parallel Grippers: A Review. In: Artificial Intelligence Review, Springer, Heidelberg (2020). https://doi.org/10.1007/s10462-020-09888-5
7. Rusu, R. B., Blodow, N., Marton, Z. C., Beetz, M.: Close-range Scene Segmentation and Reconstruction of 3D Point Cloud Maps for Mobile Manipulation in Domestic Environments. In: 2009 IEEE/RSJ International Conference on Intelligent Robots and Systems, pp. 1–6. USA, St. Louis (2009). https://doi.org/10.1109/IROS.2009.5354683
8. Qu, L., He, S., Zhang, J., Tian, J., Tang, Y., Yang, Q.: RGBD Salient Object Detection via Deep Fusion. IEEE transactions on image processing **26**(5), 2274–2285 (2017). https://doi.org/10.1109/TIP.2017.2682981
9. Qi, C. R., Chen, X., Litany, O., Guibas, L. J.: ImVoteNet: Boosting 3D Object Detection in Point Clouds with Image Votes. In: 2020 IEEE/CVF Conference on Computer Vision and Pattern Recognition (CVPR), pp. 4403–4412, Seattle, WA, USA, (2020). https://doi.org/10.1109/CVPR42600.2020.00446
10. Yang, Z., Sun, Y., Liu, S.,Jia, J.: 3DSSD: Point-based 3D Single Stage Object Detector. In: 2020 IEEE/CVF Conference on Computer Vision and Pattern Recognition (CVPR), pp. 11037–11045, Seattle, WA, USA, (2020). https://doi.org/10.1109/CVPR42600.2020.01105
11. Han, L., Zheng, T., Xu, L., Fang, L.: OccuSeg: Occupancy-aware 3D Instance Segmentation. In: IEEE/CVF Conference on Computer Vision and Pattern Recognition (CVPR), pp. 2937–2946. IEEE, Seattle, WA, USA (2020). https://doi.org/10.1109/CVPR42600.2020.00301
12. Song, C., Song, J., Huang, Q.: Hybridpose: 6d object pose estimation under hybrid representations. In: IEEE/CVF Conference on Computer Vision and Pattern Recognition (CVPR), pp. 431–440. IEEE, Seattle, WA, USA (2020). https://doi.org/10.1109/CVPR42600.2020.00051
13. Zeng, A., Song , S., Nießner, M., Fisher, M., Xiao , J., Funkhouser, T.: 3DMatch: Learning Local Geometric Descriptors from RGB-D Reconstructions. In: IEEE Conference on Computer Vision and Pattern Recognition (CVPR), pp. 199–208. IEEE, Honolulu, HI, USA (2017). https://doi.org/10.1109/CVPR.2017.29
14. Tian, Z., Shen, C., Chen, H., He, T.: Robust 6d object pose estimation by learning rgb-d features. In: IEEE International Conference on Robotics and Automation (ICRA), pp. 6218–6224, IEEE, Paris, France (2020). https://doi.org/10.1109/ICRA40945.2020.9197555
15. Pereira, N., Alexandre, L.A.: MaskedFusion: Mask-based 6D Object Pose Estimation. Preprint (2019). arXiv:1911.07771
16. Gonzalez, M., Kacete, A., Murienne, A., Marchand, E.: Yoloff: you only learn offsets for robust 6dof object pose estimation. Preprint (2020). arXiv :2002.00911

17. Wang, C., Xu, D., Zhu, Y., Martín-Martín, R., Lu, C., Fei-Fei, L.,Savarese, S.: Densefusion: 6d object pose estimation by iterative dense fusion. In: IEEE/CVF Conference on Computer Vision and Pattern Recognition (CVPR), pp. 3338–3347, IEEE, Long Beach, CA, USA (2020). https://doi.org/10.1109/CVPR.2019.00346
18. Bohg, J., Morales, A., Asfour, T., Kragic, D.: Data-Driven Grasp Synthesis–A Survey. IEEE Transactions on Robotics **30**(2), 289–309 (2014). https://doi.org/10.1109/TRO.2013.2289018
19. Mahler, J., Matl, M., Satish, V., Danielczuk, M., DeRose, B., McKinley, S., Goldberg, K.: Learning ambidextrous robot grasping policies. In: Science Robotics, Vol. 4, Issue 26 (2019). https://doi.org/10.1126/scirobotics.aau4984
20. Zeng, A., Song, S., Yu, K.-T., Donlon, E., Hogan, F. R., Bauzá, M., Ma, D., Taylor, O., Liu, M., Romo, E., Fazeli, N., Alet, F., Chavan-Dafle, N., Holladay, R., Morona, I., Nair, P. Q., Green, D., Taylor, I., Liu, W., Funkhouser, T., Rodriguez, A.: Robotic Pick-and-Place of Novel Objects in Clutter with Multi-Affordance Grasping and Cross-Domain Image Matching, In: IEEE International Conference on Robotics and Automation (ICRA), pp. 3750–3757, IEEE, Brisbane, Australia (2018). https://doi.org/10.1109/ICRA.2018.8461044
21. Zakka, K., Zeng, A., Lee J., Song, S.: Form2Fit: Learning Shape Priors for Generalizable Assembly from Disassembly, In: IEEE International Conference on Robotics and Automation (ICRA), Paris, France, pp. 9404–9410, (2020) https://doi.org/10.1109/ICRA40945.2020.9196733

Robot Programming

Playback Robot Programming Framework for Fiber Spraying Processes

Edgar Schmidt and Dominik Henrich

Abstract

Robot-based automation is still not widespread in small and medium-sized enterprises, since programming industrial robots is usually costly and only feasible by experts. This disadvantages can be resolved by using intuitive robot programming approaches like playback programming. At the same time, there are currently not automatized automatized, like fiber spraying. We present a novel approach in programming a robot system for fiber spraying processes, which extends a playback programming framework inspired by video editing concepts. The resulting framework allows the programming of also the periphery devices needed for the fiber spraying process. We evaluated the resulting programming framework to measure the intuitiveness in the use and show that the framework is not only able to program fiber spraying tasks but is also rather intuitive to use for domain experts.

Keywords

Flexible manufacturing • Small batch production • Intuitive robot programming • Playback programming • Fiber spraying process

E. Schmidt (✉) · D. Henrich
Chair for Applied Computer Science III Robotics and Embedded Systems, Universität Bayreuth, 95440 Bayreuth, Germany
E-mail: Edgar.Schmidt@uni-bayreuth.de
URL: http://robotics.uni-bayreuth.de

D. Henrich
E-mail: Dominik.Henrich@uni-bayreuth.de

1   Introduction

Robot-based automation is still not distributed in small and medium-sized enterprises (SMEs), even though a low cost and high quality production is possible with the use of robots [1]. This low spread is largely due to the small batch production, which is characterized by different phases of robot use with frequent reconfiguration of the robot system [2]. A fast reconfiguration can be achieved through a speed-up of the robot programming with the use of intuitive robot programming approaches [3, 4]. With the use of such techniques a robot system can quickly be programmed by experts in the domain of the application (*domain experts*), who have little to no programming expertise.

In the production of glass and carbon fiber composites, fiber spraying processes are one of the most commonly used production methods for short fiber-reinforced composites [5], but are only rarely automatized. However, a reduction of the production costs and an increase of the throughput is possible by means of automation of such processes.

We envision a robot programming framework for various kinds of fiber spraying processes, which can easily be programmed by experts in the domain of spraying processes who have little to no programming experience. For this, we need a framework with which we can program fiber spraying tasks and which is intuitive to use.

In this paper we present our programming framework for fiber spraying processes based on the automation approach presented in our work [6], which extends the framework presented in our work [3] for the use case fiber spraying processes. Section 2 gives an overview of the related work regarding the automation of fiber spraying processes and intuitive programming of robots. In Sect. 3, we describe our extensions to the playback programming framework for fiber spraying processes. In Sect. 4, we evaluate the intuitiveness of our framework with a user study. At the end, Sect. 5 summarizes and concludes the paper.

2   Related Work

Most of the commercially available fiber spraying systems are manually operated systems, so that they can be used like a spray paint gun. When used, these manually operated systems expose the domain expert to polluted air during the spraying, the quality of the produced composite depends heavily on the expertise of the operator while at the same time the composite is hardly reproducible and has a high error rate. An automation approach puts away these disadvantages, as seen in an off-line programming approach using graphical programming and virtual reality [7]. This off-line approach is only of limited suitability for small batch production in SMEs since off-line programming requires special training in robot programming.

In the wide variety of robot programming frameworks, only few are intuitive in use. One point is, that textual programming is only suitable for use with robotic experts [8], so that most of the frameworks of robotic manufacturers cannot be used by domain experts.

Commonly used industrial robot programming frameworks are usually based on a certain form of graphical representation for the user interface [9], but many of the frameworks only use one form of graphical representation like icon-based or wizard-based. There also exists frameworks which combine a robot programming paradigm with more than one type of graphical representation, as they utilize a data-flow representation, include a 3D simulation and provide icon based editing [3][10]. As several intuitive representation concepts are combined in these frameworks, they allow domain experts to program complex tasks easily.

The playback robot programming paradigm is of good suitability for an intuitive robot programming framework, as it can be used easily by domain experts who have no knowledge about programming itself. In playback robot programming, the robot is manually guided through the domain expert, and the trajectory is been recorded and stored by the programming system. After that, the trajectory can be played back. This approach is already used for different applications, like deburring [2], spray painting [11], welding [12], contour following [13], and also workpiece assembling with dual arms [14]. Since this approach exploits the process specific knowledge of the operator directly, it is suitable as robot programming paradigm for fiber spraying processes.

Summarized, there does not exist a robot programming framework for the programming of fiber spraying processes, which on the one side utilizes an intuitive robot programming paradigm like playback programming, and on the other side uses an easy to understand and intuitive to use graphical user interface.

3   Programming Framework for Fiber Spraying Processes

In the following subsections we will describe our approach. First, we explain our use case fiber spraying process, then we define important terms. After that, we explain the general structure of the programming framework and describe the user interface in general. At last, we give a closer insight into the programming of the periphery for the fiber spraying processes. Since this framework is based on previous work [3], we put a focus into the extension of the programming framework.

3.1   Fiber Spraying Process

In a *fiber spraying process* for the manufacturing of fiber composite ceramics, continuous oxidic fiber bundles (rovings) are chopped to uniform length immediately before being ejected out of the cutting unit and into a slurry spray, by which they are entrained. Due to the angle and the distance between the tool head (cutting unit and spray gun) and the mold surface the bundles are infiltrated during flight with the slurry. The infiltrated fiber bundles, with slurry being inside the fiber bundles as well as between the fiber bundles reach the mold surface and are orientated randomly (Fig. 1).

Fig. 1 Schematic illustration of a manual fiber spraying process

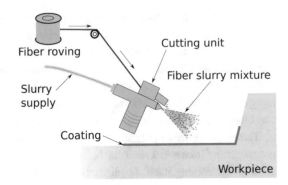

The most important requirement for this process is that the spray gun mounted on the robot toolhead should have a constant distance and orientation to the mold surface. This can be supported by the use of an external axis like a turntable, which allows the spray gun to reach every surface by rotating the mold. Since the spray pattern of the spray gun has also direct influence on the quality of the produced composite and the used fiber bundles are expensive, the material flow and the spray pattern has also to be programmable with high accuracy through the use of pressure regulators. In case of [6], five pressure values have to be programmed.

3.2   Definitions

Playback Programming, also called Walk-Through programming [12] or Programming by Guiding [15], describes a programming approach which consists of two phases: the programming phase and the playback phase. In the programming phase, the robot can be guided either kinesthetically or by a teach-pendant. While guiding, the playback programming framework records the robot joints motions and stores it. In the playback phase, the stored motions can be played back like it was recorded. We call the concatenation of motions in a playback program a *sequence*. A sequence of *n*-dimensional joint configurations with timestamps for each joint configuration, where *n* is the amount of joints of the robot, is further called a *trajectory*.

Besides the term *actor* in [3], we additional define the term *periphery device*: A periphery device is a special case of an actor, characterized in the programming framework by his own *periphery sequence*, which consists of device-specific commands. Examples for periphery devices used in our framework are a turntable which acts as a external axis and pressure regulators which define the material flow and the spray pattern.

A command in the periphery sequence of a turntable is called a *turntable motion*. These motions are specified by a duration, given in seconds [s] and a rotatory distance in degree [°] in the rotatory axis of the turntable. A command in the periphery sequence of the pressure regulators is called a *pressure configuration*. A pressure configuration specifies a duration, given in seconds [s], and five pressure values in bar [bar]. The flat and round jet are defining the spray pattern of the spray gun, while the slurry value defines the amount of slurry given into the process. The pilot air defines the de-/activation of the actual material flow through the spray gun and the cutter value defines the cutting unit speed.

3.3   General Structure

The programming framewok is designed as a client-server application using TCP/IP communication, whereby the graphical user interface acts as the server and the robot act as a client. When a client tries to connect to the server, a registration dialog will pop up. If the periphery devices are controlled centralized though the controller of the robot, the periphery is registered together with the robot. The framework allows to register more than one robot with periphery devices at the same time and the server application can be used via a tablet or a personal computer.

After registration of the robot, the programming of a task can take place immediately. The user first records a robot sequence or loads an already existing sequence from the storage. This sequence can be reviewed or edited at any time before its execution. The periphery can also be programmed right after the registration, but is usually programmed after a trajectory is available. When the programming of the fiber spraying task is finished and the playback phase is started, the sequences are processed into machine code by an interpreter. This machine code is sent to the robot controller and is executed for play back.

3.4   User Interface

The main part of the interaction takes places through the graphical user interface. The design is based on guidelines for the design of operating concepts from the ISO-9241 standard, and the work of [16, 17]. The user interface is further designed to be operated via a touch screen.

The user interface is divided into three parts: The control bar at the top, the timeline area at the bottom left, which shows the data-flow of our current program, and the simulation window at the bottom right. Furthermore, it has an edit mode and a playback mode, which separate both phases of the playback programming paradigm: The edit mode is used for the record phase and the playback mode is used to play back. The different modes with their interfaces are shown in Fig. 2.

The control bar acts as a control area, which contains all functions to control the different actors. These functions are divided into three groups in edit mode. The first group are file operations, the second and third group are extensions to the playback programming paradigm like editing of sequences and synchronizing between actors. In play mode, functions for start, pause and stop are provided.

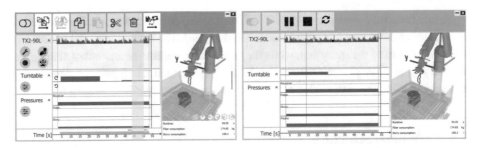

Fig. 2 The graphical user interface of our programming framework, showing the topbar, timelines and 3D-simulation in the edit mode where the actual programming takes place (left) and the playback mode, where the execution of the program is supervised (right)

The timeline area consists of label areas and graphical representations of sequences. In a label area the name of the actor is shown and additional control buttons related to the actor are positioned, like the de-/activation of the hand guidance and the recording of a robot or the programming menu in case of periphery devices. The graphical representation of a sequence is shown as a velocity graph for a robot or turntable. The turntable timeline includes an additional indicator for the direction of the rotation. For the pressure regulators, the pressure values in [bar] are shown. The start and the end of the pressure configurations is indicated, as it hints the material flow of the program. All timelines can be folded resulting in a space-saving but less informative representation, so that a better overview over the registered actors is given but only binary information is displayed.

In the simulation window, a model of the robot is shown with the exhaust unit and the turntable. Further, a 3D representation of a trajectory is also shown if one is available. The timelines and the simulation window are connected so that the simulation window is showing the configuration of the robot when selecting a part of the timeline. During playback mode, the simulation window shows the current position of the physical actor. Below the simulation window, statistical values are displayed, which show the duration of the whole program and the material consumption of slurry and fiber of the program.

3.5 Programming of Periphery Devices

By clicking on the programming button of a periphery device, a window replacing the simulation window is shown. This programming window is designed based on the definition of single commands, which are selected with a combo box. The periphery sequence is shown as a preview and the current command is highlighted in color. The different input parameter can be set through spin boxes. Additionally, timestamps can be defined by selecting from the timeline. Already programmed commands can be edited or deleted. When a command is added to the periphery sequence, an empty run until the next configuration is added automatically, so that the sequence is well defined over the whole program time.

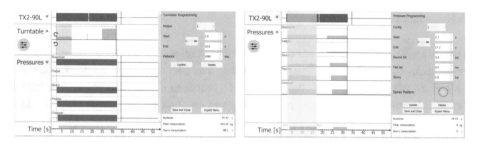

Fig. 3 Two periphery devices have to be programmed in a fiber spraying process. While the turntable acts as external axis (left), the pressure regulators control the actual material flow in the program

For the programming of a turntable motion a start and an end timestamp has to be defined, from which the duration of the motion is calculated. With the input of the moved distance for this time period the motion is fully defined. The corresponding velocity will be calculated from the input and is shown in the timeline. For the use of large scale motions like constant coverage of a cylindrical mold, the adjustment of a velocity for a motion and the setting of a endless running mode of the turntable is achieved with the help of an expert menu. The programming window with the timeline is shown in Fig. 3 left.

Besides the definition of timestamps, the pressure configuration programming takes place through the input of the round jet, flat jet and the slurry. The resulting spray pattern is visualized below the input fields and provides additional information about the expected spray pattern when using the defined configuration. The pilot air and the cutter value can be set in a separate expert menu, as their values only differ for special use cases in the given process. Also, the delay time between the activation of the pressures can be set in the expert menu, which is mostly only needed when the spray gun is changed. The base timeline shows a green bar when every component is running, indicating that the pressure regulator system is spraying at this time interval. In contrast to the aforementioned indicators in the timeline, the green bar indicates that all components are ready to spray. The input is converted into single configurations for the activation of every pressure value, so that actually five pressure configurations are inserted into the sequence when the input is processed. The programming window with timeline is shown in the right subfigure of Fig. 3.

4   Experiments

In this section we will first explain our experiment setup in general. After this, we describe the outcomes of our user study in which we try to measure the intuitiveness of our system when programming fiber spraying tasks.

4.1   Experiment Setup

The participants of this study were 10 persons from the research field of robotics and material engineering, divided in two groups of same size. The average age of the participants was 26.9 years, with 24 years the youngest and 33 years the oldest participant. One half of the participants has expertise in programming and one half has little to no experience in programming. Three of the participants have experience with fiber spraying processes and can be called domain experts. All of the domain experts have also little to no programming knowledge.

As hardware setup for our experiments we use the cell proposed in [6], which utilizes a six degree of freedom robot. Further we use a tablet that provides touch input as input device. The turntable is mounted in the center of the cell providing a rotation around the z-Axis of the robot. The pressure regulators are connected via EtherCAT with the robot control while the turntable is commanded through serial port. A smooth hand guidance is possible through the use of a hybrid force controller, which combines the advantage of a P-controller in fine movement with the advantage of a N-controller for large-scale motions [18].

4.2   Evaluation of Intuitiveness

We rated the intuitiveness in the use of our programming framework with the MINERIC toolkit [19]. For this, the participants had to program a fiber spraying task with the goal to produce a plate. We separated the task into three sub tasks. Task 1 was to program an initial run, which was a meander trajectory there and back over the mold. Task 2 was to edit the trajectory from task 1 so that the robot executes two runs over the mold. Task 3 was to program the periphery devices so that the mold rotated between the runs and for every run other pressures configurations had to be set. Each participant was introduced in the interface before a task was explained. Example solutions are shown in Fig. 4.

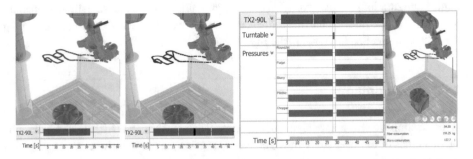

Fig. 4 Possible solutions for the sub tasks of the fiber spraying task which was used to measure the intuitiveness of our framework

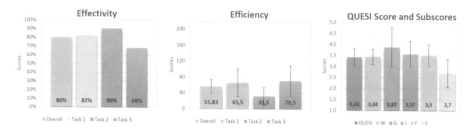

Fig. 5 Outcomes of the user study. The left chart shows the results of the effectivity. Higher values are better. The middle chart shows the results of the efficiency. Lower values are better. The right chart shows the QUESI scores and subscores representing the satisfaction. Higher values are better

The effectivity score is shown in the left chart of Fig. 5. It is measured in percent where a higher score indicates an effective use. The overall rating is 80%, which indicates a rather effective framework to use. Furthermore, the subscores show that task 2 was very effective in use, but the programming of the periphery could be more effective. The efficiency score is shown in the middle chart of Fig. 5. Its value ranges from 0 to 220, where lower scores indicate a better efficiency. The overall rating of 55.83 is translatable to little to some effort when using the framework. The increase of the efficiency for task 3 may be explained with the fact that the periphery programming is harder to understand, while the editing is easy to understand. The QUESI score and its subscores are shown in the right chart of Fig. 5. The QUESI score is measured with five subscores on a scale from 1 to 5, where 5 means that the participants are fully satisfied with the framework and 1 means that they are not satisfied [20]. An overall QUESI score of 3.42 shows that the participants were rather satisfied, but there can be further improvements. The subscores are *subjective mental workload* (W), *perceived achievement of goals* (G), *perceived effort of learning* (L), *familiarity* (F), and *perceived error rate* (E). The score of 2.7 for perceived error rate indicates, that in the process of using the framework some errors were made. Compared to our work in [17], the use of the framework received good ratings although the framework provides now more functionalities and is used to program a more difficult application.

As of the remarkable difference in effectivity and efficiency in task 3, we discuss our observations while the participants were completing this task. Although the current periphery configuration was highlighted, a previous programmed configuration was often mistakenly edited instead of a new one and so the programming was corrupted. Due to the lack of an Undo button, this error could not be corrected by the user and he*she had to start this task again. Also, the option to transfer timeline ranges from selection was used delayed or not at all, resulting in a strenuous text input. These facts caused that task 3 was overall more strenuous for the user than the other two, resulting in a not so effective use. We assume further, that these problems frustrated the user and lead to a also not so well rating in the efficiency.

5   Conclusion and Future Work

In this paper, we presented a framework for the programming of fiber spraying processes, which utilizes a framework inspired by video concepts. We described the new extensions to the framework, with which a fiber spraying task can be programmed, especially periphery devices in form of a turntable and pressure regulator. As of the outcomes of the user study, the framework is rather intuitive to use but the periphery programming has to be further improved with the the observations made in the user study.

Since our framework is only capable of editing parts of a trajectory in their temporal course, a new trajectory has to be programmed by hand guidance every time when only minor details in the spatial course of the sequence have to be changed. This disadvantage could be compensated with the opportunity to edit the spatial course of a trajectory manually or automatically. Also, a CAD model of the mold could be integrated in the simulation window and this model could be used for further optimization in the programming process, like giving feedback when the mold is actual been splashed or overspray is detected.

Acknowledgements This work has partly been supported by the European Regional Development Fund (ERDF) under the project "Roadmap zur flexiblen Produktion individueller Produkte" (Roadmap flexPro). The authors would like to thank Pascal Ruppert and David Harrer for their valuable help in the implementation of the prototype system.

References

1. Perzylo A., et al.: SMErobotics Smart Robots for Flexible Manufacturing. In: IEEE Robotics and Automation Magazine. pp. 78–90 (2019)
2. Dietz T., Schneider, U., Barho, M., Oberer-Treitz, S., Drust, M., Hollmann, R., Hägele, M.: Programming System for Efficient Use of Industrial Robots for Deburring in SME Environments. In: ROBOTIK 2012, VDE, 1–6 (2012)
3. Riedl M., Henrich D.: A Fast Robot Playback Programming System Using Video Editing Concepts. In: Tagungsband des 4. Kongresses Montage Handhabung Industrieroboter, Springer Vieweg, 259–268 (2019)
4. Schraft R. D., Meyer C.: The need for an intuitive teaching method for small and medium enterprises. In: VDI BERICHTE (2006)
5. Witte, E., Mathes, W., Sauer, M., Kühnel, M.: Composites-Marktbericht 2018: Marktentwicklungen. Ausblicke und Herausforderungen. Carbon Composites e.V. und Industrievereinigung Verstärkte Kunststoffe e.V, Trends (2018)

6. Schmidt E., Winkelbauer J., Puchas G., Henrich D., Krenkel W.: Robot-Based Fiber Spray Process for Small Batch Production. In: Annals of Scientific Society for Assembly, Handling and Industrial Robotics (2020)
7. Aßhoff C.: Herstellung von Tailored Composites. Fraunhofer WKI, (2016). https://www.wki.fraunhofer.de/content/dam/wki/de/documents/Mediathek/themen/hofzet/HOFZET_Faserspritzen_2016-10.pdf
8. Orendt E. M., Riedl M., Henrich D.: Programmierung von Robotern. In: Handbuch Mensch-Roboter-Kollaboration, 243-270 (2019)
9. Rossano G. F., Martinez. C., Hedelind M., Murphy S., Fuhlbrigge T. A.: Easy robot programming concepts: An industrial perspective. In: IEEE International Conference on Automation Science and Engineering (CASE), 1119-1126 (2013)
10. TracePen DE [online]: https://wandelbots.com/de/, last access: February 1st 2021
11. Ferraguti F., Landi C. T., Secchi C., Fantuzzi C., Nolli M., Pesamosca M.: Walkthrough programming for industrial applications. In: 27th International Conference on Flexible Automation and Intelligent Manufacturing, FAIM2017, 31–38 (2017)
12. Ang Jr. M. H., Lin W., Lim S. Y.: A walk-through programmed robot for welding in shipyards. In: Industrial Robot: An International Journal, 377-388 (1999)
13. Bascetta L., Ferretti G., Magnani G., Rocco P.: Walk-through programming for robotic manipulators based on admittance control. In: Robotica, 31(7) (2013)
14. Park, C., Park, K., Park D., Kyung J.-H.: Dual Arm Manipulator and Its Easy Teaching System. In: Proceedings of 2009 IEEE International Symposium on Assembly and Manufacturing, 242–247 (2009)
15. Lozano-Pérez, T.: Robot Programming. Proceedings of the IEEE **71**(7), 821–841 (1983)
16. Kraft M., Rickert M.: How to teach your robot in 5 minutes: Applying UX paradigms to human-robot-interaction. 26th IEEE International Symposium on Robot and Human Interactive Communication (RO-MAN), 942–949 (2017)
17. Colceriu C., Riedl M., Henrich D., Brell-Cokcan S., Nitsch V.: User-Centered Design of an Intuitive Robot Playback Programming System. Annals of Scientific Society for Assembly, Handling and Industrial Robotics, 193–203 (2020)
18. Stolka, P., Henrich, D.: A hybrid force following controller for multi-scale motions. IFAC Proceedings Volumes **247–252**, (2003)
19. Orendt E. M., Fichtner M., Henrich D.: MINERIC toolkit: Measuring instruments to evaluate robustness and intuitiveness of robot programming concepts. 26th IEEE International Symposium on Robot and Human Interactive Communication (RO-MAN), 1379–1386 (2017)
20. Naumann A., Hurtienne J.: Benchmarks for intuitive interaction with mobile devices. Proceedings of the 12th international conference on Human computer interaction with mobile devices and services (MobileHCI '10), 401–402 (2010)

LiDAR-Based Localization for Formation Control of Multi-Robot Systems

Tobias Recker, Bin Zhou, Marvin Stüde, Mark Wielitzka, Tobias Ortmaier and Annika Raatz

Abstract

Controlling the formation of several mobile robots allows for the connection of these robots to a larger virtual unit. This enables the group of mobile robots to carry out tasks that a single robot could not perform. In order to control all robots like a unit, a formation controller is required, the accuracy of which determines the performance of the group. As shown in various publications and our previous work, the accuracy and control performance of this controller depends heavily on the quality of the localization of the individual robots in the formation, which itself depends on the ability of the robots to locate themselves within a map. Other errors are caused by inaccuracies in the map. To avoid any errors related to the map or external sensors, we plan to calculate the relative positions and velocities directly from the LiDAR data. To do this, we designed an algorithm which uses the LiDAR data to detect the outline of individual robots. Based on this detection, we estimate the robots pose and combine this estimate with the odometry to improve the accuracy. Lastly, we perform a qualitative evaluation of the algorithm using a Faro laser tracker in a realistic indoor environment, showing benefits in localization accuracy for environments with a low density of landmarks.

T. Recker (✉) · B. Zhou · A. Raatz
Leibniz University Hannover, Institute of Assembly Technonolgy, An der Universität 2, 30823 Garbsen, Germany
E-mail: Recker@match.uni-hannover.de
URL: https://www.match.uni-hannover.de

M. Stüde · M. Wielitzka · T. Ortmaier
Leibniz University Hannover, Institute of Mechatronic Systems, At the Universit at 1, 30823 Garbsen, Germany

© The Author(s) 2022
T. Schüppstuhl et al. (eds.), *Annals of Scientific Society for Assembly,*
Handling and Industrial Robotics 2021,
https://doi.org/10.1007/978-3-030-74032-0_30

Keywords

Multi-robot systems • Formation control • Relative localization

1   Introduction

Robot formations are often used to extend a single robot's capabilities or break down complex tasks into simpler subtasks Feng et al. [1]. In this way, robot formations increase the flexibility of the overall robotic system and improve its ability to do new tasks. A group of robots is required to maintain their formation even under the influence of disturbances to function as a unit. This is done by the formation control. Depending on the control law, the formation control requires the robot's positions to be known either regarding a static map or relative to the other robots. This localization is either done using external sensors (external camera or laser tracker) Irwansyah et al. [2] or internal (on-board) sensors (internal cameras, LiDAR, ultrasonic, ultrawideband ...) Nguyen et al. [3],Li et al. [4],Güler et al. [5],Bisson et al. [6] together with knowledge of the environment (map). There are also solutions that combine both principles, for example, an external transmitter sends a signal that is then received by internal sensors. Examples of this are triangulation methods based on time of flight measurements Choi et al. [7] or the recognition of artificially placed landmarks with known positions Zhang et al. [8]. While external localization is generally more precise, it does require more equipment and limits the working area. This is why we will focus on internal sensors in this work.

When calculating two robots' relative positions using internal sensors, it is common to localize both robots within the map and then subtract the absolute positions. The main drawback of this method is the introduction of several sources of error. Firstly, the localization within a given (perfect) map contains a rather significant error, which alone amounts to an inaccuracy of several mm Chan et al. [9],Gang et al. [10]. Secondly, the map does not provide a perfect representation of the real environment, which reduces accuracy even further. Also, individual robots in a robot formation can obstruct the line of sight of other robots to important features in the map, making localization more difficult. To counteract this type of error, researchers have tried to estimate the relative position of the other robots directly Wasik et al. [11],Franchi et al. [12],Teixido et al. [13],Huang et al. [14],Rashid and Abdulrazaaq [15]. For example, this is done by detecting the outline of other robots in the laser scanner data Wasik et al. [11]. For this purpose, the robot's contour is approximated by simple shapes such as rectangles or circles. These shapes are then searched for in the measurement data of the laser scanners Teixido et al. [13]. If a sufficient number of points can be detected on the robot's outline, the position of the robot can be determined. If not, the position can be estimated based on the last known position by using an extended Kalman filter (EKF) taking the robot's velocity either from odometry or derivation of the previous positions Huang et al. [14].

While there has been much research in this area, we could not find a solution perfectly fitted to our needs. While researchers like Huang et al. Huang et al. [14] were able to localize robots relative to each other, the pose estimation performance is insufficient. In general, either the frequency of the estimation Wasik et al. [11] or the accuracy Huang et al. [14],Franchi et al. [12],Koch et al. [16] was insufficient.[1] In other cases, the algorithms used require features (markers) Teixido et al. [13], a special robot shape Howard et al. [18] or additional sensors Bisson et al. [6] not present on our hardware.

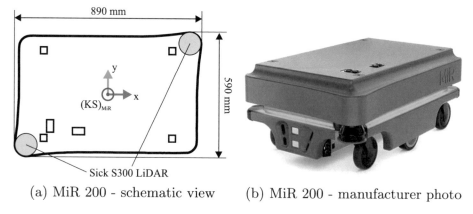

(a) MiR 200 - schematic view   (b) MiR 200 - manufacturer photo

Fig. 1 Dimensions of the MiR 200 [mobile-industrial-robots.com]

2   System Overview

For our use case, we want to determine the position and orientation (pose) of several MiR 200 robots. The MiR 200 is a rectangular mobile robot with a footprint of 890 mm by 580 mm (see Fig. 1). It is equipped with two sick s300 laser scanners on opposing corners of the robot, giving it a 360-degree detection radius without any blind spots. In this publication, we want to use both of these scanners to directly estimate the position of all robots in our formation without the need to subtract error-prone absolute positions. For this, we are proposing the algorithm shown in Fig. 2. In the first step, we fuse the sensor data from the front and rear laser scanner. We then use the map to remove every data point caused by an obstacle. Afterward, we use Density-Based Spatial Clustering of Applications with Noise (DBSCAN) to detect the potential robots in the remaining data. The next step is to check if

---

[1]In another publication Recker et al. [17] we have already investigated the theoretical accuracy of our formation control and developed mechanical compensation units to compensate for the lateral errors within the formation. Due to the mechanical design of the compensation units, they can only compensate errors of individual robots up to 100 mm.

the potential robots are one of our MiR 200 robots. We do this by checking if the shape of the respective point-cloud matches the shape of one of our robots. The last step is to compute the position and orientation and validate it.

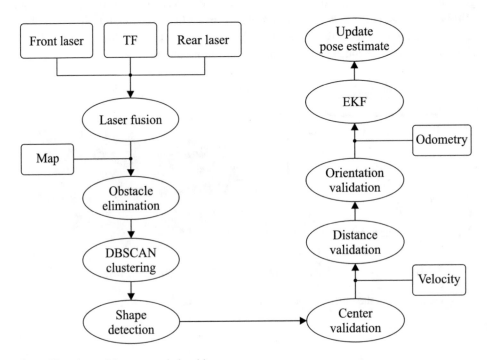

Fig. 2 Flowchart of the proposed algorithm

3   Algorithm

As described above, our algorithm starts by removing known objects (obstacles) from the measurement data. For this, we use a standard adaptive Monte Carlo Localization (AMCL) algorithm to estimate the pose of every robot within our map. Next, we remove every data point within close proximity (0.1 m) of a known obstacle within the map.[2] This process is shown in Fig. 3. Note that this step is not mandatory for our algorithm to work and can be skipped if no map is available or if localization in the map is impossible.

The next step is to detect clusters of points, which could be a robot. For this, we use an implementation of DBSCAN clustering (based on Ester et al. [19]) to assign every point

---

[2]The safety PLC of the MiR 200 will trigger an emergency stop if it detects an object within 0.1 m. Accordingly, we except our robots to be at least 0.1 m away from every object.

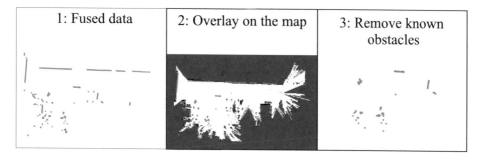

Fig. 3 Eliminating obstacles from the data based on the map

to a group of points and remove noise from the data. As our laser scanner has an angular resolution of about 0.4 degrees and the smallest side of our robot is 0.58 m long, we expect the scanners to detect at least 16 points on our robot - given the robot is less than 5 m away. Accordingly, every cluster consisting of less than 16 points can be disregarded. The step of clustering is shown in Fig. 4.

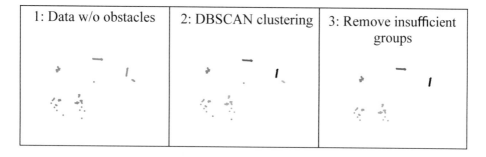

Fig. 4 DCSCAN clustering to remove scattered points

The third step is to remove clusters that do not fit the shape or size of a robot (see Fig. 5). As our robot is rectangular, only straight lines or 90-degree corners are valid shapes. In addition, we require the length of the point clusters to match the length of one of our robots sides with a maximum deviation of 20%. To do this detection, we expand on work from Zhang et al. Zhang et al. [20] by adjusting the geometric parameters to our robot and changing the orientation estimation to a multi-stage process. Instead of trying all possible orientations between 0 and 90 degrees with a fixed step size, we make several passes with decreasing step size. For the next stage, we only reduce the step size for the search space between the two angles, which previously yielded the highest correspondence between measured values and robot shape.

Applying all the previous steps, we are left with two similar groups of points. As booth groups form a line, there are four theoretical robot positions for each line (see Fig. 6). Out of

1: Cluster	2: Shape detection	3: Remove invalid shapes
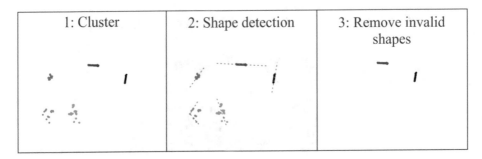		

Fig. 5 Shape detection to remove (dynamic) obstacles with the wrong shape

these four positions, we eliminate the two with the closest distance to the measuring robot. If a robot were at one of those two positions, the structure of the robot would obstruct the line of sight to the detected line. Next, we compare the length of the point cluster to the length of our robot's front and side. This step is quite prone to errors. Because of the noise of the sensors, and because of the clustering, the length of the visible side cannot always be determined correctly. Therefore, the last step is to compare the new position estimate with the previous one. By comparing the two estimates and the velocity readings from the odometry, the most likely new position can be determined. In this way, other objects with a similar shape to the robot can also be excluded.

1: Possible robot poses	2: Center Validation	3: Side detection
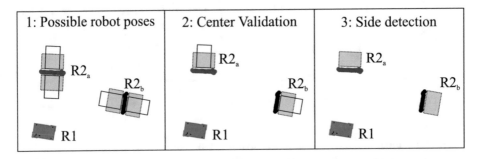		

Fig. 6 Center validation to eliminate impossible robot positions

Remark: The last step shown here necessarily requires the exchange of data between the individual participants of the formation. However, in many cases, this exchange is not possible or desired. Given that situation, it would also be possible to get the velocities of the other robots by differentiating their last known positions (we also validated this in our experiments). Since in this case, the accuracy will most likely decrease in some situations (also see Sec. 4), we, therefore, recommend using the real velocity data if available.

Since the measurement frequency of LiDAR systems is usually lower than the measurement frequency of the wheel encoders (in our 12 Hz vs. 50 Hz), we suggest a sensor

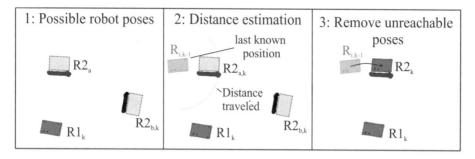

Fig. 7 Travel distance estimation to eliminate poses too far apart

fusion using an EKF. Our implementation is based on the `robot_pose_ekf`[3] package using a plugin made for GPS signals to update the prediction with the localization from our algorithm. As sensor data fusion with an EKF has already been covered extensively in the literature Khatib et al. [21]Eman and Ramdane [22]Huang et al. [14], we do not want to elaborate on it any further in this publication.

4   Evaluation

As mentioned before, we are using the MiR 200 industrial robot for the evaluation of our algorithm. In the first experiment (scenario I) we used two of these robots in our laboratory. To get an impression of the achievable accuracy, in this experiment, we only move one of the two robots and measure its position via the LiDAR of the other robot. We then compare the position estimated from our algorithm with a standard AMCL implementation. For AMCL, we have used the existing implementation in the ROS navigation stack and parameterized it manually to the best of our knowledge. In this scenario, we will not transfer any data between the robots, so the velocity of the robot has to be derived from previous position data. For reference, we use an external laser tracker to get the "true" relative position and distance. Our Faro Vantage laser tracker provides 3-DOF position measurements with an accuracy of under 0.1 mm and a frequency of 1000 Hz.

For the second experiment, we perform a simulation. This is necessary because we want to investigate the behavior of the algorithm compared to AMCL in an environment with fewer features. The simulation environment Gazebo also allows us to measure the true position of all robots in real-time, which would not be possible in reality (due to a lack of sensors). In the simulation, we will therefore be able to move both robots simultaneously and still get their true positions. Besides, we now transfer the odometry data between both robots. For the S300 laser scanners we are assuming a standard deviation of the noise of 29 mm for distances under $3\,m$ and 1% of the distance above $3\,m$ (according to the datasheet). The

---

[3]https://github.com/ros-planning/robot_pose_ekf.

encoders are modeled with a noise equal to 1% of the current velocity. Again we will try to estimate the pose of both robots using AMCL and the direct LiDAR-based detection we described in this article.

When comparing the measured trajectories in Fig. 9 (a), it becomes apparent that for the scenario I conventional distance estimation via AMCL is superior to the direct distance measurement. Figure 8 also confirms that in this case, both the average and maximum deviation are lower, and the signal is less noisy. We assume that due to the very good map and the high density of recognizable features in the surrounding area, localization via AMCL is better suited to determine the relative position in this and similar cases. By combining odometry and LiDAR, the AMCL can also compensate most of the noise for an individual robot. Since, in this scenario, we did not transfer the odometry data between robots, we can not use the EKF to reduce the noise in the LiDAR data for our algorithm. As a result, we recognize that applying the presented algorithm does not make sense under the condition of a good map and many features in the environment. Therefore, in the following, we test our algorithm under the assumption that the map is known poorly, or only a few features are available for localization via AMCL.

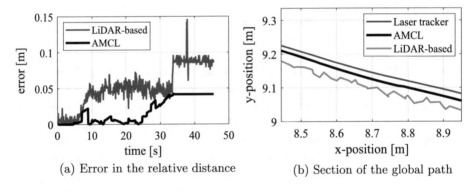

(a) Error in the relative distance   (b) Section of the global path

Fig. 8 Scenario I: One real robot moving - one real robot recording

In the second scenario, our robots are starting close to three landmarks in an otherwise empty environment. The robots then move along an oval trajectory around those landmarks. You can see this in Fig. 9 (b). Since the AMCL has significantly fewer features available, the localization and thus the determination of the relative position is less accurate. This is especially apparent in Fig. 10 (a), where starting at 30 s, two of the landmarks are partially covered by the second robot. Since the localization via AMCL is based exclusively on odometry without a line-of-sight connection to known features, the error temporarily doubles.

However, this experiment also shows a weakness in our proposed method: If the angle between the two robots is a multiple of 90 degrees (e.g., the robots are moving side by side), only one side of the other robot is detected. This is also visible in Fig. 10 (18 sec to 25 sec).

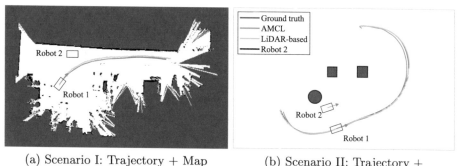

(a) Scenario I: Trajectory + Map   (b) Scenario II: Trajectory + Environment

Fig. 9 Global trajectories for scenario I and II

Since no corner can be identified in the contour, the determination of the start- and endpoint of the robot is subject to greater uncertainty. Also, in some cases, there is a slight delay between measured and real position, which also has a negative influence on the error. This delay occurs when the computer of our robot is busy with another process, and the necessary calculations are not completed in time. In general, we can still calculate a new position with every new LiDAR measurement, which averages to about 12,8 Hz on an Intel Core i7 8700k with 16 Gb of RAM using a Preempt-RT Linux Kernel (5.9.1-rt20).[4] Overall our method does provide some benefits, especially in environments with only a few landmarks.

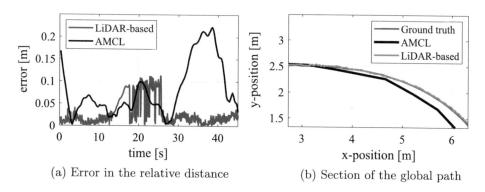

(a) Error in the relative distance   (b) Section of the global path

Fig. 10 Scenario II: Both simulated robots are moving - robot one measures robot 2

---

[4]In normal operation the processor is not fully utilized by the algorithm. Only in the case that a real time process occupies the processor the computing time cannot be kept.

5   Conclusion and Future Work

In this paper, we introduced an improved method for the estimation of relative position for multi-robot formation control. We then compared the performance and applicability of this method to another standard method for relative robot localization. Although our algorithm does not always achieve better results, there is a clear area of application. Additionally, in all cases we studied, we did not require any form of communication or map for the algorithm to function. In the future, we plan on improving the algorithm regarding computing time. This could be done, for example, by first estimating the new robot position and, based on this estimation, removing distant objects from the measured data. Also, we will analyse more edge cases like two robots getting extremely close to each other. Another interesting aspect is the combination/fusion of different localization methods. As our method shows advantages in certain scenarios and AMCL in others, we plan to combine both algorithms to generate a more robust localization.

References

1. Feng, Z., et al.: An overview of collaborative robotic manipulation in multi-robot systems. Annual Reviews in Control - **49**, 113–127 (2020)
2. Irwansyah, A., Ibraheem, O.W., Hagemeyer, J., et al.: FPGA-based multi-robot tracking. Journal of Parallel and Distributed Computing **107**, 146–161 (2017)
3. Nguyen, T. H., Kim, D. H., Lee C. H. et al.: Mobile Robot Localization and Path Planning in a Picking Robot System Using Kinect Camera in Partially Known Environment. In: International Conference on Advanced Engineering Theory and Applications, pp 686–701 (2016)
4. Li, X., Du, S., Li, G. et al.: Integrate Point-Cloud Segmentation with 3D LiDAR Scan-Matching for Mobile Robot Localization and Mapping, Sensors 2020 20(1), 1–23 (2020)
5. Güler, S., Abdelkader, M., Shamma, J. S.: Infrastructure-free Multi-robot Localization with Ultrawideband Sensors. In: American Control Conference (ACC), pp. 13–18 (2019)
6. Bisson, J., Michaud, F., Létourneau, D.: Relative positioning of mobile robots using ultrasounds. Intelligent Robots and Systems **1783–1788**, (2003)
7. Choi, W., Li, Y., Park, J. et al: Efficient localization of multiple robots in a wide space. In International Conference on Information and Automation, pp. 83–86 (2010)
8. Zhang, H., Zhang, L., Dai, J.: Landmark-Based Localization for Indoor Mobile Robots with Stereo Vision. International Conference on Intelligent System Design and Engineering Application, pp. 700–702 (2012)
9. Chan S., Wu P., Fu L.: Robust 2D Indoor Localization Through Laser SLAM and Visual SLAM Fusion. IEEE International Conference on Systems, Man, and Cybernetics (SMC), pp. 1263–1268 (2018)
10. Gang, P. et al.: An Improved AMCL Algorithm Based on Laser Scanning Match in a Complex and Unstructured Environment. Complexity, pp. 1–11 (2018)
11. Wasik, A. et al.: Lidar-Based Relative Position Estimation and Tracking for Multi-Robot Systems. In: Robot 2015: Second Iberian Robotics Conference, pp. 3–16 (2015)
12. Franchi, A., Oriolo, G., Stegagno, P.: Mutual localization in a multi-robot system with anonymous relative position measures, In; IEEE/RSJ International Conference on Intelligent Robots and Systems, pp. 3974–3980 (2009)

13. Teixido, M., Pallejà, T., Font, D., et al.: Two-Dimensional Radial Laser Scanning for Circular Marker Detection and External Mobile Robot Tracking. Sensors **12**, 16482–97 (2012)
14. Huang, G.P., Trawny, N., Mourikis, A.I., et al.: Observability-based consistent EKF estimators for multi-robot cooperative localization. Auton Robot **30**, 99–122 (2011)
15. Rashid, A., Abdulrazaaq, B.: RP Lidar Sensor for Multi-Robot Localization using Leader Follower Algorithm. The Iraqi Journal of Electrical and Electronic Engineering, pp. 21–32 (2019)
16. Koch, P., May, S., Schmidpeter, M., et al.: Multi-Robot Localization and Mapping Based on Signed Distance Functions. J Intell Robot Syst **83**, 409–428 (2016)
17. Recker T.,Heinrich, M., Raatz, A.: A Comparison of Different Approaches for Formation Control of Nonholonomic Mobile Robots regarding Object Transport, CIRPe – 8th CIRP Global Web Conference on Flexible Mass Customisation, (in publication) (2020)
18. Howard, A., Mataric, M.J., Sukhatme, G. S.: Putting the 'I' in 'team': an ego-centric approach to cooperative localization. In: IEEE International Conference on Robotics and Automation, pp. 868–874 (2003)
19. Ester, M., Kriegel, H.P., Sander, J. et al.:A density-based algorithm for discovering clusters in large spatial databases with noise, Second International Conference on Knowledge Discovery and Data Mining, pp. 226–231 (1996)
20. Zhang, X., Xu, W., Dong, C., et al.: Efficient L-shape fitting for vehicle detection using laser scanners. IEEE Intelligent Vehicles Symposium (IV) **54–59**, (2017)
21. Khatib, E. I. Al., Jaradat, M. A., Abdel-Hafez M. et al.: Multiple sensor fusion for mobile robot localization and navigation using the Extended Kalman Filter. In: 10th International Symposium on Mechatronics and its Applications (ISMA), pp. 1–5 (2015)
22. Eman, A., Ramdane, H.: Mobile Robot Localization Using Extended Kalman Filter. In: 3rd International Conference on Computer Applications & Information Security (ICCAIS), pp. 1–5 (2020)

Playback Robot Programming with Loop Increments

Michael Riedl and Dominik Henrich

Abstract

Playback robot programming is fast and easy to use for non-experts, because the robot only needs to be manually guided. However, it is only capable of replaying the trajectory exactly as it was taught. We present the concept of loop increments for playback programmed robots to allow the user to teach tasks like palletizing or stacking without having to explicitly guide the robot through each trajectory. Only the base trajectory for one repetition needs to be program med. After each loop iteration, the user-defined increment is added to the incremental configurations, e.g. to the pick or place configurations. To achieve this, two methods of defining the loop increments are shown. Afterwards, linear, Gaussian, and cosine blending functions in combination with the point and interval method are introduced for weighting the increments and as a foundation for the adaption algorithm. The evaluation showed, that the cosine blending function with the interval method best fits the needs of our programming system.

Keywords

Intuitive robot programming · Programming system · Trajectory adapting

M. Riedl (✉) · D. Henrich
Chair for Applied Computer Science III, Robotics and Embedded Systems, Universität Bayreuth, 95440 Bayreuth, Germany
E-mail: Michael.Riedl@uni-bayreuth.de
URL: https://robotics.uni-bayreuth.de

D. Henrich
E-mail: Dominik.Henrich@uni-bayreuth.de

T. Schüppstuhl et al. (eds.), *Annals of Scientific Society for Assembly, Handling and Industrial Robotics 2021*,
https://doi.org/10.1007/978-3-030-74032-0_31

1   Introduction

Programming robots using the playback programming approach is rather easy to understand and fast to use for the user. However, when it comes to repetitive tasks where after each iteration small changes to the trajectory are needed, the playback programming approach reaches its limits. This is the case with tasks such as stacking and palletizing. Every iteration of the task needs to be programmed explicitly when using the classic playback programming approach. In a previous work, we already presented an approach that allows the user to define sensor-based loops within a playback-programmed robot program, so that repetitive tasks that always play back exactly same trajectory can be programmed. The novel was, that the number of repetitions of the loop is defined by sensor input, e.g. camera images [1]. In this paper, we present an extension to the concept of loops for playback-programmed robot programs called loop increments. With these loop increments, we allow the user to program repetitive tasks with small changes to the trajectory after each iteration without the need to program all trajectories explicitly. Only the base trajectory, i.e. the trajectory of the first iteration of the loop, needs to be programmed and then the increments need to be defined. An example depalletizing and stacking task is shown in Fig. 1. This is a major extension to the playback programming approach, because until now, playback-programmed robot only replayed trajectory exactly as it was programmed and no adaption of the programs was possible during execution.

In the rest of the paper, we show how increments are defined within the existing programming system. Afterwards, the adapting of the trajectory with the help of blending functions

Fig. 1 **a** Example task consisting of depalletizing (left robot) and stacking (right robot) that can be programmed with loop increments by demonstrating only one pick and place trajectory. **b** Timeline representation in our programming system of the above task

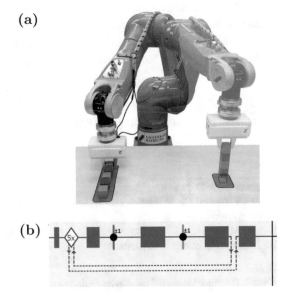

during playback of the robot program is presented. Finally, we compare the different blending functions and evaluate, which blending function best fits our programming system.

2   Related Work

The playback programming approach for industrial robots is one of the oldest programming paradigms. It was described for the first time in [2]. It consists of two phases, the programming phase where the user manually guides the robot through the desired trajectory, and the playback phase where the robot exactly reproduces the taught trajectory. Because no knowledge of textual programming in general and no robot programming skills in particular are needed, it is suitable for non-experts that have only knowledge about the task that should be fulfilled, but not about programming itself. There are several application areas that use this kind of programming technique. A lot of applications are in the field of surface treatment, e.g. spray painting [3], deburring [4], welding [5], gluing [6], contour following [7], or fiber spraying [8]. There are also applications in pick-and-place scenarios, e.g. workpiece assembly [9], moving workpieces from one machine to another [1], or placing of workpieces [10]. When it comes to pick-and-place tasks, there is the problem of palletizing or stacking objects. Until now, this was only solvable with the playback programming approach by guiding each trajectory explicitly by hand, even though a stacking or palletizing task has a lot of similar trajectories that only need a slight adjustment of the trajectory. This is because of the paradigm, that a playback programmed task is played back exactly the same as it was taught by the user.

This problem should be solved with the concept of loop increments for playback programmed trajectories that we present in this work. We have given a robot trajectory, a start and end configuration of a loop within this trajectory, and certain incremental configurations within this loop. After each loop iteration, the corresponding loop increment of the incremental configuration shall be applied to it, so that afterwards these incremental configurations are shifted and tilted according to the applied increment. The wanted algorithm should now adapt the original trajectory to a new, similar trajectory that runs through the incremented configurations.

To achieve this, we need to take a look at existing algorithms to adapt trajectories. There are various approaches in the field of programming by demonstration. Most of them are using multiple demonstrations of trajectories to learn a generalized behavior and derive a new trajectory out of the learned behavior [11–13]. Some other approaches utilize only one demonstration of the trajectory (one-shot), but still try to generalize behavior and try to derive a new trajectory during execution [14, 15].

These algorithms are not suitable for adapting playback programmed trajectories, because they are mostly used to learn and generalize from one ore multiple demonstrations to completely different trajectories. In our case, we want to get a trajectory that is still similar to the original one, but runs through the incremented configurations. The proportional editing function of Blender [16] does something similar, only on three dimensional meshes. Since a trajectory in our case is similar to a mesh, we have oriented ourselves on Blender when designing and implementing the adaption algorithm.

3   Loop Increments for Playback Robot Programs

In this work, we present the concept of loop increments for the playback programming approach. These increments should work as follows: First of all, the base trajectory needs to be programmed by hand guiding the robot through the desired trajectory. Afterwards the user defines the repetitive part of the trajectory as a loop, such as described in [1]. Our novel approach extends the programming of a loop, so that the repetitive trajectories may now be adapted after each loop iteration by a predefined loop increment at runtime.

Within the loop, the user defines certain configurations as incremental configuration that are changed during each iteration of the loop. For each nesting level of loops around these incremental configurations, the user needs to define an additional increment. E.g., if there are two loops around a incremental configuration, two different increments need to be defined for this position - one for the inner loop and one for the outer loop. These increments are the transformations that is applied to the incremental configuration after each loop iteration. If the user wants to stack objects, the increment would be a trajectory in negative z direction of the tool center point coordinate system. Both, translation and rotation, can be defined within an increment by the user. The definition of increments within our programming system is described in Sect. 3.1.

When executing the trajectory, an algorithm is needed that adapts the original playback-programmed trajectory in a way, so that the incremented configurations are reached during playback. To achieve this, a adapting method with suitable blending functions is needed. This algorithm and the different blending functions are described in Sect. 3.2.

3.1   Definition of Increments Within the Programming System

Before we can define the increments, we first need to introduce variables that we are going to use during calculation of the increment and the adapted trajectory. For each incremental configuration, we need the initial pose P_0 and the pose after once applying the increment P_1. From these two poses the incremental transformation D can be calculated. P_0, P_1, and D are all homogeneous 4×4 transformations matrices with $P_1 = P_0 \cdot D$. With the help of this equation, D can be calculated as $D = P_0^{-1} \cdot P_1$. Therefore, only P_0 and P_1 need to be given to define D. With these definitions, we are now able to show how loop increments can be defined in our playback robot programming system.

Since our programming system is designed to be easy and fast to use by both, non-experts and experts in the field of robot programming [1], we allow two ways of defining

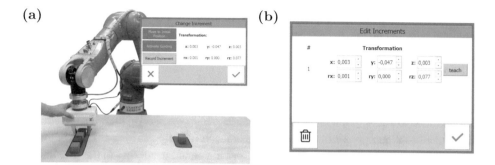

Fig. 2 **a** A photo of a non-expert teaching an increment by manually guiding the robot with a screenshot of the feedback for the user. **b** The user interface for the expert to define the transformation

loop increments. The first step for both ways is to define the incremental configurations within an existing trajectory. This is done by selecting the position within a loop in the timeline representation of the robot program and by activating the loop increment feature of our programming system. With this step, the incremental position P_0 that should be adapted after each loop iteration is selected.

The non-expert way of defining the increment is to manually guide the robot to P_1. To allow the user to do this, the robot is automatically moved to the initial pose P_0. Then it is set to manual guiding mode and the user moves it to P_1. During guiding, the user gets feedback from the programming system in the form that the current transformation D is shown as Cartesian translation and Euler angles in the user interface (Fig. 2a). Once the user is satisfied with the increment, it has to be confirmed and the definition process is over. It is still possible for the user to use the expert menu to adapt D manually afterwards.

If the user has expert knowledge about the process and wants to define D exactly, the user interface offers the possibility to enter the transformation D as Cartesian translation and Euler angle rotation. So the user may insert the transformation in x, y, and z direction and the rotation around the x-, y-, and z-axes (Fig. 2b). It should be noted that the non-expert way defines D implicitly by defining P_1 while the expert way defines D directly within the user interface.

After showing the different methods of defining increments within our programming system, the following subsection shows how the original trajectory is adapted after each loop iteration.

3.2 Trajectory Blending During Playback

The last section described two ways of defining the increments. Now, we will explain how the actual adapting of the original trajectory after each loop iteration works. To achieve this, we first define a trajectory T as a sequence of $n \in \mathbb{N}$ joint configurations q_t with $0 \leq t \leq n$

and $q_t \in \mathbb{R}^d$ for robots with $d \in \mathbb{N}$ axes. Additionally, since a trajectory also includes temporal information, we define that the duration between two joint configurations q_t and q_{t+1} is isochronous for all adjacent joint configurations in T and the duration is always a fixed number of milliseconds.

We now introduce two methods, to determine to which configurations the increment is applied to 100%. The first method only applies the increment to the defined incremental configuration to 100% (*point method*), whereas the second method applies the increment to 100% to all configurations around the incremental configuration as long as the robot stands still (*interval method*). Both methods weight the increment according to the proximity to the next and previous incremental configuration using sigmoid functions. The further away from the incremental configuration the current configuration is, the lower is the impact of the increment to the blending function. The following part showing how the adapted trajectory is calculated works for both, the point method, and the interval method.

Before we are able to adapt each position of T according to the increments D, we first need to define a blending function that calculates the weighting of D for each configuration within the trajectory. This blending functions need to return a value between 0 and 1 that defines to which extent the corresponding increment needs to be added to the current configuration. A blending function is defined as $f_{\text{blend}} : i, j, x \rightarrow w$ with $i, j, x \in \{0, \ldots, n\}$ and $w \in [0; 1]$ while i is the index of the incremental configuration, j is the index of the last configuration that should be affected by the increment, x is the index of the current configuration that should be adapted, and w is the weighting of the increment at configuration i for the configuration x. All blending functions need to fulfill two constraints, the first one is $f_{\text{blend}}(i, j, i) = 1$ and the second one $f_{\text{blend}}(i, j, j) = 0$. We compare three different blending functions: linear, Gaussian, and Cosine in combination with the point and the interval method. In Sect. 4 we show, which blending function fits best.

To define the linear blending function, we use $f(x) = m \cdot x + t$ as base formula. With the two constraints that should be fulfilled, we get Eq. 1 for the linear blending function.

$$f_{\text{blend, linear}}(i, j, x) = \frac{1}{i - j}(x - j) \tag{1}$$

For the Gaussian blending function we use $f(x) = a \cdot \exp\left(-\frac{(x-b)^2}{2c^2}\right)$ as base formula. For the first constraint we get $a = 1$ and $b = i$. This type of formula is not capable of fulfilling the second constraint, because $\forall x \in \mathbb{R} : f(x) > 0$. So we soften this constraint to $f(i, j, j) \approx 0$. To fulfill this constraint, we need to find a suitable c that results in a small enough but not too small exponent. If the exponent becomes too small, $f(x)$ will have a too big gradient and the blending function will be too steep. A steeper blending function generates a worse adapted trajectory (See Sect. 4). For our purposes, $c = \frac{i-j}{d}$ with $d = 3.5$ is the best trade-off between the second constraint and a gradient that is not too big. A bigger d results in $f(i, j, j)$ getting closer to 0, but is also steeper. A smaller d results in $f(i, j, j)$ being not so steep but also further away from 0. Example values for $f(i, j, j)$ with different d are: $d = 3 \rightarrow f(i, j, j) \approx 0.0111$, $d = 3.5 \rightarrow f(i, j, j) \approx 0.0021$, and

$d = 4 \rightarrow f(i, j, j) \approx 0.0003$. Equation 2 shows the Gaussian blending function that we use in our programming system.

$$f_{\text{blend, gaussian}}(i, j, x) = \exp\left(-\frac{(x - i)^2}{2 \cdot \left(\frac{i-j}{3.5}\right)^2}\right) \tag{2}$$

The last blending function we use is a cosine blending function. To define it, we use the base formula $f(x) = a \cdot \cos(b + c \cdot x) + d$. In addition to the two constraints it should range between 0 and 1. To fulfill this, we need to set $a = 0.5$ and $d = 0.5$. For the other two constraints $\cos(b + ci) = 1 \rightarrow b + ci = 0$ and $\cos(b + cj) = 0 \rightarrow b + cj = \pi$ need to be fulfilled. This results in $b = -\frac{\pi}{j-i} \cdot i$ and $c = \frac{\pi}{j-i}$. The simplified cosine blending function is shown in Eq. 3.

$$f_{\text{blend, cosine}}(i, j, x) = 0.5 \cdot \cos\left(\frac{\pi}{j - i} \cdot (x - i)\right) + 0.5 \tag{3}$$

With the defined blending functions, we are now able to adapt the configurations. After each loop iteration, each joint configuration $q_t \in T$ needs to be recalculated. In general, each joint configuration is affected by two increments, increment D_l belonging to the incremental configuration q_l left of q_t and increment D_r belonging to the incremental configuration q_r right of q_t. With one of the three Eqs. 1, 2, or 3, the weightings for each increment are calculated as $w_l = f_{\text{blend}, b}(l, r, c)$ and $w_r = f_{\text{blend}, b}(r, l, c)$ with $b \in \{\text{linear, gaussian, cosine}\}$. In the next step, all D_x with $x \in \{l, r\}$ are divided into their Cartesian translations t_x and Euler angles r_x. Then the translatory $(t_{\text{shift}, x})$ and rotatory $(r_{\text{shift}, x})$ shifts are calculated as $t_{\text{shift}, x} = w_x \cdot t_x$ and $r_{\text{shift}, x} = w_x \cdot r_x$. Now all translatory and rotatory shifts are converted back into 4×4 homogeneous transformation matrices $D_{\text{shift}, x}$. With the help of the robot specific forward kinematics $f_{\text{forward}} : \mathbb{R}^d \rightarrow \mathbb{R}^{4 \times 4}$ that transforms joint configurations into position and orientation of the tool-center-point in homogeneous Cartesian coordinates and the robot specific inverse kinematics $f_{\text{inverse}} : \mathbb{R}^{4 \times 4} \rightarrow \mathbb{R}^d$ that transforms the tool-center-point into joint configurations, we are able to calculate the adapted joint configuration q_c' with the weighted increments $D_{\text{shift}, l}$ and $D_{\text{shift}, r}$:

$$q_c' = f_{\text{inverse}}\left(f_{\text{forward}}(q_t) \cdot D_{\text{shift}, l} \cdot D_{\text{shift}, r}\right) \tag{4}$$

Equation 4 is used to calculate the adapted configuration within a loop after each loop iteration. The resulting trajectory is then played back, so that the repetitive task with slightly different configurations is executed. Section 4 will compare and evaluate the three blending functions shown in Eqs. 1 to 3 with both, the point and interval method.

4 Comparison and Evaluation of Blending Functions

After describing how the trajectories are adapted with the help of the loop increments and the blending functions, we will evaluate this method in this section. In Subsection 4.1, we are going to compare the three different blending functions in combination with both, the point method and the interval method. Afterwards, in Subsection 4.1, we show which of the six combinations of blending function and weighting method is the best for the playback programming approach.

4.1 Comparison

Figure 3 shows the six different combinations of the three blending functions and both, the point method and the interval method for an example trajectory with 343 configurations. This is the trajectory that was taught by manual guiding to fulfill the example depalletizing and stacking task of Fig. 1. Within the graphs, the loop start, the loop end, and the two positions where increments are added to the configurations are depicted. Furthermore, the intervals where the robot is stopped and therefore where the interval method interferes are marked, so that the difference between the point method in Fig. 3a and interval method in Fig. 3b can be seen.

If we compare the different blending functions, we can see that both, the linear and the cosine blending function are point symmetrical within an interval between two increments. However, the Gaussian blending function has a shorter interval where it weights the increment with a high weight and a wider interval where the increments are weighted with a

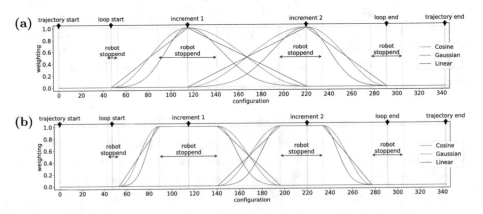

Fig. 3 Comparison of the different blending functions for the trajectory from the task in Fig. 1. Chart **a** shows the result for the blending functions the *point method*, chart **b** with the *interval method*. The red graphs show the cosine, the green graphs the Gaussian and the blue graphs the linear blending functions

low weight compared to linear and cosine. Furthermore, the Gaussian blending function has the disadvantage that it is not continuous at the last configuration that shall be affected by the increment. This is because the Gaussian blending function never reaches the value 0 as already stated in the derivation of Eq. 2. Additionally, if we compare the linear and cosine blending function, we see that the linear function is not smooth at the loop start, loop end and the increments, whereas the cosine blending function is smooth over the whole trajectory. Therefore the cosine blending function is to be preferred over the other two blending functions, regardless of whether the point method or the interval method is chosen. To determine, which of the two weighting methods is better, we are going to evaluate an example trajectory in the next section.

4.2   Evaluation

To determine if the point method or the interval method is better, we evaluate the six combinations of blending function and weighting methods. Figure 4 shows the three dimensional trajectory for the example task shown in Fig. 1.

When using the point method, it is recognizable that regardless of the chosen blending function, the trajectory makes a dip shortly before the green position. If these trajectories are played back, this dip would result in a collision of the robot with the stacked objects and therefore a failure to fulfill the task. This dip is explainable with the graphs of Fig. 3, because we have a interval at increment two where the robot does not move, but the point method already results in the blending function to decrease the weight of the increment. This results in the green position being dragged in z-direction downwards, because the increment is not applied fully to the position anymore.

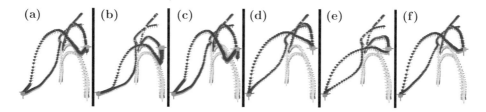

Fig. 4 Simulation of the blending functions for the task in Fig. 1. In each screenshot the original trajectory is drawn in light blue and the adapted trajectory in black with the increment in x-direction for the depalletizing position (red cube in the figure) and the increment in z-direction for the stacking position (green cube in the figure). Subfigures (**a**), (**b**), and (**c**) show the adapted trajectories for the *point method*. Subfigures (**d**), (**e**), and (**f**) show the adapted trajectories for the *interval method*. The cosine blending function is shown in (**a**) and (**d**), the Gaussian blending function in (**b**) and (**a**), and the linear blending function in (**c**) and (**f**)

To avoid this kind of error, we can use the interval method that starts using the blending function to apply the blending function only when the robot is moving. The result of this method for the three dimensional trajectory is depicted in (d), (e), and (f) of Fig. 4. It can be seen that the trajectories do not have the dip anymore and therefore all result in a successful execution of the task.

Because of the outcomes of the comparison of the different blending function and the evaluation of the point method and the interval method, we chose the cosine blending function with the interval method as our preferred trajectory adapted algorithm. It showed the best results in our simulation and is the best with regard to mathematical aspects such as continuity and smoothness. The implementation in our playback programming system showed promising results and different tasks increment tasks were programmed and executed successfully.

5   Conclusion

In this paper we presented an approach to adapt playback programmed trajectories after each loop iteration during runtime, so that tasks like stacking and palletizing can be programmed with our playback programming system. To achieve this, we first showed two ways of defining loop increments, both for non-experts and experts. Afterwards we showed an algorithm that iterates through all joint configurations after each loop iteration and applies the corresponding loop increments with a weighting factor to each configuration, so that the adapted trajectory reaches the next position on the pallet or stacks the next item on top of the previous one. In the evaluation, we compared different blending functions and two weighting methods and showed that the cosine blending function with the interval method is the best for the playback programming approach. This is a next step into the direction of making the playback programming approach more versatile and powerful, because up to now, trajectories could only be played back exactly as they were taught by the user.

Future work may include finding and evaluating other blending functions to further improve the trajectory adaption or using this concept for applications in the field of surface treatment.

References

1. Riedl M., Henrich D.: A Fast Robot Playback Programming System Using Video Editing Concepts. In: Tagungsband des 4. Kongresses Montage Handhabung Industrieroboter, 259–268 (2019)
2. Lozano-Perez, T.: Robot Programming. Proceedings of the IEEE **71**(7), 821–841 (1983)
3. Ferraguti F., Landi C. T., Secchi C., Fantuzzi C., Nolli M., Pesamosca M.: Walk-through programming for industrial applications. 27th International Conference on Flexible Automation and Intelligent Manufacturing, FAIM2017, 31–38 (2017)

4. Dietz, T., Schneider, U., Barho, M., Oberer-Treitz, S., Drust, M., Hollmann, R., Hägele, M.: Programming System for Efficient Use of Industrial Robots for Deburring in SME Environments, ROBOTIK2012. VDE **1–6**, (2012)
5. Ang M. H., Lin W., Lim S.Y.: A walk-through programmed robot for welding in shipyards. In: Industrial Robot: An International Journal, 377–388 (1999)
6. Meyer C., Hollmann R., Parlitz C., Hägele M.: Programming by Demonstration for Assistive Systems - Intuitive Programming of Welding and Gluing Trajectories. In: it-Information Technology, 238–246 (2007)
7. Bascetta, L., Ferretti, G., Magnani, G., Rocco, P.: Walk-through programming for robotic manipulators based on admittance control. Robotica **31**(7), (2013)
8. Schmidt E., Winkelbauer J., Puchas G., Henrich D., Krenkel W.: Robot-Based Fiber Spray Process for Small Batch Production. In: Annals of Scientific Society for Assembly, Handling and Industrial Robotics. 295–304 (2020)
9. Park C., Park K., Park D., Kyung J.-H.: Dual Arm Manipulator and Its Easy Teaching System. In: Proceedings of 2009 IEEE International Symposium on Assembly and Manufacturing, 242–247 (2009)
10. Landi C. T., Ferraguti F., Secchi C., Fantuzzi C.: Tool compensation in walk-through programming for admittance controlled robots. IECON 2016 - 42nd Annual Conference of the IEEE Industrial Electronics Society, 5335–5340 (2016)
11. Nicolescu, M., Mataric, M.: Natural methods for robot task learning: instructive demonstrations, generalization and practice. International Joint Conference on Autonomous Agents and Multiagent Systems AAMAS **241–248**, (2003)
12. Cederborg, T., Li, M., Baranes, A., Oudeyer, P.-Y.: Incremental Local Online Gaussian Mixture Regression for Imitation Learning of Multiple Tasks. IEEE/RSJ International Conference on Intelligent Robots and Systems **267–274**, (2010)
13. Calinon S., Li Z., Alizadeh T., Tsagarakis N. G., Caldwell D. G.: Statistical dynamical systems for skills acquisition in humanoids. 12th IEEE-RAS International Conference on Humanoid Robots, 323–329 (2012)
14. Wu, Y.: Demiris Y: Towards One Shot Learning by Imitation for Humanoid Robots. IEEE International Conference on Robotics and Automation **2889–2894**, (2010)
15. Groth C., Henrich D.: One-Shot Robot Programming by Demonstration using an Online Oriented Particles Simulation. Proceedings of the 2014 IEEE International Conference on Robotics and Biomimetics, 154–160 (2014)
16. Blender Documentation Team: Blender 2.79 Manual. https://docs.blender.org/manual/en/2.79/editors/3dview/object/editing/transform/control/proportional_edit.html, last access: September 17th 2020

Web-Based Platform for Planning and Configuration of Robot-Based Automation Solutions: A Retrospective View on the Research Project ROBOTOP

Eike Schäffer, Philipp Gönnheimer, Daniel Kupzik,
Matthias Brossog, Sven Coutandin, Jörg Franke
and Jürgen Fleischer

Abstract

Automation solutions in production represent a sensible and long-term cost-effective alternative to manual work, especially for physically strenuous or dangerous activities. However, especially for small companies, automation solutions are associated with a considerable initial complexity and a high effort in planning and implementation. The ROBOTOP project, a consortium of industrial companies and research institutes has therefore developed a flexible web platform for the simplified, modular planning and configuration of robot-based automation solutions for frequent tasks. In this paper, an overview of the project's scientific findings and the resulting platform is given. Therefore, challenges due to the scope of knowledge-based engineering configurators like the acquisition of necessary data, its description, and the graphical representation are outlined. Insights are given into the platform's functions and its technical separation into different Microservices such as Best Practice selection, configuration, simulation, AML-data-exchange and spec-sheet generator with the focus on the configuration. Finally, the user experience and potentials are highlighted.

Keywords

Configuration · Planning · Robot-based automation solution · Web-based platform · ROBOTOP

E. Schäffer · M. Brossog · J. Franke
Institute for Factory Automation and Production Systems (FAPS), Friedrich-Alexander-University Erlangen-Nürnberg (FAU), Erlangen, Germany

P. Gönnheimer · D. Kupzik (✉) · S. Coutandin · J. Fleischer
Karlsruhe Institute of Technology (KIT), Wbk Institute of Production Science, Karlsruhe, Germany
e-mail: daniel.kupzik@kit.edu

© The Author(s) 2022
T. Schüppstuhl et al. (eds.), *Annals of Scientific Society for Assembly, Handling and Industrial Robotics 2021*,
https://doi.org/10.1007/978-3-030-74032-0_32

1   Introduction

Despite the availability of numerous software tools for digital planning and virtual commissioning, there are only few efficient and usable, ready-to-use platforms for the concept planning of robot-based automation solutions (RAS). One of the reasons for this is that the concept planning is strongly based on the empirical knowledge of the respective system integrators and their employees. In order to digitalize these process steps, a structured knowledge acquisition and preparation is required. Structured methods are mandatory to prepare knowledge in a scalable, standardized and user centered manner. On the other hand, the platform and the knowledge within must be provided in an accessible, usable and scalable way. Thus, a suitable software architecture (e.g. based on microservices) and the appropriate, scalable technologies (e.g. configurators, 3D web viewers and simulation) are utilized.

2   State of the Art

2.1   System Integration of Automation Solutions

For the planning and commissioning of RAS, an interdisciplinary approach is essential [13]. RAS are technically highly interconnected and complex investment goods. Therefore, the cost of the RAS is strongly influenced through the early planning steps. For this reason, expert knowledge and Best Practice (BP) examples, regarding reasonable realizable physical functions and interaction with the product as well as first basic path planning for position and range evaluation of RAS, are necessary. Thereby, BP [14, 21] are modularized and digitalized, successfully implemented RAS.

In order to steer the planning in a sensible direction, most robot manufacturers currently recommend experienced system integrators for their own systems [9, 30]. Usually, potential robotics users contact these integrators initially to clarify the feasibility of their own ideas and to inform themselves which tasks can be automated in an economically beneficial manner. This is a hurdle, especially if the potential user has little experience in working with RAS, because extensive information and coordination effort is required with poor chances of implementation for the user and signing of a sales contract (order) for the system integrator.

Firstly, they have to define technical specifications for the task. Based on these, the system integrator will derive the requirements to the robotic components. After generating several solutions, one will be chosen, finalized and implemented. All steps after defining technical specifications have to be conducted by the system integrator. One main disadvantage of this is that the process, plant and product knowledge of the user cannot be directly used for the design of the RAS.

In order to digitalize and automate a process like system integration, it is firstly necessary to understand the essential process steps and their results. The system integration

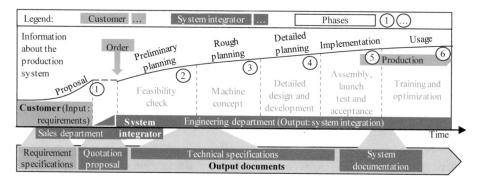

Fig. 1 Phases of the system integration process and relevant output documents [24]

process is therefore divided into five phases before serial production of the system starts in the customer's production facility (Fig. 1). These are used in the following as a scope for the functionalities of the ROBOTOP platform as well as the targeted configuration results.

2.2 Knowledge-Based Engineering Configurators (KBEC) as Tools for Automating the Planning Process of RAS

Knowledge-based configurators which can be developed table-, statement-, rule- or constraint-based [12], are a subgroup within expert systems and Artificial intelligence (AI) [10] and offer an approach to automate knowledge work such as engineering tasks via structured knowledge reuse [28]. Mostly, configurators are used for the individualization of predefined products such as clothes or automobiles [2] rather than new and dynamic engineering tasks. Therefore, we introduced the idea of a real knowledge-based engineering configurator (KBEC) [24], with which new RAS can be created based on BP [21] as well as customized and validated through constraints [15]. KBEC based on constraints [5, 7] provide advantages in transferability, scalability, and maintainability. Constraint models offer a solution description instead of case-specific result description whereas a product configurator (PC) provides existing product knowledge, as mentioned in [2] or [4]. Summarizing, the term engineering configurator is frequently used within literature but implies in most cases a PC e.g. in [4] or [6]. In general, a PC gives automatic access to different existing product variants [2]. Due to the large solution space of RAS planning, the functional scope of a KBEC is less predefined compared to a PC [15, 24], which requires additional development methods.

Existing approaches are either not based on the product to be produced and do not consider interdependencies between product and production system or are too generic or focused on deviating production areas in order to adequately represent the details of specific RAS [6, 11].

2.3 Digitalization of the Engineering Industry by Using Web-Based Platforms and Need for Action for Current Industry

For years, web-based platforms for the configuration of products have successfully offered users the opportunity to find personalized solutions tailored to their own needs. The configuration platforms hereby have the advantage of a significantly lowered inhibition threshold of the customer compared to their conventional equivalents, in the case of the automobile or furniture configurators for the individual configuration in the respective car dealer or furniture store. Systems for configuration and quotation of complex products in the B2B field are called Configure, Price, Quote (CPQ) software [8]. A prominent example for this within the mechanical engineering sector is the company Tacton, which promises high efficiency gains with its application-specific configuration platforms [27].

Production systems as the goods to be configured in the industrial environment lead to much greater challenges than in the customer market. Especially the high complexity of the industrial context with production equipment consisting of numerous components from various globally distributed manufacturers poses a major challenge for web-based configuration platforms.

3 Overcoming Barriers of Web-Based Engineering Configurators

3.1 Need for Action for a Knowledge-Based, Engineering Configurator Platform (ROBOTOP) for Planning of RAS

Increasing demands on the quality and repeatability of manufacturing processes as well as rising labor costs and a growing shortage of qualified personnel favor the use of robots in industrial production. However, the selection and configuration of robotic applications pose a great challenge, especially for small and medium-sized enterprises (SME), as in most cases they do not have adequate tools or specialized experts.

In this context, SMEs are characterized above all by the problem that they are not sufficiently familiar with the solution process of configuring RAS. Trial and error procedures with trying out and testing individual component combinations are both costly and time-consuming. Moreover, many and potentially the best solution variants remain undiscovered by such a procedure, as they are not known to the inexperienced user. SMEs have the opportunity to use the services of experienced system integrators who are aware of both the procedure as well as possible solution combinations. However, the use of these services is also expensive and characterized by a higher barrier than an internet-based configurator, the greatest challenge of which is to break down the complexity of all interactions. In order to master this complexity, the objective of the approach described in this paper is to develop a web-based platform for the configuration of RAS.

3.2 Challenges Within Data Acquisition and Preparation for Knowledge-Based Engineering Configurators Within a Web Platform

The following findings and results are based on the experiences of the last 3.5 years within the research project ROBOTOP [22]. BP based configuration strongly depends on the accessibility of examples the user can identify with as well as suitable components for customization [3]. Therefore, a sufficient number of already existing RAS must be collected and represented on the platform. In these steps, several challenges need to be overcome:

1. **Willingness to share knowledge**: Users and owners of already implemented robot-based automation solutions are usually not willing to fully represent their applications on a platform. First of all, they do not want to present their intellectual property and knowledge in manufacturing on a publicly available platform. Secondly, they want to avoid the effort for documenting and preparing the applications in a user friendly and standardized way. Furthermore, companies that already have robotic applications would not be the main profiteers of the platform.

2. **Missing templates and data models for knowledge acquisition and preparation**: To determine which data is needed, e.g. for BP and how it should be prepared, available templates and data models are missing. These include information about which data is required for the software functions as well as how this data should be prepared for the technologies used, such as the configurator and 3D visualization.

3. **High effort of knowledge acquisition and preparation**: The obtained BP must be implemented in an interactive 3D environment for a good user-experience. The presentation will be based on CAD data. As the models for a component are usually provided by the manufacturer, he also owns the rights to those models. For the publication of the components' 3D representation, it would either have to be replaced by similar components with self-designed CAD representations or the component manufacturers would have to join with the operator of the web platform. The second solution would require a certain size and market penetration by the platform. Therefore, CAD data acquisition is a "chicken or the egg" type of problem.

4. **Availability of easy-to-use tools for data acquisition**: For data acquisition for the knowledge base, usually only expert tools with poor usability are available which lowers the quota of users who feed their applications back to the platform.

4 Platform Architecture and Lessons Learned

4.1 Software Architecture of ROBOTOP

To reduce the number of combination options and to secure the functionality of the configured target RAS, a BP based approach is ideal for KBEC. Due to the existing uncertainty and in order to specify the architecture based on concrete empirical values, various sub-prototypes were created [15, 18–20, 24]. Through the prototypes a large interdependency becomes obvious, e.g. the BP preparation is necessary for the click-prototype and the click-prototype to validate BP in the overall context [24, 26]. ROBOTOP is based on the scalable and flexible idea of a microservice architecture, firstly introduced in 2006 through amazon [29]. In doing so, the ROBOTOP functions are divided into several independent microservices (MS), such as BP selection, configuration, simulation, AML-data-exchange [1, 14] and spec-sheet generator [21]. The MS can communicate via standardized interfaces. Each MS can be developed by using independent technologies [16]. Thus, frameworks and new technologies can be used as appropriate and functions can be dynamically exchanged and extended. This enables an easy implementation of multi-user configuration of RAS and even more flexible user interfaces via micro front end approaches [17].

In order to solve the problem of overly complex user interfaces, a target group specific front end concept for ROBOTOP was developed based on a user-centered approach [26]. Parallel to this, a specific procedure for the development of user-centered user interfaces for KBECs was developed [23]. Thereby, user-centered front-ends could partly perform tasks of a sales person.

In order to develop consistent back-end semantics as a general data model for the individual microservices, a modeling methodology based on a division of labor was introduced as well [20]. This was implemented using the AutomationML Editor, but can be applied to all modeling tools based on potent modeling languages.

4.2   Constraint and Best Practice Acquisition for KBECs

For purpose of KBEC, we have developed firstly a generic constraint acquisition method and secondly a BP specific knowledge acquisition method.

(1) **Generic approach for constraint acquisition**

For a general development of KBEC, especially for the structured division of labor preparation of engineering knowledge in the form of constraints, the "150% topology" including a modelling methodology was introduced [15]. For specific identification and selection of knowledge sources a six-step method was created, including a generic table with potential knowledge sources for constraint acquisition [24].

(2) **Seven step Best Practice knowledge acquisition method**

1. **Defining the required BP data for the 3D-configurator**: Especially the definition of the required data is an agile and cyclical process. This is an interplay between the development of the click-prototype [23], the 3D-models [19] and needed data in an idealistic configurator [15] and the available real data.

2. **Developing concepts to convince partners to provide BP**: In order to convince companies to provide the data, various approaches were explored. In particular, the provision of BP for sales support was well received.

3. **Finding and convincing system integrators**: The first step of every acquisition was to identify a company which is interested in listing their existing RAS on the web platform. To this end, technical questionnaires were sent to contact persons at suitable companies after an initial phone conference.

4. **BP data acquisition**: In this questionnaire, the company can categorize itself according to its size and its used automation systems and existing tools for the planning of RAS. To collect the BP information, the following subjects are requested:

   - Short description of the application.
   - Implemented degree of automation.
   - Motivation for the automation.
   - Approximate cost for planning and implementation.

   In the further course of filling in the questionnaire, a more detailed description of the task and solution is obtained. In the course of the research project, the input data from the company was validated in a meeting with the contact person at that company and an automation expert from the ROBOTOP team. To ensure a realistic representation on the web platform, the application was observed during operation and a simulation. This way, peculiarities of the implementations can be copied to the BP catalogue if necessary. Finally, measurements are made for a true-to-scale display of the application and photographs are taken as a guide for the implementation in the CAD model.

5. **Postprocessing of the collected data**: An abstraction of the acquired applications was executed as the first step of the preparation. The aim was to reduce complexity by removing or replacing unnecessary or highly specific components and to avoid the publication of confidential information of the company which provided the example. For example, the workpiece will be altered to one with similar handling and machining properties without being identifiable as the original workpiece.

6. **Preparation of BP data as 3D-model**: From the simplified representation of the BP, a 3D scene and an attribute table are derived. For the 3D representation of existing applications, CAD models of the components were assembled. These full-models of the application were rendered for the preview pictures in the BP overview.

7. **Web platform integration of BP**: For the manipulation of the example, the individual components of the models were transferred to the configurator database together with their placement information for each scene. The platform user can manipulate copies of the Best Practices during the configuration process.

5   Validation of the Iterative Development of ROBOTOP Based on Concepts Implementation and Platform Facts

Within ROBOTOP, 34 BP solutions were acquired and prepared to be an input in the web-based configuration. The process to be run through by the user on the platform from the initial identification of a suitable solution to the user-specific adaptation is divided into four steps: BP selection, configuration, simulation and finally overview and specification. For the recording and assignment of user requirements, 118 specific rules were defined to cover interactions.

In the BP selection, the user can choose between eleven different processes in the application area, from assembly to drilling up to testing processes. Here, the BP solutions that do not fit the user's target application are filtered out first. The user can then filter out other unsuitable BP by entering information about the workpiece. Information includes workpiece shape and dimensions interacting with the BP gripper and weight information interacting with the robot's payload.

In the subsequent configuration, the user then has the opportunity to adapt the selected BP to his specific application by changing the arrangement and positioning of the individual components. In addition, the configured application can be simulated and provided to the user with detailed specifications.

Initially, in order to integrate existing robotic applications for the BP based configuration, a BP template was developed Figs. 2 and 3, left) as well as the flow for a user-centered configuration. These served as a guideline for the development of the ROBOTOP Platform Figs. 2 and 3, right).

Fig. 2 Best Practice Selection: comparison of the click-prototype [23, 26] (left) and the ROBOTOP platform (right) to show the iterative development of the platform

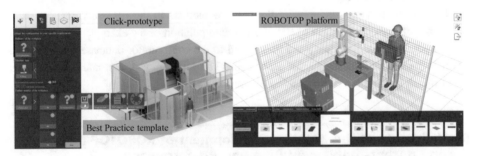

Fig. 3 Customization configuration: comparison of the click-prototype [23, 26] (left) and the ROBOTOP platform (right) to show the iterative development of the platform

6 Conclusion and Outlook

Within ROBOTOP we have shown how the agile development of a knowledge-based engineering configurator (KBEC) platform can be achieved. To this end, we have divided the platform functions such as Best Practice selection, configuration, simulation, AML-data-exchange and spec-sheet generator into different microservices. In addition, the requirements and possible functions of the platform were developed top-down using agile sub-prototypes. In parallel, bottom-up knowledge of Best Practices was gathered. In the process, challenges and possible solutions were identified.

For future research, there is a lack of user-centered tools for data preparation as well as other methods and tools to further scale-up the data acquisition as a primary bottleneck of knowledge-based platforms. BPMN in combination with process engines could be used for process control of data acquisition as well as for coordinating individual user-centric data acquisition tools. BPMN in combination with process engines could be used for the management of data acquisition processes [25] as well as for the coordination of specialized developed user-centered data acquisition microservices [17, 23].

Acknowledgements The research project ROBOTOP (01MA17009E, 01MA17009D) is funded by the German Federal Ministry of Economic Affairs and Energy (BMWi). ROBOTOP is part of the technology program "PAiCE Digitale Technologien für die Wirtschaft" and is guided by the DLR Project Management Agency in Cologne.

References

1. Bartelt, M., Stecken, J., Kuhlenkötter, B.: Microservice zur Erzeugung von digitalen Zwillingen. Seamless Convergence of Automation & IT : Automation: Baden-Baden. 679–690 (2018)
2. Blažek, P., Kolb, M., Streichsbier, C., et al.: The evolutionary process of product configurators. In: Bellemare J, Carrier S, Nielsen K et al (eds) Managing complexity. In: Proceedings of the 8th World Conference on Mass Customization, Personalization, and Co-Creation. Springer International Publishing, pp. 161–172 (2017)
3. Dackweiler, M., Krause, M., Coutandin, S., et al.: Konfiguration Von Robotiklösungen. VDI-Z **8**, 62–65 (2019)

4. Felfernig, A., Hotz, L., Bagley, C., et al.: Knowledge-based configuration: From research to business cases. Morgan Kaufmann, Waltham, MA (2014)
5. Fleischanderl, G., Friedrich, G.E., Haselböck, A., et al.: Configuring large systems using generative constraint satisfaction. IEEE Intell. Syst. **13**, 59–68 (1998). https://doi.org/10.1109/5254.708434
6. Gönnheimer, P., Kimmig, A., Ehrmann, C., et al.: Concept for the configuration of Turnkey production systems. Procedia CIRP **86**, 234–238 (2019). https://doi.org/10.1016/j.procir.2020.01.047
7. Hofstedt, P.: Constraints. In: Görz G, Schneeberger J, Schmid U (eds) Handbuch der künstlichen Intelligenz, 5., überarb. und aktualisierte Aufl. Oldenbourg, München, pp. 205–233 (2014)
8. Jordan, M., Auth, G., Jokisch, O.: Knowledge-based systems for the Configure Price Quote (CPQ) process—a case study in the IT solution business. J. Appl. Knowl. Manage. (2020)
9. KUKA AG KUKA System Partners. www.kuka.com/en-se/industries/system-partners
10. Koch, M.: Wissensbasierte Unterstützung der Angebotsbearbeitung in der Investitionsgüterindustrie. Zugl.: Erlangen-Nürnberg, Univ Diss Fertigungstechnik - Erlangen, 41 Hanser, München (1995)
11. Novak, P., Kadera, P., Wimmer, M.: Model-based engineering and virtual commissioning of cyber-physical manufacturing systems—transportation system case study. In: 2017 22nd IEEE International Conference on Emerging Technologies and Factory Automation (ETFA). IEEE, pp. 1–4 (2017)
12. Puppe, F.: Einführung in Expertensysteme, Zweite Springer, Studienreihe Informatik (1991)
13. Reinhart, G., Maǧaña, F.A., Zwicker, C.: Industrieroboter: Planung, Integration, Trends. Vogel Commun. (2018)
14. Schäffer, E., Bartelt, M., Pownuk, T., et al.: Configurators as the basis for the transfer of knowledge and standardized communication in the context of robotics. Procedia CIRP **72**, 310–315 (2018a). https://doi.org/10.1016/j.procir.2018.03.190
15. Schäffer, E., Fröhlig, S., Mayr, A., et al.: A method for collaborative knowledge acquisition and modeling enabling the development of a knowledge-based configurator of robot-based automation solutions. Procedia CIRP **86**, 92–97 (2019a). https://doi.org/10.1016/j.procir.2020.01.018
16. Schäffer, E., Leibinger, H., Stamm, A., et al.: Configuration based process and knowledge management by structuring the software landscape of global operating industrial enterprises with microservices. Procedia Manuf. **24**, 86–93 (2018b). https://doi.org/10.1016/j.promfg.2018.06.013
17. Schäffer, E., Mayr, A., Fuchs, J., et al.: Microservice-based architecture for engineering tools enabling a collaborative multi-user configuration of robot-based automation solutions. Procedia CIRP **86**, 86–91 (2019b). https://doi.org/10.1016/j.procir.2020.01.017
18. Schäffer, E., Mayr, A., Huber, T., et al.: Gradual tool-based optimization of engineering processes aiming at a knowledge-based configuration of robot-based automation solutions. Procedia CIRP **81**, 736–741 (2019c). https://doi.org/10.1016/j.procir.2019.03.186
19. Schäffer, E., Metzner, M., Pawlowskij, D., et al.: Seven levels of detail to structure use cases and interaction mechanism for the development of industrial virtual reality applications within the context of planning and configuration of robot-based automation solutions. Procedia CIRP (2020)
20. Schäffer, E., Penczek, L., Mayr, A., et al.: Digitalisierung im Engineering: Ein Ansatz für ein Vorgehensmodell zur durchgehenden, arbeitsteiligen Modellierung am Beispiel von AutomationML. I40M 2019: 61–66 (2019). https://doi.org/10.30844/I40M_19-1_S61-66

21. Schäffer, E., Pownuk, T., Walberer, J., et al.: System architecture and conception of a standardized robot configurator based on microservices. In: Schüppstuhl, T., Tracht, K., Franke, J. (eds.) Tagungsband des 3. Kongresses Montage Handhabung Industrieroboter. Springer Vieweg, Berlin, pp 159–166 (2018)

22. Schäffer, E., Schulz, J.P., Franke, J.: Robotiklösungen im Baukastenprinzip für den Mittelstand. BigData Insider (2019)

23. Schäffer, E., Shafiee, S., Frühwald, T., et al.: A development approach towards user-centered front-ends for knowledge-based engineering configurators: a study within planning of robot-based automation solutions. In: Proceedings of the 22st Configuration Workshop (2020)

24. Schäffer, E., Shafiee, S., Mayr, A., et al.: A strategic approach to improve the development of use-oriented knowledge-based engineering configurators (KBEC). Procedia CIRP (2020)

25. Schäffer, E., Stiehl, V., Schwab, P.K., et al.: Process-driven approach within the engineering domain by combining business process model and notation (BPMN) with process engines. Procedia CIRP (2020)

26. Schäffer, E., Thi, K., Franke, J.: Bedarf und Konzeptvorstellung des Forschungsprojekts ROBOTOP für einen webbasierten Konfigurator für den Massenmarkt (2020). https://doi.org/10.13140/RG.2.2.15152.30720

27. Tacton Systems GmbH: Smart commerce und CPQ für Industrieunternehmen (2020). https://www.tacton.com/?lang=de

28. Valavanis, K.P., Kokkinaki, A.I., Tzafestas, S.G.: Knowledge-based (expert) systems in engineering applications: a survey. J. Intell. Robot Syst. **113–145** (1994). https://doi.org/10.1007/BF01258225

29. Wolff. E.: Microservices: Grundlagen flexibler Softwarearchitekturen, 1., korrigierter Nachdruck. dpunkt.verlag, Heidelberg (2016)

30. Yaskawa, Y.: Motoman strategic automation partners. https://www.motoman.com/en-us/integration/partners

Partial Automated Multi-Pass-Welding for Thick Sheet Metal Connections

Sascha Lauer, Sebastian Rieck, Martin-Christoph Wanner
and Wilko Flügge

Abstract

The production of tubular-node-connections, which are required for the construction of offshore wind energy plants or converter platforms, is subject to high manufacturing standards. The welding process is currently carried out manually and requires a great deal of experience on the part of the welder. In this process, one or more branch member pipes are welded to a base pipe, which vary in their diameters and alignment to each other. This results in a small batch size for which no standard automation solution can be considered. The approach of a pre-defined offline path-planning is not expedient, since the weld metal forms differently with the multiple curved geometries and the desired target result cannot be achieved with an integrated compensation. The approach for automation combines the experience of a skilled welder with the accuracy of an industrial-robot. For implementation, the robot system moves along the welding contour with a 2D-profile sensor. The joint profile is recorded at defined measurement

S. Lauer (✉) · S. Rieck · M.-C. Wanner · W. Flügge
Fraunhofer Research Institution of Large Structures in Production Engineering IGP,
Albert-Einstein-Straße 30, 18059 Rostock, Germany
e-mail: sascha.lauer@igp.fraunhofer.de

S. Rieck
e-mail: sebastian.rieck@igp.fraunhofer.de

M.-C. Wanner
e-mail: martin-christoph.wanner@igp.fraunhofer.de

W. Flügge
e-mail: wilko.fluegge@igp.fraunhofer.de

W. Flügge
Chair of Production Technology, University of Rostock, Albert-Einstein-Straße 2, 18059
Rostock, Germany

© The Author(s) 2022
T. Schüppstuhl et al. (eds.), *Annals of Scientific Society for Assembly,
Handling and Industrial Robotics 2021*,
https://doi.org/10.1007/978-3-030-74032-0_33

points. Parallel to the seam cross-section, the current gradient of the geometry in relation to the horizontal plane is stored. After all the information has been generated, it is visualized for the operator in a graphical user interface. The operator can use his experience in the field of welding technology and can carry out the positioning of the weld seam in every single scan generated. The decisions on positioning are stored in the system and serve as a base for a future implementation of an automatic system for positioning welding beads on multi-curved contours.

Keywords

Multi-pass-welding · Sensor-based programming · Lightweight robots · Robot arc welding

1 Introduction

In order to reduce the manual work the developed process steps for the semi-automated welding of tubular-node connections already start with the consideration of the tolerance-afflicted pipe production. Here, production deviations of 1% in the pipe-diameter can occur. The variance in the wall thickness can be around 17.5% [1]. In order to take these inaccuracies into account and to achieve an optimal result for welding, both pipes are measured for the cutting process. The branch member pipe is cut by robot on the base of the real geometry [2, 3].

For the welding process it is first of all assumed that a constant welding parameter set is used over the entire three dimensional welding contour. This has already been determined by experimental tests. Since this way the identical weld volume is produced at any time, the cut and the seam preparation is designed with regard to this boundary condition. The main geometry is based on the AWS D1.1 [4].

If the branch member pipe has been cut, it is placed on the position to be welded. The root layer is weld by hand and then grounded. This preparation serves as a base for semi-automatic welding, which will be done by the orbital welding system (Fig. 1).

The orbital welding machine is clamped in the branch member pipe with a clamping mechanism. By means of an additional linear unit and an endless rotation axis, the working area of the robot is extended so that the tubular node connection can be processed continuously [5].

2 Description of the Developed Process

The developed process serves as a base for the automatic welding of tubular node connections. Here, a 2D-profile sensor is used to record the geometric data of weld seams, which are manually evaluated by an experienced welder and serve to plan the path of the

Fig. 1 Orbital welding system
for partial automated welding
[5]

welding process. For the evaluation of the geometric data a software was designed, which serves the welder for visualization. Based on this, the evaluated data are linked to the raw data. These links are to be used for the training of a neural network.

2.1 Solution Approach

As much data as possible must be generated to implement the solution approach. First, measurement points are determined on the intersection contour of the two pipes to be welded. Based on these points, a robot measuring program is generated, which is run with the orbital welding machine. As the number of measurement points increases, the data generation as well as the accuracy of the welding path and geometry generation increases.

At each of these points a scan of the seam geometry is generated during the measuring run. In the next process step these are transformed in a self-developed software in combination with the measuring program of the robot with regard to the horizontal plane. This is necessary to consider the influence of the earth's gravity on the formation of the weld bead.

Based on the transformed scan data, an experienced welder determines the position of the bead to be welded. Depending on the slope and seam opening angle, the welder also adjusts the torch orientation.

Depending on the selected positions and orientations, a robot welding program is generated. Parallel to this, this data is stored in the transformed scan data and saved. These data serve as a base for teaching the neural network. After the welding process, the measurement process is repeated. Now the comparison between welded and unwelded geometry can be performed and the choice of welding position and orientation can be evaluated.

2.2 Test Setup

A lightweight robot with a load capacity of 10 kg is used to carry out the measurement data generation and welding. It has a 2D-profile sensor and a welding torch attached to the hand axis. The robot controller and the profile sensor are accessed and the communication is coordinated via a higher-level control system with the self-developed software. The programs for the measuring and welding program are generated on the control system and transferred to the robot control system. The robot stops at defined measurement points and triggers the sensor to start a scan. This scan data is then transferred to the higher-level control system, where it is transformed and visualized on the software for the operator (Fig. 2).

2.3 Software/GUI

The developed software with user interface serves the operator to visualize the scanned weld seam profiles and for spatial orientation. Here, he is given a multitude of setting options, which he can also influence during manual welding. Depending on the welding

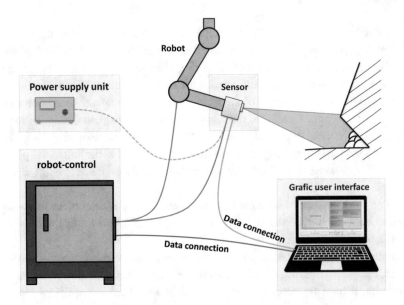

Fig. 2 Test setup for measurement process

position, the orientation of the torch can be adjusted or the respective type of movement can be changed. When these settings are adjusted, the orientation of the torch changes in relation to the scan data. Thus the visualization of the scan data in combination with the welding torch geometry is helpful as a preliminary stage of a collision control.

The following describes the areas shown in Fig. 3 and shows which changes can be made manually:

1. Load Profiles or new data
2. Slider for scans
3. Definition welding torch geometry
4. Definition torch orientation
5. Positioning weld
6. Coordinates for the robot-program
7. Visualization welding torch and Scan.

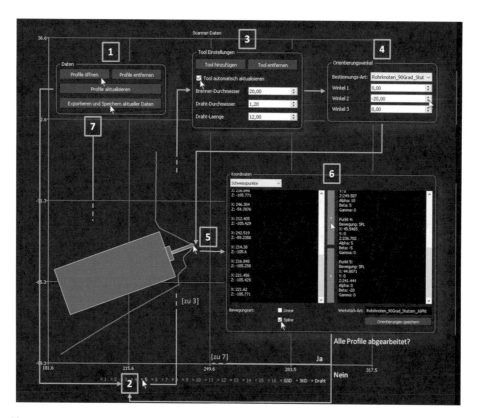

Fig. 3 Graphic user interface for positioning the weld

3 Implementation

To implement the solution, the first step is to create a gradient diagram of the intersection contour. This diagram is used to define interpolation points at which scan data of the seam geometry are generated in the subsequent process. Using this scan data, an experienced welder determines at which position and with which orientation the next weld bead is to be welded. At the same time, the system stores the welder's decisions and stores them with the generated scan data.

3.1 Planning the Welding Path

Depending on the geometry of the tubular pipe connection and the resulting intersection contour, measurement points must be defined to generate the measuring path (Fig. 4). Based on a 90° tubular node connection, the intersection contour is shown in Fig. 5. Based on the gradient, characteristic points, such as a change in gradient, minima or maxima are selected, which represent the contour as accurately as possible by means of a spline interpolation. At the defined points (red crosses in Fig. 4), measurements are made for the welding process, at which welding positions and alignments are determined by means of a graphical user interface.

By increasing the number of measurement points, both the accuracy of the trajectory and the amount of data required for training are improved.

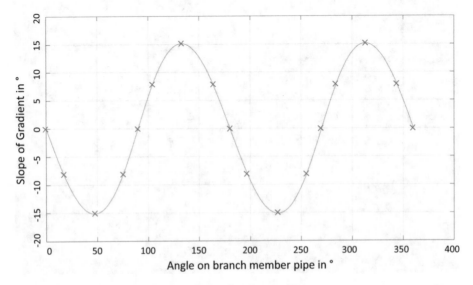

Fig. 4 Slope of gradient for a welding path at a 90° tubular connection with measuring points

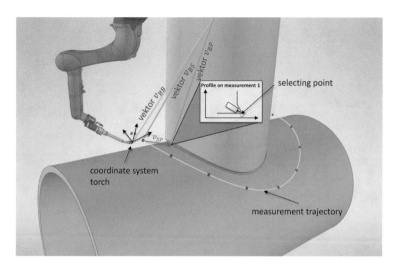

Fig. 5 Measuring run including vectors

3.2 Scan of the Welding Path

A measuring program is generated based on the generated interpolation points. The weld seam is scanned at the measurement points with a defined safety distance. For the subsequent transformation into the welding torch coordinate system, the scan data are generated in the RobRoot coordinate system, which is located in the center of the orbital welding system. After the measuring run is completed, the scan data are transferred to the software.

3.3 Human–Machine Collaboration

For the visualization, the scan data are first transformed according to their real orientation to the earth's gravitational force in order to take the influence of this force into account when positioning the next weld seam. Figure 6 shows a transformed scan with a welding torch. The red contour corresponds to the recorded scan data where 2 welding beads have already been welded. For the positioning of the third welding bead the welding position is selected in the first step. With the torch geometry stored in the software, the torch orientation is defined in the second step. Here the accessibility is already checked. By changing the wire length, minor adjustments can also be made.

Since the welds are performed in out of position and not in continuous flat position like in [6] or [7], the torch orientation along the welding contour must be adjusted in the third step as shown in Fig. 7. Welding is done in a falling position, whereas welding must be done in a dragging orientation and in a rising position in stabbing orientation.

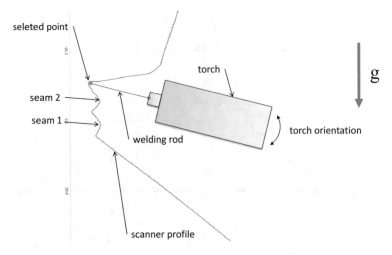

Fig. 6 Measured scanner profile for positioning next seam

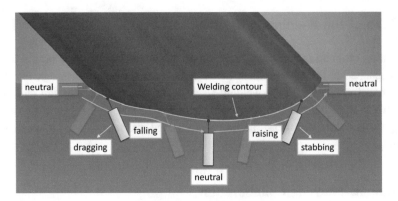

Fig. 7 Torch orientation depending on the slope of gradient

After confirming the welding position with orientation in each scan, a post-processor generates a robot welding program which is transferred to the robot controller. In the post-processor, additional welding commands, inputs and outputs and motion profiles are added.

After the performed welding, the measuring program is started again and the process is repeated with the selection of the welding positions.

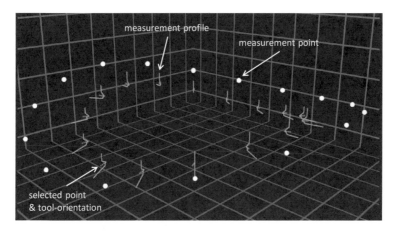

Fig. 8 Three-dimensional representation of the selected welding data sets

4 Generating Data

For control and traceability, the individual scans, the points of the measurement run and the selected welding points and their orientation are transformed into the RobRoot coordinate system in a final step and presented to the user in three dimensions as shown in Fig. 8.

Thus, with every weld bead created, new data is generated that can be used for artificial intelligence. Depending on the tubular node connection geometry, the material thickness or the welding bead to be positioned, new boundary conditions arise again and again, which the experienced welder covers in the first steps. In order to be able to access the data in the subsequent steps, they are stored systematically as shown in Fig. 9.

5 Using AI

Current research activities focus on the development of methods to determine the optimal weld seam positions automatically.

Due to constantly changing joint geometries and boundary conditions, an exclusively analytical approach is most likely not effective. The reason for this is that, as a rule, assumptions have to be made to establish the mathematical relationships and that there is insufficient generalization ability.

To solve this problem, a grey-box model is used [8, 9]. On the one hand, it contains an analytical sub-model based on a priori knowledge (white-box approach). This is used to suggest an optimal position first. On the other hand, there is a data-based sub-model from the field of machine learning (black-box approach) which serves to compensate for possible errors in the analytical part (see Fig. 10).

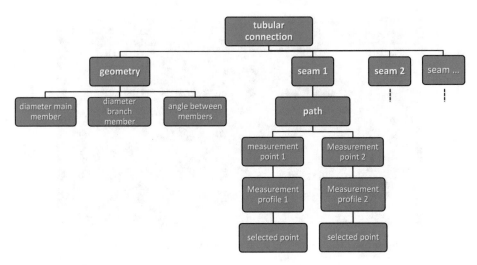

Fig. 9 Data to be stored and its hierarchy

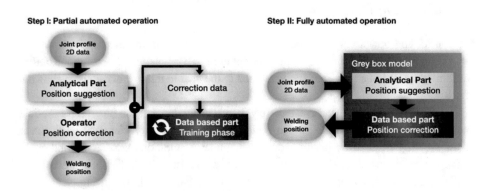

Fig. 10 Workflow to train and use the grey box model

In the analytical model part, a simplified seam geometry is assumed, the proportions or area of which depend on the selected welding parameters (wire feed, welding speed, voltage, current etc.). An algorithm fits the geometry into the point cloud of the joint scan afterwards. The supporting effect of the flanks is also taken into account here. After determining the initial position for the nth seam, the following positions are limited to a certain area around this point.

In a partially automated phase, seam positions for each scan are suggested by the analytical model part first and are corrected manually by the user if necessary. The correction data is collected in order to train the machine learning model part. Input–output data pairs are required for the training. The joint scan, the gravitational vector, the calculated cross-sectional area of the seam and the result of the analytical seam position determination represent the input data. The seam position corrected by the user, however, represents the output data. An artificial neural net is used as a model.

The duration of the partially automated phase depends on the amount of data collected. By using a new tubular node connection with a defined pipe geometry, as many new scan and positioning data as possible are generated. For example, if the branch member pipe is divided into 1° steps and about 25 single weld beads are necessary, 9000 data sets are generated. The machine learning model is trained at regular intervals and the performance of the entire grey box model is recorded as part of evaluation welds. The mean absolute deviation between the predetermined seam positions of the model and the selected positions of an experienced welder could serve as a metric for evaluating the model performance. If the deviation is in the range of 1–2 mm, it is probably possible to switch from the partially automated phase to the test phase for fully automated operation.

6   Conclusion and Outlook

The first milestone for the automation of tubular node connection production was laid with the development of a software with an user interface. This involves determining defined measurement points on the joint to be welded at which scans are generated. Based on these scans, an experienced welder generates a robot welding program. Thus, the welder is not exposed to the direct welding process but can contribute his experience to the process. Compared to the manual teaching of such a welded joint connection, the programming time is reduced from about 30 min to only 5 min. For complete automation, the decisions of the welder are stored in a database in order to build up a generally valid neural network for the most diverse tubular node connection geometries.

References

1. Deutsches Institut für Normung: DIN EN 10216–1:2014–03 Nahtlose Stahlrohre für Druck-beanspruchungen – Technische Lieferbedingungen – Teil 1:Rohre aus unlegierten Stählen mit festgelegten Eigenschaften bei Raumtemperatur, Beuth Verlag (2014)
2. Ambrosat, T., Lauer, S., Geist, M., Flügge, W.: Bestimmung komplexer Schneidkonturen als Vorbereitung zum Verschweißen von 3D Rohrstößen, in GFaI, Berlin (2018)
3. Lauer, S., Ambrosat, T., Wanner, M.C., Flügge, W.: Measurement based robot cutting as preparation for welding tubular connections, in MHI-Fachkolloquium, Bayreuth (2020)
4. Society, A.W.: AWS D1.1:Structural Welding Code-Steel, Miami: American Welding Society, LNCS Homepage (2000). https://www.springer.com/lncs. Accessed 21 Nov 2016
5. Wanner, M.C., Dryba, S., Weidemann, B., Harmel, A.: Schweißanordnung zum dauerhaften Fügen eines ersten rohrförmigen Bauteils mit einem zweiten Bauteil. Germany Patent DE 10 2015 206 044 A1 (2016)
6. Lotz, S., Wolski, U., Mückenheim, U.: Robotergestütztes Schweißen von Rohrknoten, Halle, SLV Halle (2017)

7. Zhang, Y., Lv, X., Xu, L., Jing, H., Han, Y.: A segmentation planning method based on the change rate of cross-sectional area of single V-groove for robotic multi-pass welding in intersecting pipe-pipe joint. The International Journal of Advanced Manufacturing Technology: Springer Verlag (2018)
8. Didona, D., Romano, P.: Hybrid machine learning/analytical models for performance prediction: a tutorial. In: ICPE 2015—Proceedings of the 6th ACM/SPEC International Conference on Performance Engineering (2015). https://doi.org/10.1145/2668930.2688823
9. Sohlberg, B., Jacobsen, E.: Grey box modelling—branches and experiences. IFAC Proceedings Volumes (IFAC-Papers Online). 17 (2008). https://doi.org/10.3182/20080706-5-KR-1001.01934

Author Index

Printed in the United States
by Baker & Taylor Publisher Services